Foundations in Signal Processing, Communications and Networking

Volume 22

Series editors

Holger Boche, Technische Universität München, München, Germany
Rudolf Mathar, ICT cubes, RWTH Aachen University, Aachen, Germany
Wolfgang Utschick, Technische Universität München, München, Germany

This book series presents monographs about fundamental topics and trends in signal processing, communications and networking in the field of information technology. The main focus of the series is to contribute on mathematical foundations and methodologies for the understanding, modeling and optimization of technical systems driven by information technology. Besides classical topics of signal processing, communications and networking the scope of this series includes many topics which are comparably related to information technology, network theory, and control. All monographs will share a rigorous mathematical approach to the addressed topics and an information technology related context.

More information about this series at http://www.springer.com/series/7603

Andreas Gründinger

Statistical Robust Beamforming for Broadcast Channels and Applications in Satellite Communication

Springer

Andreas Gründinger
Ergolding
Bayern, Germany

ISSN 1863-8538 ISSN 1863-8546 (electronic)
Foundations in Signal Processing, Communications and Networking
ISBN 978-3-030-29580-6 ISBN 978-3-030-29578-3 (eBook)
https://doi.org/10.1007/978-3-030-29578-3

This Springer imprint is published by the registered company Springer Nature Switzerland AG.
The registered company address is: Gewerbestrasse 11, 6330 Cham, Switzerland

To my family—thanks for always being there for me and for the smiling faces after a walk.

Preface

Reliable data access within wireless networks gains importance for an increasing number of applications in the age of autonomous vehicles and latency sensitive communication. Transmit beamforming is seen as the key technology of modern communication systems for simultaneous transmission to a growing number of receivers in the same frequency bands. While ideal beamforming spatially separates the received data at the mobile devices, practical transmit limitations and noise-prone estimation of the unknown channel result in inevitable interference and unwanted data losses. The book addresses these issues. It provides robust formulations that take into account the statistical channel properties and, thus, enable reliable data services for mobiles, e.g., phones and even cars, in future terrestrial and satellite communication (SatCom) systems that are sensitive to data losses.

Therefore, this book is for readers with interest in modern physical-layer beamforming and/or robust optimization techniques. The subject overview and classification of the average and probability restrictions and the presentation of closed-form expressions besides novel and known approximations is one strength of the work. Another strength is that the introduced approximations for the ergodic data rate and the probability for data losses provide also structural information about the original problem. These bounds especially also enable testing feasibility for given target data demands, analyzing the infinite transmit power limit, and transforming the downlink beamformer design task into an uplink filter computation and power allocation. The downlink-uplink duality framework covers unequal demands and existing dualities from literature as special cases. By applying the approximations for SatCom, where interference and channel fading are the main limitations, a fully adaptive beamformer optimization for up to a hundred receivers was possible.

Ergolding, Germany
May 2018

Andreas Gründinger

Acknowledgments

Working at the Institute for Methods of Signal Processing toward recent research in multiuser MIMO communication was an inspiring time. I thank all those who accompanied me during these exciting years.

I am deeply grateful to my adviser, Prof. Wolfgang Utschick, for his guidance from being my mentor in the master's program until now, introducing me to so many interesting areas in signal processing, and for the confidence which he always had in me. I also want to express my sincere gratitude to Michael Joham for guiding my steps as a doctoral student, the effort he has put into correcting my first publications, and the shared insights in MIMO systems. This work and the collaborations with the German Aerospace Center and the University of A Coruña would not have been possible without their encouragement and commitment.

I would also like to express my sincere gratitude to Prof. Luis Castedo Ribas for the opportunity to visit A Coruña, accepting to examine the thesis, and serve in the examination committee. This work has benefited from the collaboration with him and the discussions with his doctoral student José Pablo González Coma.

Many thanks to all the colleagues at the institute for coauthoring papers and for the valuable discussions, constructive criticisms, and friendship.

Furthermore, I would like to thank the Deutsche Forschungsgemeinschaft (DFG) for supporting the research to this work under funds Jo 724/1-1, Jo 724/1-2, and Jo 724/2-1 and Qualcomm for the European Innovation Fellowship Award in 2012.

My warmest thanks go to my family, my wife, and our beloved kids for their understanding and patience.

Introduction

Modern communication systems use physical-layer beamforming for serving a continuously growing number of mobile receivers. The aim is to exploit multipath propagation in wireless communication for simultaneous and interference-free data transmission from one multi-antenna transmitter to multiple receivers. Practical limitations for the dynamic range of the amplifiers and erroneous transmitter channel state information (CSI), e.g., due to the varying environment and noise-prone channel estimations, make interference-free transmission inefficient.

To ensure reliable data transmission, beamformers are designed either to achieve the receivers' quality of service (QoS) demands or to maximize their data rate relative to the demands. For imperfect CSI, a careful choice of the performance metric is crucial. Otherwise, transmission can be subject to frequent data loss (outages). For fast-changing channel conditions, demands in data rate shall naturally be satisfied on average, while the probability for the outage event shall be limited for slower changes.

The goal of this work is to provide conservative and efficient solutions for the so obtained statistically robust optimizations and applications to downlink satellite communication (SatCom). The bases are the theory of interference functions, convex conic programming, and Lagrangian duality. Two imperfect CSI models are distinguished the additive Gaussian error model and scalar multiplicative errors.

Multiplicative channel errors essentially enhance the noise power. This enables beamformer solutions that are tractable for SatCom with a hundred of terminals and spotbeams. For additive channel errors, the average rate maximization is analyzed by a minimization of the maximum mean square error between the transmit and receive signals. The solutions benefit from the provided uplink-downlink duality for general conic power limitations. The duality framework is sufficiently general to analyze the effects of practical power restrictions for SatCom. Outage robust beamforming also relies on conservative approximations for the probability constraints. Such formulations either restrict the channel error to reside in a predefined uncertainty set or are based on concentration inequalities. Two new formulations are analyzed in this context. The resulting solutions outperform some relevant results from the literature. Applications in mobile SatCom employed a further approximation to separate the outages due to rain fading and multipath scattering from each other.

Zusammenfassung

Moderne Funkkommunikationssysteme verwenden Beamforming im Physical-Layer, um eine steigende Zahl mobiler Endgeräte zu bedienen. Ziel ist es, die räumliche Mehrwegeausbreitung in der drahtlosen Funkkommunikation zur gleichzeitigen und interferenzfreien Übertragung an mehrere Empfänger zu nutzen. Interferenzfreie Übertragung ist in realen Systemen jedoch häufig ineffizient. Grund hierfür ist die beschränkte Linearität von Leistungsverstärkern und fehlerhafte Kanalkenntnis am Sender als Folge der sich ändernden Umgebung und unzureichender Kanalschätzung.

Zur zuverlässigen Datenübertragung werden deshalb Beamforming-Verfahren eingesetzt, die entweder den Serviceanforderungen der Empfänger genügen oder die Datenraten relativ zu diesen Anforderungen maximieren. Entscheidend ist auch die Wahl der Metrik, um regelmäßig auftretende Datenverluste (Outages) zu beschränken. Für schnell veränderliche Übertragungskanäle sind die Anforderungen an die Datenraten im Mittel zu erfüllen, während für langsam veränderliche Kanäle die Wahrscheinlichkeit des Outage-Ereignisses begrenzt wird.

Ziel der Arbeit sind konservative Abschätzungen und effiziente Lösungen für die resultierenden Optimierungsprobleme und deren Anwendung in der Satellitenkommunikation. Mathematische Grundlagen für die entwickelten Algorithmen sind die Theorie der Interferenz Funktionen, konisch konvexe Optimierung und Lagrange Dualität. Zudem werden zwei Kanalmodelle unterschieden: das Standardmodell mit additivem Gauß'schen Fehler und ein Modell mit skalaren multiplikativen Fehlern.

Multiplikative Kanalfehler steigern hauptsächlich die Rauschleistung. Dies ermöglicht duale Fixpunktverfahren zur Berechnung des Beamforming für über hundert Antennen und Empfänger in der Satellitenkommunikation. Für additive Kanalfehler wird die Maximierung der mittleren Datenraten durch eine Minimierung des maximalen mittleren quadratischen Fehlers zwischen den Sende- und Empfangssignalen approximiert. Lösungen dieses Problems profitieren von der eingeführten Uplink-Downlink Dualität mit konischen Nebenbedingungen. Damit werden die Effekte praxisnaher Leistungsbeschränkungen auf das Sendeverhalten analysiert. Bei Outage begrenzten Verfahren sind ebenfalls konservative Näherungen für die Wahrscheinlichkeitsnebenbedingungen erforderlich. Solche

Näherungen beschränken entweder Kanalfehler auf einen definierten Bereich oder basieren auf stochastischen Ungleichungen. Die Arbeit analysiert in diesem Zusammenhang zwei neue Formulierungen. Die daraus resultierenden Lösungen übertreffen dabei relevante Verfahren aus der Literatur. Bei Anwendungen für mobile Satellitenkommunikation werden neben den Kanalfehlern durch die Mehrwegeausbreitung noch Schwankungen in der Atmosphäre mit berücksichtigt.

Contents

Acronyms

The following acronyms and abbreviations are used throughout this work:

ACS	Alternating convex search
BB	Branch and bound
BC	Broadcast channel
CDF	Cumulative distribution function
CSI	Channel state information
dB	Decibel
DPC	Dirty paper coding
EVD	Eigenvalue decomposition
FDD	Frequency division duplex
FSL	Free space loss
GEO	Geostationary earth orbit
GP	Geometric program
IFC	Interference channel
i.i.d.	Identically and independent distributed
KKT	Karush–Kuhn–Tucker
LMI	Linear matrix inequality
LoS	Line of sight
LP	Linear program
LPM	Lorentz positive map
MAC	Multiple access channel
MF	Matched filter
MIMO	Multiple-input multiple-output
ML	Maximum likelihood
MMSE	Minimum mean square error
MRT	Maximum ratio transmission
MSE	Mean square error
PCLI	Probabilistically constrained linear inequality
PDF	Probability density function
PG	Projected gradient

PtP	Point-to-point
QCP	Quadratically constrained program
QF	Quantile function
QoS	Quality of service
RB	Rate balancing
RF	Radio frequency
RZF	Regularized zero-forcing
SatCom	Satellite communication
SCS	Sequential convex search
SDC	Semidefinite cone
SDP	Semidefinite program
SDR	Semidefinite relaxation
SINR	Signal-to-interference-plus-noise ratio
SIR	Signal-to-interference ratio
SMSE	Sum mean square error
SNR	Signal-to-noise ratio
SOC	Second-order cone
SOCP	Second-order cone program
SQ	Scalar quantization
TDD	Time division duplex
THP	Tomlinson–Harashima precoding
VQ	Vector quantization
WSMSE	Weighted sum mean square error
ZF	Zero-forcing

Nomenclature

The basic symbols and operators used throughout this work are listed below. Generally, boldface lower case is used for column vectors and upper case for matrices. Other symbols, functions, or operations are defined at the required position within the text.

$\lvert \cdot \rvert$	Absolute value
$\lVert \cdot \rVert_2$	Euclidean norm, i.e., L^2 norm
$\lVert \cdot \rVert_\mathrm{F}$	Frobenius norm
$\otimes$	Kronecker product
$(\cdot)^*$	Complex conjugate
$(\cdot)^\mathrm{T}$	Transpose
$(\cdot)^\mathrm{H}$	Complex conjugate transpose
$(\cdot)^{-1}$; $(\cdot)^\dagger$	Inverse, pseudo-inverse
$(\cdot)_{\setminus\{k\}}$	All columns except for the k-th one
$\mathrm{Re}(\cdot)$	Real part
$\mathrm{Im}(\cdot)$	Imaginary part
$\mathrm{tr}(\cdot)$	Trace
$\mathrm{diag}(\cdot)$	Diagonal matrix with scalars as diagonal entries
$\mathrm{bdiag}(\cdot)$	Block diagonal matrix with matrices as diagonal entries
$\mathrm{vec}(\cdot)$	Column-stacking operation
$\det(\cdot)$	Determinant operation
$\mathrm{rank}\{\cdot\}$	Rank of a vector or matrix
$\mathrm{range}\{\cdot\}$	Range space of a linear map
$\mathrm{null}\{\cdot\}$	Kernel or nullspace of a linear map
j	Imaginary number, i.e., $\sqrt{-1}$
e	Euler's number, i.e., $\mathrm{e} \approx 2.71828$
γ	Euler–Mascheroni constant, i.e., $\gamma \approx 0.57722$

$\mathbf{0}$	Zero or all-zero vector/matrix of appropriate dimension
$\mathbf{1}$	One or all-ones vector/matrix of appropriate dimension
$\mathbf{e}_i$	Canonical unit-norm vector with one at the i-th position
$\mathbf{I}$	Identity matrix (operation)
$\exp(\cdot)$	Exponential function
$\ln(\cdot)$	Natural logarithm
$\log_2(\cdot)$	Binary logarithm
$\mathrm{E}_1(\cdot)$	Exponential integral function $\mathrm{E}_1(x) = \int_x^\infty \frac{\exp(-t)}{t}\,\mathrm{d}t$
$\Gamma(\cdot)$	Gamma function, i.e., $\Gamma(x) = \int_0^\infty t^{x-1}\mathrm{e}^{-t}\,\mathrm{d}t$
$\mathrm{E}[\cdot]$	Expectation operation
$\Pr(\cdot)$	Probability operation
$f_z(\cdot)$	Probability density function of a random variable z
$F_z(\cdot)$	Cumulative distribution function of z, i.e., $F_z(x) = \Pr(z \leq x)$
min	Minimize (also minimum of a set of two or more scalars)
max	Maximize (also maximum of a set of two or more scalars)
arg min	Minimizing decision variables of the problem
arg max	Maximizing decision variables of the problem
s.t.	Subject to the constraints
$\mathbb{R}$, $\mathbb{R}_+$	Real and nonnegative numbers (with scalar inequalities $>, \geq, <, \leq$)
$\mathbb{C}$	Complex numbers
$\mathbb{R}_+^N$	Vectors with nonnegative entries, $\mathbb{R}_+ = \{\boldsymbol{x} \in \mathbb{R}^N : x_i \geq 0, i = 1, \ldots, N\}$
$\mathcal{L}^N$	Lorentz cone, second-order cone, $\mathcal{L}^N = \{(y, \boldsymbol{x}) \in \mathbb{R}_+ \times \mathbb{C}^N : \|\boldsymbol{x}\|_2 \leq y\}$
$\mathcal{H}^N$	Hermitian matrices, $\mathcal{H}^N = \{\boldsymbol{A} \in \mathbb{C}^{N\times N} : \boldsymbol{A} = \boldsymbol{A}^{\mathrm{H}}\}$
$\mathcal{H}_+^N$	Positive semidefinite matrices, $\mathcal{H}_+^N = \{\boldsymbol{A} \in \mathbb{C}^{N\times N} : \boldsymbol{x}^{\mathrm{H}}\boldsymbol{A}\boldsymbol{x} \geq 0, \boldsymbol{x} \in \mathbb{C}^N\}$
$\boldsymbol{x} \geq \boldsymbol{y}$ $(\boldsymbol{x} > \boldsymbol{y})$	Elementwise (strict) inequality, $x_i \geq y_i$ $(x_i > y_i)$, $i = 1, \ldots, N$
$\boldsymbol{x} \leq \boldsymbol{y}$ $(\boldsymbol{x} < \boldsymbol{y})$	Elementwise (strict) inequality, $x_i \leq y_i$ $(x_i < y_i)$, $i = 1, \ldots, N$
$\boldsymbol{X} \succeq \boldsymbol{Y}$ $(\boldsymbol{X} \succ \boldsymbol{Y})$	$\boldsymbol{X} - \boldsymbol{Y}$ is positive semidefinite (definite)
$\boldsymbol{X} \preceq \boldsymbol{Y}$ $(\boldsymbol{X} \prec \boldsymbol{Y})$	$\boldsymbol{Y} - \boldsymbol{X}$ is positive semidefinite (definite)
$\mathcal{N}$	Real Gaussian distribution
$\mathcal{N}_\mathbb{C}$	Circularly symmetric complex Gaussian distribution
$\mathcal{U}$	Uniform distribution
$\mathcal{X}_n^2$	(Non-)central chi-squared distribution of degree n
$\mathcal{E}$	Exponential distribution

The notations $\boldsymbol{x} \geq \mathbf{0}$ and $\boldsymbol{x} \in \mathbb{R}_+^N$ as well as $\boldsymbol{X} \succeq \mathbf{0}$ and $\boldsymbol{X} \in \mathcal{H}_+^N$ are equivalent and define a nonnegative vector $\boldsymbol{x} \in \mathbb{R}^N$ and a semidefinite matrix $\boldsymbol{X} \in \mathcal{H}^N$, respectively.

List of Figures

List of Tables

Chapter 1
Multi-User Downlink Communication

Simultaneous wireless transmission from a single multi-antenna transmitter, e.g., a base station or a satellite, to K receivers—the users—is a standard model for terrestrial and Satellite communication (SatCom) [1, 2]. This multi-user downlink model is also known as a Broadcast channel (BC) (e.g., see [1]). The transmitter forms its transmit signal from independent data that are simultaneously conveyed to the users [3]. Therefore, a terminal's received signal not only includes the intended data signal, but also interfering signals that are destined to other terminals.[1]

The interference can be taken into account in the encoding process of the data signals using non-linear Dirty paper coding (DPC)—named according to Costa's work [4]. This encoding scheme has shown to achieve the complete Shannon capacity region of the Gaussian BC. Extensions to multi-antennas at the transmitter and receivers were subsequently derived in the last decade. In terms of the sum capacity, i.e., the throughput of the BC, this result was shown by Viswanath et al. [5] for a multi-antenna transmitter. The proof that the complete capacity region of the Multiple-input multiple-output (MIMO) BC is reached via DPC was provided by Weingarten et al. [6].

The problem with DPC is that perfect transmitter Channel state information (CSI) is crucial for its application [7, Section V.D]. In particular, it requires non-causal knowledge of the block fading channels to incorporate the caused interference for the code design.[2] If the information of the block fading channel states is erroneous, DPC leads to unpredictable data loss. Then, interference is induced where it should be "pre-canceled" from the code design. Other drawbacks of DPC are its complexity

[1]This is in contrast to multicast setups, where all users receive the same information.

[2]Lately, a DPC type scheme was proposed for block fading single antenna channels [8]. However, as extensions to multi-antenna setups are missing, this approach is beyond the scope of this work.

A. Gründinger, *Statistical Robust Beamforming for Broadcast Channels and Applications in Satellite Communication*, Foundations in Signal Processing, Communications and Networking 22, https://doi.org/10.1007/978-3-030-29578-3_1

for the code design and the involved computations for physical layer precoding, which makes it intractable when scaling the system dimensions or physically separating the location of the transmitting antennas.

An approach towards pre-canceling interference are robust versions of the Tomlinson–Harashima precoding (THP) scheme. Such a version is available for the sum MSE minimization [9]. However, the complexity for THP is still too high to implement it in Quality of service (QoS) and delay sensitive precoder designs with an increasing number of receivers. Moreover, transmission with THP becomes unreliable for an increasing variance of the channel errors.

For these reasons, there is ongoing research on linear schemes that do not explicitly pre-cancel the interference, e.g., multi-user transmit beamforming [10]. Typical examples are the QoS and balancing optimizations for unicast [11] and multicast transmission [12, 13], but also the classical weighted sum rate maximization [14] and general utility maximizations [15–17]. Even though linear beamforming may be suboptimal for these problems [18], dependent on the quality of the channels' estimates, it is commonly applied to exploit the transmit degrees of freedom. Other examples where linear schemes are still under research are interference channels (e.g., see [19]), when excessively increasing the number of base station antennas in favor for a simplified filter design (e.g., see [20–22]) and for relay systems [23, 24].

This work particularly details linear beamforming techniques for the Gaussian vector BC that are robust against impairments of the transmitter's information about the channel and the actual channel state. Only imperfect knowledge of the block-fading channels is usually available for mobile wireless communication, e.g., due to estimation errors in the pilot-based training and delayed and limited feedback from the mobile devices to the base station. Herein, the transmitter is only aware of a channel estimate, the statistics of the fading channels, and the resulting error model. Examples for erroneous channel information models are the topic of Chap. 2.

The task of statistically robust beamforming requires general knowledge in the fields of communications engineering, stochastic modeling and estimation, and (convex) optimization theory. The choices for the stochastic channel model and the performance metric of the communication scenario have a big influence on the available approximations and optimization techniques for the beamformer computation. Even the analysis of a performance metric, e.g., the ergodic data rate or the data rate with limited outage probability, can be difficult if the channel model follows an involved distribution. That is the reason why research towards stochastically robust designs has two goals. For performance analysis, researchers calculate numerically tractable closed-form expressions for the achievable data rate with statistical knowledge of the vector channels, e.g., see [25–28] for ergodic rates and [29–32] for the outage probability. For the beamformer design, researchers applied robust approximations for the ergodic rate (e.g., see [33, 34]) and outage probability metrics (e.g., see [35, 36]) to employ convex optimizations.[3]

[3]Further literature on multi-user downlink beamformer optimizations with ergodic rates and the related MSE or outage probability metrics is provided in the contributions section, i.e., Sect. 1.5, and at the beginning of Chaps. 3 and 4 or Chap. 5, respectively.

Worst-case robust beamforming is related to the stochastically robust beamformer design task, but it avoids the stochastic channel modeling by a deterministic uncertainty region constraint. This type of robust beamforming became popular through the works by Voroboyev et al. [37] and Shahbazpanahi et al. [38], both guided by Gershman and Zhi-Quan Luo, and the publications of Li and Stoica [39] in 2003. Statistically robust beamforming techniques based on ergodic performance metrics and outage limitations gained interest in competition with the worst-case approach.[4] Since then, the number of publications on "robust beamforming" for signal processing and wireless communications is steadily growing (see Fig. 1.1).

Nowadays, statistically robust beamforming approaches are published for various objectives, e.g., signal power maximization [42], throughput maximization [17], sum MSE minimization [43], power minimization [44], as well as robustness maximization [45]. The herein studied beamformer optimizations are statistically robust versions of the classical QoS power minimization and Rate balancing (RB)

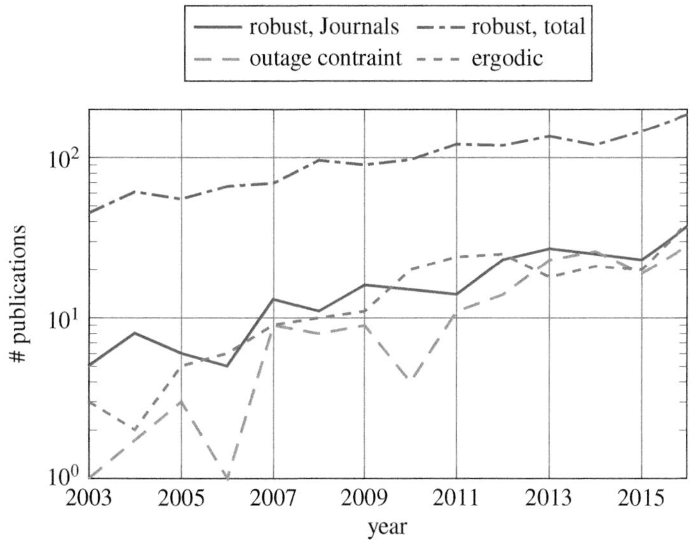

Fig. 1.1 Annual number of publications at selected Journals for the search on "robust beamforming" and total publications for searching "outage constraint beamforming" and "ergodic beamforming" at IEEEXplore from 2003 to 2016[5]

[4]See the book by Li and Stoica [40], the book by Dietrich [41], and Chaps. 3–5 for more early references on robust beamforming for wireless communications.

[5]The source for the plot is IEEEXplore and available online via ieeexplore.ieee.org. The following journals have been selected: *IEEE Transactions on Signal Processing, IEEE Transactions on Communications, IEEE Transactions on Wireless Communications, IEEE Transactions on Vehicular Technology, IEEE Selected Topics in Signal Processing, IEEE Selected Topics in Communications, IEEE Signal Processing Letters, IEEE Communications Letters.*

optimization in Sect. 1.2, which were also considered by [36, 46–51]. These QoS constrained formulations are preferred when either communication services have to meet some minimum demands or the throughput has to be shared such that all users achieve data rates proportional to their demands. Moreover, these optimizations are better tractable for linear beamforming than throughput maximization, which is already non-convex for perfect transmitter CSI (e.g., see [16]). The focused QoS and balancing optimizations can be solved via convex programming tools and up to a hundred receivers and transmit antennas dependent on the fading scenario.

For these reasons, the robust QoS based power minimization and balancing optimizations are predestined for modern SatCom, where reliability and connectivity are crucial (e.g., see [52]) and a large number of terminals have to be served simultaneously in the same frequency band. Due to strengthened investigations for increasing the frequency reuse, adaptive physical layer beamforming and power allocation are seen as a key technology to coordinate the interference at neighboring terminals and deal with the harmful channel fading [53–58].[6]

Chapter 6 contributes to these investigations by extending the ergodic and outage constrained beamformer designs for the SatCom specific channel fading, which combines the atmospheric attenuation and local scattering effects around the terminals [59]. The results show that interference management with ergodic rate bounds substantially outperforms Zero-forcing (ZF)-beamforming schemes that completely suppress the interference. For appropriately shaped channel fading, these bounds even enable a fully adaptive interference coordination for more than a hundred spotbeams that are focused to Europe. When the ergodic rate is unavailable for physical layer beamforming due to the fading, the MSE metric provides a tractable approximation [56, 60–63]. With respect to this metric, the focus is on practicable constraints for the high power Radio frequency (RF) amplifiers serving the antennas. These constraints substantially decrease the dynamic power range of the amplifiers compared to the standard sum power requirement. However, they guarantee almost the same performance if the terminals are uniformly placed in their respective areas and experience similar fading statistics.

When the block fading channels remain constant during the transmission phase, outage constraints on the data rates are the measure of interest (e.g., see [64]). The outage limits are usually small for SatCom and separated with respect to the unknown attenuation and scattering. This decreases the performance, but provides the degree of freedom to trade-off the outages between these error types. Moreover, it enables for convex deterministic approximations of the outage probability requirements, such that an analysis for a SatCom scenario with seven neighboring terminals becomes tractable.

The next sections present basic modeling aspects and optimization strategies for perfect channel information in the vector BC. These results are the basis for understanding the robust beamforming strategies with imperfect CSI.

---

[6]Interference becomes severe if the frequency reuse is one, i.e., the same bands are used in all cells.

1.1 Gaussian Vector Broadcast Channel Model

Restricting to superposition coding and modeling self-interference as additive noise for the K-user Gaussian BC with an N-antenna transmitter, the error free achievable data rate r_k for receiver k is given by the mutual information (e.g., see [1, 3])

$$\mathrm{I}(\boldsymbol{x}_k; y_k) = \mathrm{h}(y_k) - \mathrm{h}(y_k|\boldsymbol{x}_k). \tag{1.1}$$

Formulations for the differential entropies $\mathrm{h}(y_k)$ and $\mathrm{h}(y_k|\boldsymbol{x}_k)$ of the received signal $y_k \in \mathbb{C}$ and the received signal conditioned on the transmit data signal $\boldsymbol{x}_k \in \mathbb{C}^N$ depend on the individual and conditional distributions for $y_1, \ldots, y_K$ and $\boldsymbol{x}_1, \ldots, \boldsymbol{x}_K$.

The transmit signals shall be zero-mean circularly symmetric complex Gaussian. This input distribution is capacity achieving for the Point-to-point (PtP) channel or the Gaussian BC with DPC, but it is also a standard assumption for linear transmit beamforming based on Shannon rates (e.g., see [1]). These signals $\boldsymbol{x}_k = \boldsymbol{t}_k s_k$ are formed by the mutually independent zero-mean Gaussian data signals $s_k \sim \mathcal{N}_{\mathbb{C}}(0, 1)$ and the beamformers $\boldsymbol{t}_k \in \mathbb{C}^N$, $k = 1, \ldots, K$.[7] Their superposition is sent over the narrow-band block-fading channels $\boldsymbol{h}_k \in \mathbb{C}^N$. The received signal of user k reads as

$$y_k = \boldsymbol{h}_k^{\mathrm{H}} \sum_{i=1}^{K} \boldsymbol{t}_i s_i + n_k, \tag{1.2}$$

where $n_k \sim \mathcal{N}_{\mathbb{C}}(0, \sigma_k^2)$ denotes the experienced zero-mean additive Gaussian noise with variance σ_k^2 that is mutually independent to the sent data signals and the noise at other users.

For this BC model, the achievable data rate in (1.1) reads as (e.g., see [1])

$$r_k = \log_2(1 + \mathrm{SINR}_k), \tag{1.3}$$

where the Signal-to-interference-plus-noise ratio (SINR) of user k is given by

$$\mathrm{SINR}_k = \frac{|\boldsymbol{h}_k^{\mathrm{H}} \boldsymbol{t}_k|^2}{\sum_{i \neq k} |\boldsymbol{h}_k^{\mathrm{H}} \boldsymbol{t}_i|^2 + \sigma_k^2}. \tag{1.4}$$

A strongly related performance metric is the MSE $\mathrm{E}[|s_k - \hat{s}_k|^2]$ between s_k and the signal estimate $\hat{s}_k = f_k y_k$, with receive filter $f_k \in \mathbb{C}$. It reads as

[7] Alternatively, the inputs may also be modeled as complex circularly symmetric Gaussian vectors $\boldsymbol{x}_k \sim \mathcal{N}_{\mathbb{C}}(0, \boldsymbol{Q}_k)$, $k = 1, \ldots, K$. However, the optimal transmit covariance matrices $\boldsymbol{Q}_k$ of the intended problems are rank-one if sufficiently accurate CSI is available at the transmitter [18].

$$\mathrm{MSE}_k = 1 - 2\,\mathrm{Re}(f_k \boldsymbol{h}_k^{\mathrm{H}} \boldsymbol{t}_k) + \sum_{i=1}^{K} |f_k|^2 |\boldsymbol{h}_k^{\mathrm{H}} \boldsymbol{t}_i|^2 + |f_k|^2 \sigma_k^2 \tag{1.5}$$

and is minimized by the Minimum mean square error (MMSE) filter function

$$f_{\mathrm{MMSE},k} = \frac{\boldsymbol{t}_k^{\mathrm{H}} \boldsymbol{h}_k}{\sum_{i=1}^{K} |\boldsymbol{h}_k^{\mathrm{H}} \boldsymbol{t}_i|^2 + \sigma_k^2}. \tag{1.6}$$

The corresponding minimum value, i.e., $\mathrm{MMSE}_k = \min_{f_k} \mathrm{MSE}_k \in (0, 1]$, reads as

$$\mathrm{MMSE}_k = 1 - \frac{|\boldsymbol{h}_k^{\mathrm{H}} \boldsymbol{t}_k|^2}{\sum_{i=1}^{K} |\boldsymbol{h}_k^{\mathrm{H}} \boldsymbol{t}_i|^2 + \sigma_k^2} \tag{1.7}$$

and relates to the rate (1.3) and the SINR (1.4) by (e.g., [65])[8]

$$r_k = -\log_2(\mathrm{MMSE}_k), \tag{1.8a}$$

$$\mathrm{MMSE}_k = (1 + \mathrm{SINR}_k)^{-1}, \tag{1.8b}$$

respectively. Obviously, the rate, SINR, and MMSE are all continuous functions in the precoder $\boldsymbol{t} = \mathrm{vec}([\boldsymbol{t}_1, \ldots, \boldsymbol{t}_K])$. To highlight this dependency, these metrics are alternatively denoted by $r_k(\boldsymbol{t})$, $\mathrm{SINR}_k(\boldsymbol{t})$, and $\mathrm{MMSE}_k(\boldsymbol{t})$.

An important property is that the SINR and the rate monotonically decrease in $\delta > 0$ when scaling the precoder with $(1+\delta)^{-1}$, while the MMSE increases with this scaling for the precoder. This is seen by inserting $\boldsymbol{t}' = (1+\delta)^{-1}\boldsymbol{t}$ into the SINR (1.4) instead of $\boldsymbol{t}$, which finally results in the increased noise power $\sigma_k^2(1+\delta)^2$. The same result holds when scaling the channel $\boldsymbol{h}_k$ by this factor. This noise enhancement is a key property for imperfect transmitter CSI.

1.2 Quality-of-Service Optimization and Rate Balancing

The weighted sum throughput maximization is a common goal for data communication, but requires either spectrum or time sharing to guarantee a proportional partitioning of the data rate according to the users' service demands. This complicates the codebook design and the scheduling and, thereby, increases the signaling overhead for the data transmission. Furthermore, if the scheduling layer shall ensure the proportional partitioning on average, the data services must be insensitive to

[8] The advantageous use of the MSE and the MMSE compared to the rate and the SINR is especially for QoS feasibility and imperfect transmitter CSI as will be detailed in Sect. 1.2 and Chap. 4.

delays and dynamic supply changes. Therefore, QoS and balancing optimizations are employed instead when connectivity and short delays are important for a stable communication [66, Chapter 9].

1.2.1 Quality-of-Service Optimization Problem

An abstract formulation for the rate based QoS optimization with perfect CSI reads as

$$\min_{p\geq 0,\, \boldsymbol{t}} \; p \quad \text{s.t.} \quad p^{-1}\boldsymbol{t} \in \mathcal{P}, \quad \boldsymbol{r}(\boldsymbol{t}) \in \mathcal{R}. \tag{1.9}$$

The set $\mathcal{R} \subseteq \mathbb{R}_{+}^{K} \setminus \{0\}$ describes the target QoS region for the achievable rate vector $\boldsymbol{r}(\boldsymbol{t}) = [r_1(\boldsymbol{t}), \ldots, r_K(\boldsymbol{t})]^{\mathrm{T}}$ that comprises the data rates (1.3). Its Pareto frontier describes a lower bound for the target rates and it shall be unbounded above, that is, for every $\boldsymbol{r} \in \mathcal{R}$ the inequality $\boldsymbol{r}' \geq \boldsymbol{r} \geq 0$ implies $\boldsymbol{r}' \in \mathcal{R}$. This work particularly considers a target QoS region with minimal requirements $\rho_k \in \mathbb{R}_{++}$ on the user's rates, i.e.,

$$\mathcal{R} = \left\{ \boldsymbol{r} \in \mathbb{R}_{+}^{K} : r_k \geq \rho_k,\; k = 1, \ldots, K \right\}. \tag{1.10}$$

Other QoS regions from the literature either lower bound the SINRs (cf. [67]), e.g.,

$$\mathrm{SINR}_k = 2^{r_k} - 1 \geq \vartheta_k, \qquad k = 1, \ldots, K,$$

or upper bound the corresponding MMSEs as (cf. [68])

$$\mathrm{MMSE}_k = 2^{-r_k} \leq \varepsilon_k, \qquad k = 1, \ldots, K.$$

The compact convex set $\mathcal{P} \subset \mathbb{C}^{NK}$ has non-empty interior $\mathrm{int}(\mathcal{P}) \subset \mathbb{C}^{NK}$ and upper bounds the transmit power that is induced by $\boldsymbol{t}$. The set shall be symmetric around $0 \in \mathrm{int}(\mathcal{P})$, particularly, $\boldsymbol{t} \in \mathcal{P}$ imposes $\boldsymbol{t}' \in \mathcal{P}$ if $\boldsymbol{t}_k' = \mathrm{e}^{\mathrm{j}\phi_k}\boldsymbol{t}_k$ with $\phi_k \in [0, 2\pi)$, $k = 1, \ldots, K$. A basic example is the single sum power constraint set

$$\mathcal{P} = \left\{ \boldsymbol{t} \in \mathbb{C}^{NK} : \|\boldsymbol{t}\|_2^2 \leq P \right\}. \tag{1.11}$$

With (1.11) and (1.10), (1.9) is equivalent to the standard SINR constrained power minimization with targets $\vartheta_k = 2^{\rho_k} - 1$. This linear beamforming problem was already solved by [10, 67, 69–73]. A brief summary of the solution methods is provided later in this section. For the other representations of $\mathcal{P}$ in Sect. 1.4, (1.9) can be formulated as a min–max problem [74].

Two properties for the QoS problem (1.9) follow immediately from the monotonicity of the achievable rate (1.3) and the definitions of the target region $\mathcal{R}$ and the set $\mathcal{P}$ (see Sect. A.1). First, the objective is strictly positive for all feasible $\boldsymbol{t}$.[9] Second, the optimum is monotonic in a joint scaling of the rate requirement $\boldsymbol{r}(\boldsymbol{t}) \in \mathcal{R}$. It even approaches infinity if no feasible precoder $\boldsymbol{t}$ is found, because p is unbounded above. Therefore, feasibility tests are indispensable for (1.9).

A further important statement is due to the monotonicity properties of the rate (1.3) in a scaling of the individual beamformers. Namely, all constraints are satisfied with equality at the optimum when $\mathcal{R}$ consists of the per-receiver rate constraints (1.10) and $\mathcal{P}$ only includes the sum power requirement (1.11). This is a direct consequence of the discussions about resource allocation by [11, 75].

1.2.2 Rate Balancing Problem

In contrast to scaling the precoder, the rates are scaled in the balancing formulation:

$$\max_{\rho,\, \boldsymbol{t}} \ \rho \quad \text{s.t.} \quad \boldsymbol{t} \in \mathcal{P}, \quad \rho^{-1}\boldsymbol{r}(\boldsymbol{t}) \in \mathcal{R}. \tag{1.12}$$

This RB optimization is equivalent to a weighted sum rate maximization if the target region $\mathcal{R} \subseteq \mathbb{R}_+^K \setminus \{0\}$ only contained a lower bound for the positively weighted sum of the rates. For strictly positive rate targets, (1.12) can be rewritten into the familiar max–min problem formulation (cf. [76, Eq. (6)], [77, Eq. (1)], or [78, Eq. (2)])

$$\max_{\boldsymbol{t}} \min_{k} \frac{r_k(\boldsymbol{t})}{\rho_k} \quad \text{s.t.} \quad \boldsymbol{t} \in \mathcal{P}.$$

This RB formulation is equal to the common SINR balancing problem in [67, 73] if the rate targets ρ_k are equal and $\mathcal{P}$ consists only of a sum power constraint. Otherwise, SINR balancing with fixed targets only approximates the RB optimization since the rates are logarithmic functions of the SINRs (e.g., see [79]). Furthermore, the max–min RB formulation also allows for a zero objective, which is an undefined limit case for (1.12). Equivalence between the two problems holds despite this fact, because the optimum is provably non-zero as long as the channels are non-zero.[10]

A basic property of (1.12) is that its maximum is monotonic in a scaling of the precoder $\boldsymbol{t} \in \mathcal{P}$ (see Sect. A.1). This is analogous to the monotonicity of the QoS optimum in the scaling of the rates. Furthermore, the ratios r_k/ρ_k are balanced at the optimum if $\mathcal{P}$ only includes a sum power constraint (e.g., see [76]). For other power constraints, this condition only holds for those solutions that feature minimum sum power. Then, there can exist further solutions $\boldsymbol{t} \in \mathcal{P}$ with unbalanced rates.

---

[9] A precoder $\boldsymbol{t}$ is feasible if there is a $p \geq 0$ such that $p^{-1}\boldsymbol{t} \in \mathcal{P}$ and $\boldsymbol{r}(\boldsymbol{t}) \in \mathcal{R}$.

[10] See Sect. A.1 for more detailed discussions to the basic properties of (1.12).

1.2.3 Relation Between the Problems

Similar to the connection between the sum power minimization with SINR constraints and the SINR balancing optimization with a sum power constraint [73, Section V.A], also the rate based QoS optimization (1.9) and balancing formulation (1.12) are inverse to each other. Mathematically, this relation reads as

$$p(\rho(p_0)) = p_0$$
$$\rho(p(\rho_0)) = \rho_0,$$

where $p(\rho)$ is the optimum of (1.9) with the scaled rate constraint of (1.12), i.e.,

$$p(\rho) = \min\left\{p \in \mathbb{R}_+ : p^{-1}\boldsymbol{t} \in \mathcal{P},\ \rho^{-1}\boldsymbol{r}(\boldsymbol{t}) \in \mathcal{R},\ \boldsymbol{t} \in \mathbb{C}^{NK}\right\}, \tag{1.13}$$

and $\rho(p)$ denotes the optimum of (1.12) with the precoder scaling as in (1.9), i.e.,

$$\rho(p) = \min\left\{\rho \in \mathbb{R}_+ : p^{-1}\boldsymbol{t} \in \mathcal{P},\ \rho^{-1}\boldsymbol{r}(\boldsymbol{t}) \in \mathcal{R},\ \boldsymbol{t} \in \mathbb{C}^{NK}\right\}. \tag{1.14}$$

The proof in Sect. A.1 is by contradiction and only requires the monotonicity properties of the QoS and balancing optimizations (cf. [73, Proof of Theorem 3]).

This inverse relationship between the optimal values of the QoS and RB optimizations shows one way to solve either of the problems in (1.9) and (1.12) with the help of the other. For example, (1.12) can be solved via a series of QoS optimizations until the optimum $p(\rho)$ equals one. Since $p(\rho)$ is continuously monotonically increasing in ρ, a bisection procedure can be employed for the line search. The resulting precoder for the QoS optimization with $p(\rho) = 1$ is also a solution for the balancing optimization.

This procedure is computationally complex due to the computational burden for each QoS optimization. Therefore, direct algorithmic solutions are preferred.

1.3 Solutions for Perfect Transmitter Channel Knowledge

Solutions for QoS and RB optimization can be divided into two categories [73]:

1. using fixed point methods based on uplink–downlink dualities or
2. rewriting the constraints into conic (convex) formulations for the beamformers.

The dual uplink problem reformulation aims at a reduction of the decision variables. The beamformers become scaled receive equalizers, whose solutions are well established. Thus, only the scaling variables—the uplink power allocation—remain for the optimization. This enables fixed point methods for the initial precoder computation, which otherwise requires a conic (convex) problem reformulation and convex general purpose solvers in the downlink domain. A summary for these strategies is appended.

1.3.1 Uplink–Downlink Duality and Uplink Power Allocation

The concept of QoS based beamformer designs using (SINR) duality was already employed by [10] and [80]. It has become popular because of [81, 82], which both show that exactly the same SINR is achievable in the downlink and the uplink if and only if the sum transmit powers are equal in the two systems. This has enabled a transformation of the original downlink problems into tractable uplink formulations. The role of the transmitter and receivers is reversed in the dual uplink, that is, transmission is from K users to an N-antenna base station, which employs unit-norm equalizers $\boldsymbol{u}_k \in \mathbb{C}^N$, $k = 1, \ldots, K$ to separate the data streams. The dual uplink SINR reads as

$$\mathrm{SINR}_k^{(\mathrm{ul})} = \frac{|\boldsymbol{h}_k^{\mathrm{H}} \boldsymbol{u}_k|^2 \sigma_k^{-2} \lambda_k}{\sum_{i \neq k} |\boldsymbol{h}_i^{\mathrm{H}} \boldsymbol{u}_k|^2 \sigma_i^{-2} \lambda_i + \|\boldsymbol{u}_k\|_2^2}, \tag{1.15}$$

where $\lambda_k \geq 0$ is the transmit power allocated to data signal k, $k = 1, \ldots, K$. Therewith, the SINR balancing optimization and the QoS power minimization become a power allocation (e.g., see [67, 83]) and an equalizer design. Its solution is the normalized MMSE equalizer $\boldsymbol{u}_k = \boldsymbol{u}_{k,\mathrm{MMSE}} / \|\boldsymbol{u}_{k,\mathrm{MMSE}}\|_2$ with

$$\boldsymbol{u}_{k,\mathrm{MMSE}} = \left(\mathbf{I} + \sum_{i=1}^{K} \lambda_i \boldsymbol{h}_i \boldsymbol{h}_i^{\mathrm{H}}\right)^{-1} \boldsymbol{h}_k \sqrt{\lambda_k}. \tag{1.16}$$

Power allocations for the related Signal-to-interference ratio (SIR) balancing (without noise) have already been investigated by Zander [84], via solving a principle eigenvector problem.[11] In contrast, Yates [75] has taken the noise into account for the power allocation in SINR based QoS optimization. He has reformulated the constraints $\mathrm{SINR}_k^{(\mathrm{ul})} \geq \vartheta_k$, $k = 1, \ldots, K$, to the abstract vector constraint

$$\boldsymbol{\lambda} \geq \boldsymbol{I}(\boldsymbol{\lambda}; \boldsymbol{\vartheta}), \quad \boldsymbol{\vartheta} = [\vartheta_1, \ldots, \vartheta_K]^{\mathrm{T}} \in \mathbb{R}_+^K, \tag{1.17}$$

where $\boldsymbol{I} : \mathbb{R}_+^K \to \mathbb{R}_+^K$ is a continuous monotone map [85, Section 5.4]:

$$\begin{aligned} &\boldsymbol{I}(\boldsymbol{\lambda}; \boldsymbol{\vartheta}) > \mathbf{0} && \text{for} \quad \boldsymbol{\lambda} \geq \mathbf{0}, && (\textit{positivity}) \\ &\boldsymbol{I}(\boldsymbol{\lambda}; \boldsymbol{\vartheta}) \geq \boldsymbol{I}(\boldsymbol{\lambda}'; \boldsymbol{\vartheta}) && \text{for} \quad \boldsymbol{\lambda} \geq \boldsymbol{\lambda}' \geq \mathbf{0}. && (\textit{monotonicity}) \end{aligned}$$

[11] Detailed reviews on this topic are also provided by Schubert and Boche [11, 67, 83].

These properties enable a fixed point search for solving the QoS problem. In particular, the sequence $\boldsymbol{\lambda}^{(n)}$ that is induced by the vector fixed point iteration[12]

$$\boldsymbol{\lambda}^{(n+1)} = \boldsymbol{I}(\boldsymbol{\lambda}^{(n)}; \boldsymbol{\vartheta}) \tag{1.18}$$

converges under various conditions [86, Chapter 2], e.g., weak contraction. Yates explicitly examined the convergence when the mapping additionally satisfies

$$\alpha \boldsymbol{I}(\boldsymbol{\lambda}; \boldsymbol{\vartheta}) > \boldsymbol{I}(\alpha\boldsymbol{\lambda}; \boldsymbol{\vartheta}) \qquad \text{for} \quad \boldsymbol{\lambda} \geq \mathbf{0}, \quad \alpha > 1. \qquad (\textit{sublinearity})$$

Then, the following result is a consequence of [87, Corollary 1].

Theorem 1.1 ([75, Theorems 1 and 2]) *Let $\boldsymbol{I} : \mathbb{R}_+^K \rightarrow \mathbb{R}_+^K$ be a positive, sublinear, monotone, and continuous map. Then, the sequence $\boldsymbol{\lambda}^{(n)}$ induced by* (1.18) *globally converges to the unique fixed point $\boldsymbol{\lambda}^\star$ if existing. Otherwise, it diverges and its entries grow unbounded.*

Convergence to the unique minimizer is even ensured by an asynchronous fixed point update, e.g., one entry per iteration [75, Theorem 4]. Furthermore, the sequence in (1.18) increases monotonically when starting from the all-zero vector and it decreases when starting from $\boldsymbol{\lambda}^{(0)} > \boldsymbol{\lambda}^\star$ [75, Lemma 1 and 2]. Due to these two properties, the unique fixed point $\boldsymbol{\lambda}^\star$ has minimum one norm among all feasible points $\boldsymbol{\lambda} \geq \mathbf{I}(\boldsymbol{\lambda}; \boldsymbol{\vartheta})$. This shows the importance of the positive, sublinear, monotonic maps for QoS power minimization.

With the uplink SINRs as QoS measures, the entries of $\boldsymbol{I}(\boldsymbol{\lambda}; \boldsymbol{\vartheta})$ read as

$$I_k(\boldsymbol{\lambda}; \vartheta_k) = \vartheta_k \lambda_k / \, \mathrm{SINR}_k^{(\mathrm{ul})}, \quad k = 1, \ldots, K. \tag{1.19}$$

Since $I_k : \mathbb{R}_+^K \rightarrow \mathbb{R}_{++}$ is inherently the (effective) interference plus noise scaled by the target and $\boldsymbol{I}$ comprises these functions, these functions are herein denoted as *(individual) interference functions*.

Definition 1.1 ([75, Definition]) A (vector) interference function $\boldsymbol{I} : \mathbb{R}_+^K \rightarrow \mathbb{R}_+^K$, $\boldsymbol{I}(\boldsymbol{\lambda}; \boldsymbol{\vartheta})$ is called standard if it is a positive, monotonic, and sublinear map in $\boldsymbol{\lambda}$.

For fixed equalizers, the monotone map evenbecomes an affine function

$$\boldsymbol{I}(\boldsymbol{\lambda}; \boldsymbol{\vartheta}) = \boldsymbol{M}\boldsymbol{\lambda} + \boldsymbol{n}, \tag{1.20}$$

which is clearly standard as $[\boldsymbol{M}]_{k,i} \geq 0$ for $i, k = 1, \ldots, K$ and $\boldsymbol{n} > \mathbf{0}$. The interference function based on (1.19) is also standard when inserting the MMSE equalizers, which minimize the individual interference functions (see Sect. A.2).

[12] This fixed point map is also known as *Picard iteration* [86, Secion 1.2]. This reference also shows other iteration procedures with superior convergence that focus more on stability and contraction. The restriction to (1.18) for this work is due to its simplicity and variability for power allocation.

A standard interference function for the individual rate constraints $r_k^{(\mathrm{ul})} \geq \rho_k$, using the dual uplink data rate $r_k^{(\mathrm{ul})} = \log_2(1 + \mathrm{SINR}_k^{(\mathrm{ul})})$ instead of (1.3), reads as

$$I_k(\boldsymbol{\lambda}; \rho_k) = \rho_k \lambda_k / r_k^{(\mathrm{ul})}, \quad k = 1, \ldots, K. \tag{1.21}$$

Section A.2 shows this for the case of an arbitrary fixed equalizer, where the uplink SINR is written as the ratio of λ_k over some positively affine function in $\boldsymbol{\lambda}$. Since $I_k(\boldsymbol{\lambda}; \rho_k)$ is concave increasing (see Lemma A.3), also its minimum over the equalizers is monotonically increasing and concave in $\boldsymbol{\lambda}$. Hence, $\boldsymbol{I}(\boldsymbol{\lambda}; \boldsymbol{\rho})$ is also standard when inserting the MMSE equalizers (1.16) into the uplink rates.

A counterexample is the linear vector function for SIRs balancing from [88] that reads as (1.20) but with $\boldsymbol{n} = \mathbf{0}$. Schubert and Boche [11, 67, 83] studied the corresponding power iterations—the spectral radius computation—in detail for SINR balancing. To this end, they established the connection between the power iterations to the positive, monotone, sublinear maps from Yates [75] (e.g., see [11, Section 2.1]) by replacing the above *sublinearity* with the strict *linearity* condition [11, Definition 2.1]

$$\alpha \bar{\boldsymbol{I}}(\bar{\boldsymbol{\lambda}}; \boldsymbol{\vartheta}) = \bar{\boldsymbol{I}}(\alpha \bar{\boldsymbol{\lambda}}; \boldsymbol{\vartheta}) \qquad \text{for} \quad \bar{\boldsymbol{\lambda}} \geq \mathbf{0}, \quad \alpha \geq 0. \qquad (linearity)$$

For the above uplink SINR example, this type of interference function is obtained by extending the power vector with the noise, i.e., $\bar{\boldsymbol{\lambda}} = [\boldsymbol{\lambda}^{\mathrm{T}}, 1]^{\mathrm{T}}$, and defining the $(K + 1)$th interference term with the sum power constraint as

$$\bar{I}_{K+1}(\bar{\boldsymbol{\lambda}}; \boldsymbol{\vartheta}) = \frac{1}{P} \sum_{k=1}^{K} \frac{\vartheta_k \lambda_k}{\mathrm{SINR}_k^{(\mathrm{ul})}}.$$

Rewriting this term as an affine function in $\bar{\boldsymbol{\lambda}}$ and employing the linear formulation (1.20) for the remaining interference terms, this leads to the SINR balancing optimality criterion

$$\vartheta^{-1} \bar{\boldsymbol{\lambda}} = \bar{\boldsymbol{M}} \bar{\boldsymbol{\lambda}},$$

where $\bar{\boldsymbol{M}}$ only depends on the equalizers as remaining optimization variables. The thereof deduced repeated eigenvector computation for SINR balancing [89, Section IV.C] is also a method to construct feasible power allocations for the QoS optimization. This enables the super-linear converging fixed point search [67, 89, 90]

$$\boldsymbol{\lambda}^{(n+1)} = (\mathbf{I} - \boldsymbol{M}^{(n)})^{-1} \boldsymbol{n}^{(n)}, \tag{1.22}$$

where the update of $\boldsymbol{M}$ and $\boldsymbol{n}$ is due to an intermediate equalizer optimization.

The same authors have subsequently provided a much more comprehensive analysis for interference functions with either convex, concave, or log-convex shape [91], e.g., they extend the super-linear convergence property of (1.22) to the class of log-convex interference functions. For max–min SIR balancing, Vučić and Schubert [92] have shown convergence also for a normalized version of the fixed point mapping (1.18) when $\boldsymbol{I}$ satisfies the linearity property. Such an update is exploited in Sect. 5.4.

Alternatively to the fixed point search, the uplink power allocation for RB may be found via a convex Geometric program (GP) (e.g., see [93]). The uplink equalizer update and power allocation may be implemented in an alternating fashion. This strategy increases the computational complexity compared to the above fixed point search, but it allows to handle also sum rate requirements within $\mathcal{R}$ (e.g., see [94]). This made the GP formulation especially interesting for the MIMO downlink [95, 96].

1.3.2 Convex Problem Reformulations

Solutions for (1.9) and (1.12) were also found with the BC formulations. Bengtsson and Ottersten [71] and also Wiesel et al. [73] have shown that the exemplary sets $\mathcal{P}$ and $\mathcal{R}$ have conic convex formulations in $\boldsymbol{t}$. The solution for the QoS and the balancing optimizations is then obtained with convex interior-point solvers, e.g., SeDuMi [97] or SDPT3 [98] and the disciplined convex programming toolbox CVX [99].[13] These formulations are a basis for the robust formulations in subsequent chapters and are therefore revisited next.

A Second-order cone (SOC) representation for (1.9) reads as (cf. [73, 101, 104])

$$\begin{aligned}\min_{\boldsymbol{t},p}\; p \quad \text{s.t.} \quad & \left(\mathrm{Re}(\boldsymbol{h}_k^{\mathrm{H}}\boldsymbol{t}_k), \sqrt{2^{\rho_k}-1}\begin{bmatrix}\boldsymbol{T}_{\backslash\{k\}}^{\mathrm{H}}\boldsymbol{h}_k \\ \sigma_k\end{bmatrix}\right) \in \mathcal{L}^K, \quad k=1,\ldots,K, \\ & \left(p\sqrt{P}, \boldsymbol{t}\right) \in \mathcal{L}^{KN}\end{aligned} \tag{1.23}$$

where $\boldsymbol{T}_{\backslash\{k\}} = [\boldsymbol{t}_1,\ldots,\boldsymbol{t}_{k-1},\boldsymbol{t}_{k+1},\ldots,\boldsymbol{t}_K]$. The latter constraint represents the power limitation $p\sqrt{P} \geq \|\boldsymbol{t}\|_2$ and the previous constraints are the rewritten rate requirements

$$\mathrm{Re}\{\boldsymbol{h}_k^{\mathrm{H}}\boldsymbol{t}_k\} \geq \sqrt{2^{\rho_k}-1}\Big(\sum_{i\neq k}|\boldsymbol{h}_k^{\mathrm{H}}\boldsymbol{t}_i|^2+\sigma_k^2\Big)^{1/2}. \tag{1.24}$$

Equivalence to the rate constraints holds because all the approximations

$$\mathrm{Re}\{\boldsymbol{h}_k^{\mathrm{H}}\boldsymbol{t}_k\} \leq |\boldsymbol{h}_k^{\mathrm{H}}\boldsymbol{t}_k|, \quad k=1,\ldots,K \tag{1.25}$$

[13] The mathematical basis for conic optimization, the generalized inequality formulations, and the corresponding interior point methods is amongst others due to Nesterov and Nemirovski [100]. A brief tutorial on conic optimization can be found in [101], for example, and more detailed introductions and applications are provided by Boyd and Ben-Tal in their study books [102, 103].

can be satisfied with equality without changing the interference and transmit powers. Both, the power constraint and the right-hand side of (1.24), are independent with respect to exchanging either of the beamformers $\boldsymbol{t}_k$ with $\boldsymbol{t}'_k = \mathrm{e}^{\mathrm{j}\phi_k}\boldsymbol{t}_k$, $\phi_k \in [0, 2\pi)$.

The equality in approximation (1.25) cannot be ensured if the transmitter was uncertain about the channel states [101, 104, 105] or one beamformer serves a group of receivers [12]. An accepted approach in these cases is to formulate a Semidefinite program (SDP) [106, 107]. The SDP formulation of (1.9) has already been derived by Bengtsson and Ottersten [71]. Using the substitutes $\boldsymbol{Q}_k = \boldsymbol{t}_k\boldsymbol{t}_k^{\mathrm{H}}$, it reads as

$$\begin{aligned} \min_{\boldsymbol{Q},\phi\geq 0} \ \phi \quad \text{s.t.} \quad & \boldsymbol{h}_k^{\mathrm{H}}\Big(\frac{1}{2^{\rho_k}-1}\boldsymbol{Q}_k - \sum_{i\neq k}\boldsymbol{Q}_i\Big)\boldsymbol{h}_k \geq \sigma_k^2, \\ & \boldsymbol{Q}_k \in \mathcal{H}_+^N, \quad \operatorname{rank}(\boldsymbol{Q}_k) = 1, \quad k = 1,\ldots,K, \\ & \sum_{i=1}^{K} \operatorname{tr}(\boldsymbol{Q}_i) \leq \phi P, \end{aligned} \tag{1.26}$$

where $\boldsymbol{Q} = \operatorname{bdiag}(\boldsymbol{Q}_1,\ldots,\boldsymbol{Q}_K)$ and $\phi = p^2$. The initially quadratic terms in the beamformers are reformulated into linear constraints in the positive semidefinite transmit covariance matrices $\boldsymbol{Q}_k$, $k = 1,\ldots,K$, but with the drawback of the adherent non-convex rank-one constraints. Dropping these rank-one conditions,[14] the resulting SDP can be solved with the interior-point solvers of [99], for example, but on the cost of a post-processing to recast the beamformers if either of the solution covariance matrices is not rank-one [104].

The beamformers may be recast with a rank reduction scheme [108] from the transmit covariance matrices. For solutions that do not feature a rank-one solution, the dominant eigenvectors or candidates from a Gaussian process are suboptimal schemes for recasting beamformers [12, 107, 109].[15] A further post-processing power allocation then aims at compensating for the suboptimal candidate beamformers. This reconstruction scheme can still result in the dubious situation that the solver declares feasibility but no feasible beamformers are found for (1.9). Therefore, the direct beamformer optimization is preferred and a closed-form feasibility test is important.

This issue cannot arise for the RB problem (1.12), which SOC formulation reads as

$$\begin{aligned} \min_{\boldsymbol{t},\rho} \ \rho \quad \text{s.t.} \quad & \left(\operatorname{Re}(\boldsymbol{h}_k^{\mathrm{H}}\boldsymbol{t}_k), \sqrt{2^{\rho\rho_k}-1}\begin{bmatrix}\boldsymbol{T}_{\backslash\{k\}}^{\mathrm{H}}\boldsymbol{h}_k \\ \sigma_k\end{bmatrix}\right) \in \mathcal{L}^K, \quad k = 1,\ldots,K, \\ & \left(\sqrt{P}, \boldsymbol{t}\right) \in \mathcal{L}^{KN}. \end{aligned} \tag{1.27}$$

---

[14] This corresponds to a relaxation of the rank-one conditions to $\operatorname{rank}(\boldsymbol{Q}_i) \leq N$.

[15] This stochastic interpretation of the above substitution and relaxation for the quadratic terms results from $\operatorname{tr}(\boldsymbol{Q}_k\boldsymbol{R}) = \mathrm{E}_{\boldsymbol{\tau}_k}[\boldsymbol{\tau}_k^{\mathrm{H}}\boldsymbol{R}\boldsymbol{\tau}_k]$ for Gaussian vectors $\boldsymbol{\tau}_k \sim \mathcal{N}_{\mathbb{C}}(\boldsymbol{0}, \boldsymbol{Q}_k)$ (e.g., see [109]).

This RB formulation is not jointly convex in the beamformers and the rate factor ρ. Therefore, one may solve (1.27) via a line search over ρ and a QoS optimization for each candidate value. For example, a bisection over ρ can be employed that tests in each iteration whether (1.13) achieves $p(\rho) \leq 1$. Alternatively, (1.27) results in a generalized eigenvalue problem [73] for equal targets $\rho_1 = \rho_2 = \ldots = \rho_K$.

The Semidefinite cone (SDC) formulation for the RB problem reads as (cf. [71, 73])

$$\begin{aligned} \max_{\boldsymbol{Q},\rho} \ \rho \quad \text{s.t.} \quad & \boldsymbol{h}_i^{\mathrm{H}}\Big(\frac{1}{2^{\rho\rho_k}-1}\boldsymbol{Q}_i - \sum_{k\neq i}\boldsymbol{Q}_k\Big)\boldsymbol{h}_i \geq \sigma_k^2, \\ & \boldsymbol{Q}_i \in \mathcal{H}_+^N, \quad \operatorname{rank}(\boldsymbol{Q}_i) = 1, \quad i = 1,\ldots,K, \\ & \sum_{i=1}^{K} \operatorname{tr}(\boldsymbol{Q}_i) \leq P. \end{aligned} \tag{1.28}$$

Even though the constraint set becomes linear in the covariance matrices when the rank-one constraints are dropped, the interference constraints are still non-convex due to the additional variable ρ. Therefore, a sequence of QoS optimizations has to be solved for finding the solution in this case. When reconstructing the beamformers from covariance matrices which rank is larger than one, a rescaling of the beamformers and an adaption of ρ is required to ensure that the power constraint is satisfied and the proposed rates are achieved, respectively.

1.3.3 *Quality-of-Service Feasibility*

The QoS power minimization (1.9) can be infeasible if the transmitters' degree of freedom, i.e., the number of transmit antennas, is smaller than the number of served users and the users' target rates are too large. Then, there exists no precoder $\boldsymbol{t}$ that jointly satisfies all rate constraints.[16] For less users than antennas, ZF beamforming can theoretically achieve arbitrary high rate (e.g., see [110]).

If an infeasible rate tuple is targeted in the previous overloaded setup, the requested data services cannot be provided simultaneously to all users in the same frequency band due to the limiting interference. This event may be resolved by a scheduler, which allocates users to frequency bands with the available resources. To avoid an unnecessary waste of resources, e.g., by allocating one user per frequency band, a test to predict QoS infeasibility is inevitable [111]. Interior-point solvers can declare infeasibility for the above QoS problem [71] based on various convergence measures. However, their decision can be corrupted by numerical inaccuracies if the requirements are close to infeasibility.

[16] This is in contrast to the non-linear DPC case, where any finite rate targets are reachable [83].

More accurate feasibility results are obtained via the above uplink–downlink duality and the power iterations from (1.18). The global convergence and the monotonic increase for starting from the all-zero vector help to detect whether a fixed point exists below a maximum power limit. However, checking existence of a fixed point for unbounded transmit power in this way can fail. On the one hand, the number of iterations until convergence increases dramatically when the fixed point is close to the infeasibility bound [89, Figure 3]. On the other hand, the iteration (1.18) is not generally a contractive mapping [112].[17] Hence, it is unsure if the iteration truly converges to a suspected fixed point after a finite number of iterations.

Another iterative feasibility test via SIR balancing is provided in [11, Section 2.4]. With (1.20), this test can be written as the spectral radius computation

$$\vartheta^{-1}\boldsymbol{\lambda} = \boldsymbol{M}\boldsymbol{\lambda}, \tag{1.29}$$

where the coupling matrix $\boldsymbol{M}$ depends on the target SIRs and the equalizer $\boldsymbol{u}$, which itself is a function of $\boldsymbol{\lambda}$. The boundary of the SIR feasible region is characterized by the cases where the minimum spectral radius equals one and the achievable region consists of all SIR targets ϑ_k, $k = 1, \ldots, K$, where $\vartheta^{-1} < 1$ [113].

Unfortunately, this test does not circumvent numerical inaccuracies for separating feasible targets from infeasible ones when the system dimensions become large or the channels are ill-conditioned. An analytic expression of the feasibility bound for QoS power minimization with equal targets has been presented by Wiesel at al. [73, Proposition 1]. This bound only depends on the rank of the channel matrix $\boldsymbol{H} = [\boldsymbol{h}_1, \ldots, \boldsymbol{h}_K]$. For distinct targets and a single transmit antenna, Jorswieck et al. [114, Remark 1] have derived a sum MMSE bound for the feasible QoS region. These bounds are special cases of the characterization for the vector BC with proper complex signaling and regular channels by Hunger and Joham [65].

Definition 1.2 ([65, Section III]) The matrix $\boldsymbol{H} = [\boldsymbol{h}_1, \ldots, \boldsymbol{h}_K]$ is regular iff the columns $\boldsymbol{h}_i \in \mathbb{C}^N$ of any submatrix $\boldsymbol{H}_{\mathcal{I}} \in \mathbb{C}^{N\times|\mathcal{I}|}$, $i \in \mathcal{I} \subseteq \{1, \ldots, K\}$, and $|\mathcal{I}| \leq N$ are linearly independent. In other words, $\boldsymbol{H}$ must satisfy $\operatorname{rank}\{\boldsymbol{H}_{\mathcal{I}}\} = \min\{N, |\mathcal{I}|\}$ for all $\mathcal{I} \subseteq \{1, \ldots, K\}$.[18]

Lemma 1.1 ([65, Theorem III.1]) *If the channel matrix is regular, the QoS feasible set is given by the MMSE region*

$$\mathcal{F}_{\text{MMSE}} = \left\{\boldsymbol{\varepsilon} \in [0, 1]^K : \sum_{k=1}^{K} \varepsilon_k \geq K - N\right\}. \tag{1.30}$$

MSE points $\boldsymbol{\varepsilon} = [\varepsilon_1, \ldots, \varepsilon_K]^{\mathrm{T}}$ in the interior of $\mathcal{F}_{\text{MMSE}}$ are achievable with finite transmit power. Points at the boundary $\partial\mathcal{F}_{\text{MMSE}} = \{\boldsymbol{\varepsilon} \in [0, 1]^K : \sum_{k=1}^{K} \varepsilon_k = K - N\}$ require infinite power, and points strictly below $\partial\mathcal{F}_{\text{MMSE}}$ are unattainable.

---

[17] Whether (1.18) is contractive or not depends on the system realization (cf. [112, Theorem 2]).

[18] If $\exists \mathcal{I} \subseteq \{1, \ldots, K\} : \operatorname{rank}\{\boldsymbol{H}_{\mathcal{I}}\} < \min\{N, |\mathcal{I}|\}$, the matrix $\boldsymbol{H}$ is *singular* [115, Section III.C].

The authors further refined the formulation for this feasible MMSE region to non-regular channels [65, Theorem III.4] and the MIMO BC [115]. The latter region equals region (1.30). Hence, multi-stream transmission cannot increase the performance in the infinite power limit.

With the bijective mapping between rates and MMSEs (1.8), the region of feasible rate points $\boldsymbol{\rho} = [\rho_1, \ldots, \rho_K]^{\mathrm{T}}$ can be written as

$$\mathcal{F}_{\text{rate}} = \left\{ \boldsymbol{\rho} \in \mathbb{R}_+^K : \sum_{k=1}^{K} 2^{-\rho_k} \geq K - N \right\} \tag{1.31}$$

for a regular channel matrix $\boldsymbol{H}$. This representation allows an abstract description of QoS feasibility. The QoS optimization (1.9) is feasible if $\mathcal{F}_{\text{rate}} \cap \mathcal{R} \neq \emptyset$. Moreover, the optimum of (1.9) is finite if $\operatorname{int}(\mathcal{F}_{\text{rate}}) \cap \mathcal{R} \neq \emptyset$ and infinite transmit power is required if $\mathcal{F}_{\text{rate}} \cap \mathcal{R} \subseteq \partial \mathcal{F}_{\text{rate}}$. This abstract notation holds for QoS problems with general $\mathcal{R}$, even though this work mainly restricts to per-user rate constraints.

1.4 Beamformer Design with Multiple Power Constraints

The single sum power constraint (1.11) is often inadequate to model reality. In modern multi-antenna transmitters, each antenna is served by one RF amplifier and analog front-end. Transmit limitations are then due to the limited linearity of the amplifiers. Moreover, groups of antennas may be placed at different locations. A transfer of the payload between the physically separated devices is then impossible. For such systems, an appropriate model of the power limitations is crucial for the precoder design when the system shall work at its limits. If only a sum constraint were considered, the proposed rate would mismatch the actual achievable performance and the dynamic range of the amplifiers could be exceeded.

To handle a variety of power constraints within the QoS and RB optimizations, this work considers the generic convex quadratic model (cf. [63, 74, 79, 116–118]):

$$\sum_{i=1}^{K} \boldsymbol{t}_i^{\mathrm{H}} \boldsymbol{A}_{i,\ell} \boldsymbol{t}_i \leq P_\ell, \quad \ell = 1, \ldots, L. \tag{1.32}$$

The form of the matrices $\boldsymbol{A}_{i,\ell} \in \mathcal{H}_+^N$, $i = 1, \ldots, K$ depends on the application:

- Inequality (1.32) implies a *sum power* constraint if the matrices are $\boldsymbol{A}_{i,\ell} = \mathbf{I}_N$ for all $i = 1, \ldots, K$.
- For a constraint on the ℓth transmit *antenna*, the matrices are diagonal with only one non-zero element at the ℓth position, e.g., $\boldsymbol{A}_{i,\ell} = \mathbf{e}_\ell \mathbf{e}_\ell^{\mathrm{T}}$ for all $i = 1, \ldots, K$.
- *Per array constraints* result from (1.32) if the matrices are block diagonal with positive definite sub-blocks, e.g., $\boldsymbol{A}_{i,\ell} = \operatorname{bdiag}(0, \mathbf{I}_{N_\ell}, 0)$ for $i = 1, \ldots, K$.
- For *per-beamformer* constraints, $\boldsymbol{A}_{i,i} = \mathbf{I}_N$ and $\boldsymbol{A}_{i,\ell} = 0$ for $i \neq \ell$.

In either of these examples, the matrices $A_{i,\ell}$ are diagonal with elements being either one or zero. However, (1.32) also allows to model correlations for neighboring RF supply chains. Then, the matrices $A_{i,\ell}$ have non-zero off-diagonal entries.

For subsequent chapters, the compact convex set $\mathcal{P}$ is formed with (1.32), i.e.,

$$\mathcal{P} = \left\{ t \in \mathbb{C}^{KN} : \sum_{i=1}^{K} \| A_{i,\ell}^{1/2} t_i \|_2^2 \leq P_\ell,\ \ell = 1, \ldots, L \right\}, \tag{1.33}$$

where $A_{i,\ell} = A_{i,\ell}^{\mathrm{H}/2} A_{i,\ell}^{1/2}$ and $\mathrm{rank}\{\sum_{\ell=1}^{L} A_{i,\ell}\} = N$ must be satisfied to account for all transmit degrees of freedom. Other constraints for $\mathcal{P}$, e.g., one-norm or infinity-norm constraints, may be approximated by the two norm (e.g., see [119, Section 2.2]) and in this way included into the beamformer designs.

1.4.1 QoS Optimization and Balancing with Multiple Power Constraints

The SINR based QoS and balancing optimizations with (1.33) can still be solved via either a dual uplink problem formulation or a conic reformulation in the downlink. Uplink–downlink duality is now in terms of Lagrangian multiplier theory. Table 1.1 provides an overview for the available SINR dualities.[19]

Yu and Lan [74] originally derived the uplink dual problem for the SINR constrained QoS optimization with per-antenna constraints, but their derivation also fits for the above generic constraints. The resulting uplink QoS problem is a max–min optimization of the sum power $\sum_{i=1}^{K} \lambda_i$ subject to minimal requirements on the users' SINRs

Table 1.1 SINR uplink–downlink duality results for the vector broadcast channel

Year	Work	Problem	$\mathcal{P}$	Channel info.	Duality
1998	[10]	SINR, QoS	Sum power	Perfect CSI	Perron Frob.
1998	[80]	SINR bal.	Sum power	Perfect CSI	Perron Frob.
2002	[81]	Duality only	Sum power	Perfect CSI	Perron Frob.
2003	[5]	Sum capacity	Sum power	Perfect CSI	Perron Frob.
2007	[74]	SINR, QoS	Per-antenna	Perfect CSI	Lagrangian
2010	[104]	SINR, QoS	Sum power	Second moment	Lagrangian
2011	[120]	SINR bal.	Per-array	Perfect CSI	Perron Frob.
2012	[121]	SINR bal.	Per-array	Second moment	Lagrangian
2013	[118]	SINR bal.	Antenna, array	Second moment	Surrogate

[19]This table is not meant to be complete. Similar dualities may be found in other works as well.

$$\mathrm{SINR}_k^{(\mathrm{ul})} = \frac{|\boldsymbol{h}_k^{\mathrm{H}} \boldsymbol{u}_k|^2 \sigma_k^{-2} \lambda_k}{\sum_{i \neq k} |\boldsymbol{h}_i^{\mathrm{H}} \boldsymbol{u}_k|^2 \sigma_i^{-2} \lambda_i + \sum_{\ell=1}^{L} \mu_\ell \|\boldsymbol{A}_{k,\ell}^{1/2} \boldsymbol{u}_k\|_2^2} \tag{1.34}$$

with the equalizer $\boldsymbol{u}_k \in \mathbb{C}^N$ and the uplink powers $\lambda_i \in \mathbb{R}_+$, $i, k = 1, \ldots, K$. The outer maximization is with respect to the multipliers $\mu_\ell \in \mathbb{R}_+$, $\ell = 1, \ldots, L$ from the L downlink power constraints, which weight the noise portions within (1.34).

While the (inner) sum power minimization is again solved with MMSE equalizers and the previous fixed point framework, finding the dual variables μ_ℓ that maximize the minimal sum power subject to $\sum_{\ell=1}^{L} \mu_\ell P_\ell \leq 1$ requires an additional outer search. A subgradient projection update has been proposed for this task [74]. Alternatively, a semidefinite reformulation has been provided to jointly find all Lagrangian multipliers with a single SDP.

Dartmann et al. [118] have considered the downlink balancing problem for per-antenna and per-beamformer constraints, but for the "average" SINR[20], where $\boldsymbol{R}_k = \mathrm{E}[\boldsymbol{h}_k \boldsymbol{h}_k^{\mathrm{H}}] \in \mathcal{H}_+^N$ replaces $\boldsymbol{h}_k \boldsymbol{h}_k^{\mathrm{H}}$, $k = 1, \ldots, K$. Then, the duality proof of Yu and Lan [74] is no longer applicable. Therefore, the authors focus on surrogate duality for quasiconvex programming [122].[21] The surrogate dual SINRs have the same form as (1.15), but with $\boldsymbol{R}_k$ instead of $\boldsymbol{h}_k \boldsymbol{h}_k^{\mathrm{H}}$. Therefore, the SINR balancing power allocations are found as before and the SINR maximizing equalizers are the eigenvectors to the dominant eigenvalues of the matrix $\sum_{i=1}^{K} \lambda_i \sigma_i^{-2} \boldsymbol{h}_i \boldsymbol{h}_i^{\mathrm{H}} + \sum_{\ell=1}^{L} \mu_\ell \boldsymbol{A}_{i,\ell}$, given $\mu_k \in \mathbb{R}_+$, $k = 1, \ldots, K$. These multipliers are the solution for the outer minimization of the balanced SINRs. Dartmann et al. have suggested a subgradient method and a normalized fixed point update to find the solution.

Alternatively to the above solutions based on duality, convex conic constraint reformulations can be employed with (1.32). Its SOC formulation reads as

$$\left(\sqrt{P_\ell}, \boldsymbol{A}_\ell^{1/2} \boldsymbol{t}\right) \in \mathcal{L}^{NK},$$

where $\boldsymbol{A}_\ell^{1/2} = \mathrm{bdiag}(\boldsymbol{A}_{1,\ell}^{1/2}, \ldots, \boldsymbol{A}_{K,\ell}^{1/2})$, and the SDC representation reads as

$$\sum_{i=1}^{K} \mathrm{tr}(\boldsymbol{A}_{i,\ell} \boldsymbol{Q}_i) \leq P_\ell.$$

[20] We obtain these average SINRs by independently taking the mean of the nominator and the denominator if only second order channel information is available. These measures are commonly employed in the literature, even though they provide neither a lower nor an upper bound for R_k.

[21] Surrogate constrained programming is an approach to employ multiplier and duality theory for quasiconvex optimizations with multiple inequality constraints [123], where the standard Lagrangian duality theory does not generally apply. For the surrogate primal problem, the convex inequality constraints are linearly combined to a single scalar inequality constraint (cf. [122, 124]). The corresponding weights are the surrogate multipliers. This formulation allows for a simplified notion of duality.

The per-antenna version of the SOC constraint was used by Tölli et al. [79]. He has solved the SINR balancing and RB problems via a bisection, both subject to per-antenna constraints. The therein considered (MIMO) setup with per data-stream constraints is equivalent to the vector BC if the downlink receive filters are fixed.

1.5 Outline of the Chapters and the Contributions

In contrast to the perfect CSI assumption for the above literature review, this work focuses on the QoS power minimization and RB formulation for partial transmitter CSI. Only the channels' estimates $\bar{\boldsymbol{h}}_k$ and the statistics of the estimation errors are known for the beamformer design. Therefore, the receivers' rates (1.3) are stochastic variables. Their dependence on the channel realizations is highlighted by writing $r_k(\boldsymbol{t}, \boldsymbol{h}_k)$ for $r_k(\boldsymbol{t})$. The goal are robust QoS and RB beamformer designs that take the above error statistics into account.[22] Otherwise, the system performance would unpredictably degrade if we inserted the estimate $\bar{\boldsymbol{h}}_k$ into the algorithms for perfect CSI [128, 129].

The basis for robust beamforming is a channel error model that reflects reality and can be handled within the designs. Chapter 2 introduces two forms of the common Gaussian *channel knowledge model* that are considered in this work:

1. The additive error model $\boldsymbol{h}_k = \bar{\boldsymbol{h}}_k + \boldsymbol{e}_k$, e.g., which is the basic model in literature for terrestrial communication setups (e.g., [41, Chapter 2]), and
2. a multiplicative error model $\boldsymbol{h}_k = (1+\xi_k)\bar{\boldsymbol{h}}_k$, e.g., to approximate the first model.

These models mark two extreme cases with regard to the channels' covariance matrices. For very weak correlations of the channels' entries, the channels' covariance matrices are full-rank and may even be scaled identity matrices. Then, the first channel error model is appropriate. In contrast, the channel has approximately a rank-one covariance matrix for very strong correlations, which leads to the second error model. This chapter recapitulates the corresponding approximations.

With these models, this work also provides a trade-off between modeling accuracy, approximation capabilities, and computational tractability for the robust beamformer design. Tight bounds for the data rate are already required with the standard additive Gaussian error model. Without such bounds, the expressions for the achievable data rate include involved numerical integrations, which are intractable for state-of-the-art optimizations.

[22]The downlink is especially interesting due to the imposed correlations between the intended signal power and the interference within the users' SINRs. Both signal parts are distorted by the same fading channels. This is in contrast to the Multiple access channel (MAC) and Interference channel (IFC), where the distortions of the interference and the useful signal are independent (e.g., see [125–127]).

Increasing the dimensions of the vector BC, i.e., the number of antennas N and users K, is additionally challenging. Then, the complexity of the optimizations is not allowed to grow above that for perfect CSI. We achieve this with the multiplicative error model, which becomes accurate for fixed terminal SatCom.

Remark 1.1 Linear beamforming may be suboptimal dependent on the available CSI [18, 130]. For example, a scaled identity transmit covariance matrix outperforms rank-one beamforming in a PtP systems if no CSI is available. However, beamforming exploits the available degrees of freedom in multi-user systems as long as the channel direction information is sufficiently accurate.

The choice of the QoS metric for robust beamforming depends on the channels' coherence time and the quality of the estimation. For fast channel fading, the coherence time is so small that any transmitted data sequence experiences multiple channel realizations. In this case, ergodic rates are commonly employed (e.g., [126, 131–133]):

$$R_k(\boldsymbol{t}) = \mathrm{E}[r_k(\boldsymbol{t}, \boldsymbol{h}_k)], \qquad \forall k = 1, \ldots, K. \tag{1.35}$$

Closed forms for the ergodic rate are too difficult for optimizations when incorporating an additive channel error model (see Sect. 4.1). Then, the average MSEs

$$\mathrm{AMSE}_k(\boldsymbol{t}) = \mathrm{E}[\mathrm{MSE}_k(\boldsymbol{t}, \boldsymbol{h})], \qquad \forall k = 1, \ldots, K \tag{1.36}$$

and average MMSEs are tractable alternative metrics for common statistical channel knowledge models (e.g., [43, 116, 134–136]) and provide a lower bound for (1.35).

In contrast, if the coherence time is within the range of a sequence of transmitted signals in a slow-fading scenario, each data sequence experiences only one channel realization. As only an estimate for this realization and the statistics of the estimation errors are assumed to be known, data rate targets cannot be assured. However, one can restrict the probabilities for an outage event—that a receiver fails to achieve its target rate—to lie below an acceptable threshold. Denoting the threshold of receiver k by $\epsilon_k \in (0, 1)$, the outage constrained rates are defined as

$$\varrho_k(\boldsymbol{t}) = \max\{\varrho \in \mathbb{R}_+ : \Pr(r_k(\boldsymbol{t}, \boldsymbol{h}_k) \geq \varrho) \geq 1 - \epsilon_k\}, \qquad \forall k = 1, \ldots, K. \tag{1.37}$$

Relying on (1.37) instead of the rates (1.3), the QoS and RB problem formulations become chance constrained optimization problems (cf. [36, 44, 47, 51, 137, 138]).

Robust QoS and RB optimizations based on the above performance metrics are detailed in Chaps. 3–5. These chapters extend the pre-published contributions from [63, 139–147]. The beamformer design principles under study are also differentiated for the above multiplicative and standard additive (Gaussian) channel error models.

1.5.1 Contributions for Ergodic Rates with Multiplicative Channel Errors

Chapter 3 considers the QoS and RB beamformer designs with ergodic rates (1.35). General ergodic rate expressions (e.g., see [148] and [149]) are too complicated for the multi-user beamformer optimization. Therefore, the chapter's focus is on the structural less complex multiplicative channel model (cf. [139–143]).

Closed-form ergodic rate expressions are available for this transmitter CSI model. However, these terms do not preserve the logarithmic structure of (1.3), why an uplink–downlink duality and convex constraint reformulations are missing.

In order to solve the QoS and RB optimization with the available algorithms from perfect CSI, we verify the use of ZF beamforming and introduce tight lower and upper bounds for the ergodic rate.[23] Section 3.3 then addresses the QoS feasibility detection for overloaded setups—with less transmit antennas than users. We prove that the feasible ergodic rate region coincides with the feasible region (1.31) for the multiplicative channel error model. The proof provides a two-step construction scheme for strictly feasible beamformers, i.e., an approximate design with one of the tight ergodic rate bounds followed by a post-processing power allocation.

Starting from these feasible beamformers for the QoS power minimization, a Sequential convex search (SCS) (cf. [150]) guarantees convergence to a local optimizer by solving an SINR constrained QoS problem in each iteration. A similar SCS finds a local solution of the ergodic RB formulation within few iterations.

The QoS and RB optimization results of the above approximations are compared with the global optimum. A Branch and bound (BB) search similar to [16] finds the global optimum up to an acceptable accuracy (e.g., see [151]). Since the inherent computational complexity of this exhaustive search is exponential in the number of ergodic rate requirements, it is only suitable for benchmarks in small systems. Simulations show that the ergodic rate bounds are truly tight at low and high SNR regimes and that the SCS achieves the global optimum.

1.5.2 Contributions for Average MSEs with Additive Channel Errors

The motivation for QoS and balancing optimizations with the average MSE metric (1.36) in Chap. 4 is the close connection between the MSE and the mutual information [65, 152–159]. For example, a closed-form expression of the average MMSE is obtained via a single derivative of the corresponding data rate

[23] ZF beamforming for only a part of the receivers can help to reduce the computational complexity, which is especially interesting if the number of antennas N and users K are large and $N \geq K$.

formulations from [152, 160] in case of perfect receiver CSI but only covariance matrix information of the Gaussian channels at the transmitter.[24]

An Alternating convex search (ACS) of the precoder and the equalizers finds a local optimum for the MSE based QoS and balancing optimizations. In comparison to the straightforward MMSE equalizer update, the beamformer update is computational complex in the downlink. Uplink–downlink duality has proven to be an utmost useful tool to reduce this complexity. It translates the downlink precoder update to an MMSE filter calculation and a power allocation in the dual uplink that features efficient and reliable fixed point iterations. This connection was revealed by Shi et al. [68] using the SINR duality result from [5, 67]. Simplified versions of this duality for the Sum mean square error (SMSE) and per-receiver MSE were shown by Hunger et al. [161].

While these references impose a sum power constraint and perfect transmitter CSI, Chap. 4 studies the transceiver design for imperfect transmitter CSI and extends the duality to multiple power constraints. Related results for SMSE minimization with equal imperfect CSI at the transmitter and receivers have been revealed by Ding and Blostein [136] for a sum power constraint and by Bogale and Vandendorpe [116], who provided a duality relation for per-antenna and per-precoder constraints via an algorithmic argumentation. An average SMSE duality for imperfect transmitter CSI, perfect receiver CSI, and a single sum power constraint has also been shown by [162].

All these results are application examples of the uplink–downlink dualities from Sects. 4.3 and 4.5. We derive these relations for the QoS and max–min MSE optimizing precoder update concisely via Lagrangian multiplier theory for the weighted sum MSEs $\sum_{k\in\mathcal{G}_j} w_k\, \mathrm{AMSE}_k(\boldsymbol{t})$ of disjoint user subsets $\mathcal{G}_j \in \{1, \ldots, K\}$ and $j = 1, \ldots, K$. The duality takes into account the generalized power constraints of Sect. 1.4 and the imperfect transmitter CSI.[25] The dual optimizations consist of an inner power allocation (and receive equalizer design) and an outer worst-case noise search with respect to the dual variables corresponding to the primal power constraints. To solve the outer problem via a gradient projection method, the gradient is calculated in closed-form. This improves the computational efficiency compared to the solutions in [63, 63, 163, 164]. Due to strong duality, the primal reconstruction leads to an optimal precoder update.

Section 4.6 proves that the duality result is sufficiently general to cover also the weighted MSE approximation for the data rate [155]. Therewith, the suggested ACS also features a local search for RB with additive channel fading.

The duality also enables a discussion about the MSE feasibility region for imperfect CSI. The box constraint and sum constraint for the perfect CSI region (1.30) are only necessary requirements for achievable MSE targets ε_k, $k = 1, \ldots, K$. In

[24] These closed-form expressions were originally derived for the related MIMO models [157, 158].

[25] Even though the shown ACS considers only imperfect transmitter CSI, i.e., the receivers employ perfect CSI equalizers, the duality is also valid for imperfect transmitter and receivers CSI (cf. [116]).

other words, the Pareto bound of $\mathcal{F}_{\mathrm{MMSE}}$ is an outer bound for the feasible region with partial transmitter CSI. Two inner bounds are shown in Sect. 4.4.

For the balancing optimization, Sect. 4.5 especially also analyzes a paradigm for unequal MSE targets: Transmission to receivers with too high target values is switched off for stringent beamformer restrictions.[26] Thereby, different priorities for accessing the channel are established with the distinct MSE targets. The MSE balancing optimization softly switches on transmission to users with low demands when the available transmit power increases. This behavior strictly differs from SINR and rate balancing, where all the users are served, even for unequal targets (cf. [67, 166, 167]). Section 4.5 visualizes and explains this difference among RB, SINR balancing, and min–max MSE optimization. The individual steps of the suggested ACS strategy take care of this behavior, in that they allow to activate and deactivate transmission to users dependent on the objective.

Besides the effects of distinct MSE targets, the numerical results of Chap. 4 compare the MSE balancing performance for distinct power constraint sets and for decreasing the number of weighted sum MSE constraints. As a result, the performance of the average MSE measure varies only a little when employing a sum power constraint instead of per-antenna limitations. However, the dynamic power range of each antenna is tremendously increased for this modeling mismatch. When grouping the receivers' MSEs to weighted sum MSE constraints, e.g., to reduce the computational complexity, a mismatch to the per-user constraints is also visible in the receivers' performance.

1.5.3 Contributions for Outage Rate Requirements

Chance-constrained QoS and RB formulations are studied in Chap. 5, which starts with a summary on probability constrained optimization and applications for downlink transmission. For the multiplicative error model, the ϵ-outage rate (1.37) has the same logarithmic form as for perfect CSI, but with an enhanced noise power. The algorithms for QoS optimization and RB with perfect CSI also find the optima of the chance-constrained problems.

The remainder of Chap. 5 then focuses on the chance-constrained optimizations for additive channel errors. For this channel model, the known deterministic constraint reformulations [35, 36, 47, 104, 105, 138, 168–170] can be categorized into two types: (1) *uncertainty restrictions*, where the channel error is restricted to reside in a pre-defined region, and (2) *approximations with concentration inequalities*, which relate the probability for fulfilling the stochastic event $r_k(\boldsymbol{t}, \boldsymbol{h}) \geq \varrho$ with an average measure. The standard uncertainty approximation [36, 105]—with a spherical uncertainty region for the additive channel error—induces a very

[26]Examples of such stringent limitations are low power constraints and interference temperature constraints that set a threshold on the generated interference in a cognitive radio system [165].

conservative beamformer design. We propose an alternative uncertainty region in Sect. 5.5. It bounds the error that is orthogonal to the channels' mean. This promises considerable performance gains and an increased feasible target range compared to the spherical uncertainty region.

A popular concentration inequality approximation for Gaussian channel models is the Bernstein-type inequality applied by Wang et al. [171]. This deterministic approximation provides a convex constraint formulation only after a Semidefinite relaxation (SDR). Thus, it generally requires a beamformer reconstruction after solving the optimizations. As an alternative to this two-step procedure, we derive a constraint approximation for a direct beamformer optimization in Sect. 5.6, which is based on an MSE reformulation of the rate constraint $r_k(\boldsymbol{t}, \boldsymbol{h}) \geq \varrho$ and the second Bernstein-type inequality from [172]. The resulting constraint is biconvex in the beamformers and the equalizers. Thus, a (local) solution of the RB optimization is found via an ACS. This solution outperforms the uncertainty approximations if the transmit power is low or the number of receivers is large (see Sect. 5.7).

All these approximations are conservative in the sense that none of the probability requirements (1.37) is met. Thus, the QoS optimizations predicts a too large transmit power and the RB optimizations proposes much less data rate than would be achievable. The approximate QoS optimizations may even detect infeasibility when the original chance-constrained problem is feasible. While [138, 173] proposed a multi-step optimization with upscaled outage limits ϵ_k, $k = 1, \ldots, K$ to overcome this issue, we suggest a post-processing power allocation to meet the probability requirements (cf. [144]). The basis is a tractable probability evaluation, e.g., by Imhof [174]. This enables a QoS and RB based power allocation subject to the actual probability constraints (1.37) and, thus, to compensate partly for the performance losses of the approximate beamforming.

The obtained two-step optimization procedure even allows opportunistic beamformers as an input for the power allocation, e.g., the results with a multiplicative error approximation of the additive fading channels. A feasibility test detects whether the power allocation can still fulfill the original constraints. Section 5.7 shows a numerical example, where the post-processing compensates for the optimistic beamformers and, thereby, achieves a larger feasible region than for the conservative designs.

1.5.4 Applications to Satellite Communication

The high throughput demands in SatCom led to strengthened investigations for increasing the frequency reuse and serve a larger number of receivers in the same band. Interference from neighboring spotbeams and channel fading are then the main limitations for downlink communication [57].[27] To overcome these limits,

[27] Interference becomes severe for frequency reuse one as the same bands are used in all the cells.

adaptive physical layer beamforming is seen as a key technology to reliably serve terminals in neighboring regions and restrict the harmful interference [52–58, 60]. Chapter 6 extends these investigations by adapting the introduced ergodic and outage constrained beamformer designs to the SatCom model.

The robust physical layer design changes because the channels to the terminals inherently suffer from various types of fading [59]. Section 6.1 recapitulates the channel characteristics and fading impairments, which led to the SatCom specific joint additive and multiplicative channel error model from [52]. The model encounters for atmospheric attenuation and scintillation—rain fading—and local scattering around the terminals. The emphasis on the fading varies dependent on the transmit scenario.

For fixed terminals, the low rank of the channels' covariance matrices results in a joint multiplicative fading that combines the attenuation and scattering. The ergodic rate based QoS and balancing optimization from Chap. 3 and the direct deterministic reformulation of the outage constraints from Sect. 5.2 are still valid. These bounds enable a fully adaptive robust beamformer optimization of more than a hundred spotbeams that cover Europe (cf. Sect. 6.1).

When the covariance matrices of the downlink channels have a rank larger than one, e.g., for mobile receivers, the average MMSE balancing from Chap. 4 replaces the RB optimization. The MSE metric was already analyzed for SatCom, but with ZF, Maximum ratio transmission (MRT), and Regularized zero-forcing (RZF) precoding [56, 60–63]. In contrast, we directly optimize the MSEs for the multi-spotbeam SatCom system at hand. The multiplicative channel error is handled with the tight ergodic rate lower bound from Sect. 3.2 and the MMSE metric under study relates to the remaining additive channel error. This keeps the optimization robust with respect to both types of fading and guarantees the proposed performance.

Simulations with this MSE metric are performed for a mobile terminal S-band model. The optimization remains tractable for a hundred spotbeams when restricting to deterministic equalizers. The ACS with perfect CSI receivers remains tractable up to 19 neighboring receivers for multi-spotbeam SatCom. The effects for employing per-antenna constraints instead of a sum power constraint are shown for a small number of receivers. While the sum power constraint provides a good estimate for the performance, it is unable to deliver appropriate beamformers for SatCom.

The epsilon-outage rate ϱ_k (1.37) becomes the metric of interest for SatCom (e.g., see [64]) if the channels remain constant during a transmit time slot but change between the training and transmission phase. While Chap. 5 considers either multiplicative or additive channel errors, the outage constrained RB optimization for SatCom has to cope with both channel errors. To apply the available approximations, the outage probabilities are separated with respect to the multiplicative attenuation and the additive fading (see Sect. 6.5). This separation decreases the approximation accuracy, but allows to trade-off the probability bounds between these error types. Section 6.5 compares the optimized outage probability trade-off in small systems with an equal trade-off for all receivers. The results show that the equal trade-off is sufficiently accurate for SatCom services with similar outage requirements and statistics for the rain attenuation. This is required to reduce the computational complexity of the beamformer designs such that a SatCom scenario with seven neighboring spotbeams and terminals becomes tractable. The actual outage probabilities from these conservative beamformer optimizations are still far below the requirements. Therefore, a post-processing is required to increase the achievable data rates also for SatCom.

Chapter 2
Models for Incomplete Channel Knowledge

The quality of the channel knowledge is generally asymmetric at the receivers' and the transmitter's side of a vector BC. The receivers can exploit accurate knowledge about the received signals' autocorrelation and variance for equalization and decoding. Achieving a similar quality of transmitter channel knowledge for downlink wireless communication is unrealistic when the channel is subject to fading [175, Chapter 2]. Transmitter CSI is gained via pilot-based training (e.g., see [176]), either in the uplink for Time division duplex (TDD) systems or in the downlink for systems in Frequency division duplex (FDD) mode with feedback.[1] This results in errors between the channel and the transmitter's estimate. For TDD systems, the errors are due to imperfect uplink channel estimation and delayed utilization of the estimates. For the FDD model, the errors stem from estimation at the receivers and the limited feedback to the transmitter. Feedback delays additionally degrade the channel information in this scenario.

This work considers two generic models to account for the errors of the channel estimates. The standard additive (Gaussian) error model reads as[2]

$$\boldsymbol{h} = \bar{\boldsymbol{h}} + \boldsymbol{e}, \tag{2.1}$$

where $\bar{\boldsymbol{h}} \in \mathbb{C}^N$ is the channel mean or estimate and $\boldsymbol{e} \in \mathbb{C}^N$ is the additive circular symmetric Gaussian channel error, i.e., $\boldsymbol{e} \sim \mathcal{N}_{\mathbb{C}}(\boldsymbol{0}, \boldsymbol{C}_{\boldsymbol{e}})$. This is a standard model for terrestrial mobile communication in an urban environment (e.g., see [36, 43, 133, 177–179]). The error results from multi-path fading (cf. Sect. 2.1).

[1]The resources for training and transmission are standardized for current systems (cf. [128]).

[2]We neglect the receiver index k for simplicity of the channel estimation in this chapter.

A. Gründinger, *Statistical Robust Beamforming for Broadcast Channels and Applications in Satellite Communication*, Foundations in Signal Processing, Communications and Networking 22, https://doi.org/10.1007/978-3-030-29578-3_2

This assumptions distinguish (2.1) from the second multiplicative error model

$$h = (1 + \xi)\bar{h}, \tag{2.2}$$

which applies for terrestrial mobile communication in a rural environment and SatCom systems (see Chap. 6).[3] The mean represents the dominant path of the channel and the multiplicative error $\xi \in \mathbb{C}$ models the uncertainty in the fluctuating gain. For example, the dominant path may suffer from shadow fading. Then, an accurate channel estimate and feedback to the transmitter will be available after training only if the attenuation changes sufficiently slow, [181]. Otherwise, the fading and the system delays result in outdated CSI.

The exact expressions for the channel estimate and the parameters of the imposed error statistics depend on the underlying *deterministic* or *stochastic* model [1, 175]. The covered standard cases are shown next.

2.1 Additive Error Models

The abovementioned additive channel errors can result from multi-path fading and channel estimation. The quality of the channel estimate and the covariance matrix of the zero-mean (complex) Gaussian error $\boldsymbol{e} \sim \mathcal{N}_{\mathbb{C}}(\boldsymbol{0}, \boldsymbol{C}_{\boldsymbol{e}})$ depend on the imposed propagation model and the estimation process. Basic examples [36, 43, 133, 178, 179] and the influence of limited and delayed feedback [43, 182] are detailed next.[4]

A generic model of the received signal for the training reads as (e.g., see [185])

$$\boldsymbol{y}_{\text{tr}} = \boldsymbol{S}\boldsymbol{h} + \boldsymbol{n}_{\text{tr}}, \tag{2.3}$$

where $\boldsymbol{S} \in \mathbb{C}^{T \times N}$ comprises the training sequence and $\boldsymbol{n}_{\text{tr}} \in \mathbb{C}^{T}$, $\boldsymbol{n}_{\text{tr}} \sim \mathcal{N}_{\mathbb{C}}(\boldsymbol{0}, \boldsymbol{C}_{\text{tr}})$ is the training noise. If the noise is uncorrelated in space and time, $\boldsymbol{C}_{\text{tr}} \in \mathbb{C}^{T \times T}$ will be a scaled identity matrix, i.e., $\boldsymbol{C}_{\text{tr}} = \sigma_{\text{tr}}^2 \mathbf{I}$. The structure of $\boldsymbol{S}$ depends on the orthogonal training sequences and whether (2.3) comprises the received signal of either the uplink or the downlink training. For both cases, $\boldsymbol{S}$ shall be tall and have full column rank, i.e., $T \geq N$ and $\text{rank}\{\boldsymbol{S}\} = N$.

[3]SatCom considers a combination of additive and multiplicative fading, but with a low-rank covariance matrix $\boldsymbol{C}_{\boldsymbol{e}}$ [180]. Therefore, (2.2) well approximates the SatCom channel in Sect. 6.1.

[4]Other forms of limited feedback can be found in the literature. For example, instantaneous channel norm feedback and its application for downlink beamforming is discussed in [183, 184].

2.1.1 Deterministic Channel Models

Given (2.3), the channel estimate $\bar{\boldsymbol{h}}$ is a function of the observation $\boldsymbol{y}_{\mathrm{tr}}$. For example, the observation can be used to parameterize the Line of sight (LoS) path and multipath propagation of ray-tracing models [175, Chapter 2]. A joint estimation of the angles of departure and arrival, the delays, and the losses of the main paths can be performed via ESPRIT-like algorithms [186]. This approach has been investigated for rural environments [187], indoor environments [188, 189], street paths [190], and land-mobile SatCom [59, 191–193]. The channel estimation errors in (2.1) are then from uncaptured rays.

If ray-tracing models are inaccurate because the number of scatterers is large, the receivers are moving, or the reflecting surfaces are non-smooth, a deterministic finite impulse response model can be employed for the channel [175, Chapter 2]. This leads to a Maximum likelihood (ML) channel estimation based on $\boldsymbol{y}_{\mathrm{tr}}$. For Gaussian noise and known $\boldsymbol{C}_{\mathrm{tr}}$,[5] the ML estimation simplifies to the least-squares optimization [194, 195]

$$\bar{\boldsymbol{h}}_{\mathrm{ML}} = \arg\min_{\boldsymbol{h}} \|\boldsymbol{C}_{\mathrm{tr}}^{-1/2}(\boldsymbol{y}_{\mathrm{tr}} - \boldsymbol{S}\boldsymbol{h})\|_2^2. \tag{2.4}$$

The unbiased least-squares solution reads as

$$\bar{\boldsymbol{h}}_{\mathrm{ML}} = \boldsymbol{W}_{\mathrm{ML}}\boldsymbol{y}_{\mathrm{tr}} = (\boldsymbol{S}^{\mathrm{H}}\boldsymbol{C}_{\mathrm{tr}}^{-1}\boldsymbol{S})^{-1}\boldsymbol{S}^{\mathrm{H}}\boldsymbol{C}_{\mathrm{tr}}^{-1}\boldsymbol{y}_{\mathrm{tr}} = \boldsymbol{h} + (\boldsymbol{S}^{\mathrm{H}}\boldsymbol{C}_{\mathrm{tr}}^{-1}\boldsymbol{S})^{-1}\boldsymbol{S}^{\mathrm{H}}\boldsymbol{C}_{\mathrm{tr}}^{-1}\boldsymbol{n}_{\mathrm{tr}} \tag{2.5}$$

and the covariance matrix of the estimation error $\boldsymbol{e}_{\mathrm{ML}} = \boldsymbol{h} - \bar{\boldsymbol{h}}_{\mathrm{ML}}$ becomes

$$\boldsymbol{C}_{\boldsymbol{e}_{\mathrm{ML}}} = \mathrm{E}[\boldsymbol{e}_{\mathrm{ML}}\boldsymbol{e}_{\mathrm{ML}}^{\mathrm{H}}] = (\boldsymbol{S}^{\mathrm{H}}\boldsymbol{C}_{\mathrm{tr}}^{-1}\boldsymbol{S})^{-1}. \tag{2.6}$$

Least-squares channel estimation is also employed if the stochastic observation model is unknown. The even less complex correlation estimate (e.g., [185])

$$\bar{\boldsymbol{h}}_{\mathrm{ML}} = T\sigma_s^{-2}\boldsymbol{S}^{\mathrm{H}}\boldsymbol{y}_{\mathrm{tr}} = \boldsymbol{h} + T\sigma_s^{-2}\boldsymbol{S}^{\mathrm{H}}\boldsymbol{n}_{\mathrm{tr}}$$

results from (2.5) for orthogonal training sequences and white noise, in particular $\boldsymbol{S}^{\mathrm{H}}\boldsymbol{S} = T^{-1}\sigma_s^2\mathbf{I}_N$ and $\boldsymbol{C}_{\mathrm{tr}} = \sigma_{\mathrm{tr}}^2\mathbf{I}_T$. In this case, the covariance matrix of the error $\boldsymbol{e}_{\mathrm{ML}} = T\sigma_s^{-2}\boldsymbol{S}^{\mathrm{H}}\boldsymbol{n}_{\mathrm{tr}}$ is

$$\boldsymbol{C}_{\boldsymbol{e}_{\mathrm{ML}}} = T\sigma_s^{-2}\sigma_{\boldsymbol{n}}^2\mathbf{I}_N.$$

[5] If the noise covariance matrix is jointly estimated with the channel [194] the ML estimate is $\bar{\boldsymbol{h}}_{\mathrm{ML}} = (\boldsymbol{S}^{\mathrm{H}}\boldsymbol{S})^{-1}\boldsymbol{S}^{\mathrm{H}}\boldsymbol{y}_{\mathrm{tr}}$ and the error covariance is computed in the long term [41, Chapter 3].

2.1.2 Stochastic Channel Models

The ML channel estimation may lead to unwanted large estimation errors for subspaces with weak gains, because it does neither take into account a special structure of the channel nor correlations between its entries. If accurate knowledge of this long-term channel statistics is available,[6] it will be incorporated into a stochastic channel model.

Let the channel be $\boldsymbol{h} \sim \mathcal{N}_{\mathbb{C}}(\boldsymbol{m}, \boldsymbol{C}_h)$, with mean $\boldsymbol{m} = \mathrm{E}[\boldsymbol{h}]$ and covariance matrix $\boldsymbol{C}_h = \mathrm{E}[(\boldsymbol{h} - \boldsymbol{m})(\boldsymbol{h} - \boldsymbol{m})^{\mathrm{H}}]$. This model directly represents (2.1), i.e.,

$$\boldsymbol{h} \simeq \boldsymbol{m} + \boldsymbol{C}_h^{1/2}\boldsymbol{w}, \tag{2.7}$$

where $\boldsymbol{w} \sim \mathcal{N}_{\mathbb{C}}(\boldsymbol{0}, \mathbf{I}_N)$ is the *whitened* error and $\boldsymbol{C}_h^{1/2}$ the matrix square root of $\boldsymbol{C}_h = \boldsymbol{C}_h^{1/2}\boldsymbol{C}_h^{\mathrm{H}/2}$. The two most prominent examples are Rayleigh and Rician fading channels. Rayleigh fading channels are zero-mean with i.i.d. Gaussian entries, i.e., $\boldsymbol{h}_{\mathrm{Rayleigh}} \sim \mathcal{N}_{\mathbb{C}}(\boldsymbol{0}, \sigma_h^2\mathbf{I})$, and suit for non-LoS environments with a large number of scatterers. In LoS scenarios, the Rician model describes the influence of the dominant channel mean as (e.g., see [1])

$$\boldsymbol{h}_{\mathrm{Rician}} = \sqrt{\frac{\kappa}{\kappa+1}}\bar{\boldsymbol{z}} + \sqrt{\frac{\kappa}{\kappa+1}}\tilde{\boldsymbol{z}}, \tag{2.8}$$

where $\bar{\boldsymbol{z}} \in \mathbb{C}^N$ and $\tilde{\boldsymbol{z}} \sim \mathcal{N}_{\mathbb{C}}(\boldsymbol{0}, \boldsymbol{C}_z)$ are the channel components in the direct and the scattered paths, respectively, and κ marks the ratio between their average gains.

Pilot-based training and the MMSE estimation [195]

$$\bar{\boldsymbol{h}}_{\mathrm{CM}} = \arg\min_{\bar{\boldsymbol{h}}} \mathrm{E}_{\boldsymbol{h}|\boldsymbol{y}}\left[\|\boldsymbol{h} - \bar{\boldsymbol{h}}\|_2^2\right] \tag{2.9}$$

improve the channel knowledge. The solution is the conditional mean estimator $\bar{\boldsymbol{h}}_{\mathrm{CM}} = \mathrm{E}_{\boldsymbol{h}|\boldsymbol{y}}[\boldsymbol{h}]$ and the error covariance matrix reads as $\boldsymbol{C}_{\mathrm{CM}} = \mathrm{E}_{\boldsymbol{h}|\boldsymbol{y}}[(\bar{\boldsymbol{h}}_{\mathrm{CM}} - \boldsymbol{h})(\bar{\boldsymbol{h}}_{\mathrm{CM}} - \boldsymbol{h})^{\mathrm{H}}]$. For jointly Gaussian distributed $\boldsymbol{y}_{\mathrm{tr}}$ and $\boldsymbol{h}$, the MMSE estimation (2.9) results in [195]

$$\bar{\boldsymbol{h}}_{\mathrm{MMSE}} = \boldsymbol{m} + (\boldsymbol{C}_h\boldsymbol{S}^{\mathrm{H}}\boldsymbol{C}_{\mathrm{tr}}^{-1}\boldsymbol{S} + \mathbf{I})^{-1}\boldsymbol{C}_h\boldsymbol{S}^{\mathrm{H}}\boldsymbol{C}_{\mathrm{tr}}^{-1}(\boldsymbol{y}_{\mathrm{tr}} - \boldsymbol{S}\boldsymbol{m}) \tag{2.10}$$

and the error $\boldsymbol{e}_{\mathrm{MMSE}} = \boldsymbol{h} - \bar{\boldsymbol{h}}_{\mathrm{MMSE}}$ has the covariance matrix [195]

---

[6]The sample covariance matrix from the uplink training procedure can be employed for TDD systems [41, Chapter 3]. For FDD systems, the long-term estimate for the error covariance matrix can be fed back to the transmitter on a larger time scale than the channel estimates. Alternatively, the covariance matrix may be computed at the downlink transmitter using the channel covariance matrix from the uplink frequency bands [196, 197].

$$C_{e_{\text{MMSE}}} = C_h - C_h S^{\text{H}} (S C_h S^{\text{H}} + C_{\text{tr}})^{-1} S C_h. \tag{2.11}$$

For zero-mean e.g., Rayleigh fading, $\bar{h}_{\text{MMSE}}$ even becomes linear in y_{tr}, i.e.,

$$\bar{h}_{\text{LMMSE}} = W_{\text{LMMSE}} y_{\text{tr}} = (C_h S^{\text{H}} C_{\text{tr}}^{-1} S + \mathbf{I})^{-1} C_h S^{\text{H}} C_{\text{tr}}^{-1} y_{\text{tr}}, \tag{2.12}$$

which is also obtained by regularizing the objective from (2.4) with $\|C_h^{-1/2} h\|_2^2$ [41]. Since $\bar{h}_{\text{LMMSE}}$ exploits the channel's second order statistics, it achieves a better bias–variance trade-off for the estimation error than $\bar{h}_{\text{ML}}$ [41, Chapter 2]. The exact relation between the MMSE estimate and the ML estimate is provided by [185].

2.1.3 *Quantization Errors and Delays*

The above models represent the estimated channel knowledge, but miss the limited feedback for FDD systems and the system delays. For limited feedback rate, only a compressed version of the estimate is sent to the transmitter [198].[7] The quantizer maps the estimate $\bar{h}_{\text{EST}}$ to one out of a finite number of representatives [198], i.e.,

$$\bar{h}_{\text{Q}} = \bar{h}_{\text{EST}} + n_{\text{Q}}. \tag{2.13}$$

Clearly, the quantization noise $n_{\text{Q}} \in \mathbb{C}^N$ is non-Gaussian. The mapping of a Vector quantization (VQ) [129, 201] minimizes some distortion metric, e.g., the MSE [198]. The stochastic properties of n_{Q} are difficult to describe in this case. For a Scalar quantization (SQ) of $\bar{h}$, the real and imaginary parts of n_{Q} are almost uniformly distributed [43]. Since the resulting probability expressions are too complicated for the precoder design, this work restricts to a Gaussian approximation $n_{\text{Q}} \sim \mathcal{N}_{\mathbb{C}}(\mathbf{0}, C_{n_{\text{Q}}})$.

If the transmitter cannot exploit this fed back information within the same coherent time block, this will further increase the error. Jake's temporal correlation model describes the increase of the errors covariance for such delays, by modeling the channel's correlation from time slots $n - d$ (delay d) and n as (e.g., [43, 202])

$$\mathrm{E}\left[(h[n] - m)(h[n-d] - m)^{\text{H}}\right] = J_0(\omega d) C_h,$$

where $\omega = 2\pi \frac{f_{\text{doppler}}}{f_{\text{slot}}}$, with maximum Doppler frequency f_{doppler} and coherence rate f_{slot} in slots, and J_0 is the 0th order Bessel function of the first kind [203, Chapter 9].

This correlation can additionally be included into the additive channel error model if the transmitter only exploits the latest information, e.g., $\bar{h}_{\text{Q}}[n - d]$, with

[7] Either a quantized estimate or the codebook entry of its representation is fed back [199]. A summary of the available feedback approaches provides in the overview by Love et. al [200].

delay d.[8] Then, the conditional channel estimate $\mathrm{E}[\boldsymbol{h}[n]|\bar{\boldsymbol{h}}_{\mathrm{Q}}[n-d]]$ for time slot n is

$$\bar{\boldsymbol{h}}[n|n-d] = \boldsymbol{m} + J_0(\omega d)\boldsymbol{C}_{\boldsymbol{h}}\boldsymbol{S}^{\mathrm{H}}\boldsymbol{W}^{\mathrm{H}}\boldsymbol{C}_{\bar{\boldsymbol{h}}_{\mathrm{Q}}}^{-1}(\bar{\boldsymbol{h}}_{\mathrm{Q}}[n-d] - \boldsymbol{m}) \tag{2.14}$$

and the covariance matrix of $\bar{\boldsymbol{h}}_{\mathrm{Q}}[n-d] = \bar{\boldsymbol{h}}[n-d] + \boldsymbol{n}_{\mathrm{Q}}[n-d]$ becomes

$$\boldsymbol{C}_{\bar{\boldsymbol{h}}_{\mathrm{Q}}} = \boldsymbol{W}(\boldsymbol{S}\boldsymbol{C}_{\boldsymbol{h}}\boldsymbol{S}^{\mathrm{H}} + \boldsymbol{C}_{\mathrm{tr}})\boldsymbol{W}^{\mathrm{H}} + \boldsymbol{C}_{\boldsymbol{n}_{\mathrm{Q}}}.$$

Here, $\boldsymbol{W}$ stands for either of the linear filters for the ML or MMSE estimation. Using (2.14), the transmitter's error model for the delayed channel information is

$$\boldsymbol{h}[n] = \bar{\boldsymbol{h}}[n|n-d] + \boldsymbol{e}[n|n-d] \tag{2.15}$$

with the complex Gaussian error $\boldsymbol{e}[n|n-d] \sim \mathcal{N}_{\mathbb{C}}\left(\boldsymbol{0}, \boldsymbol{C}_{\boldsymbol{h}[n]|\bar{\boldsymbol{h}}_{\mathrm{Q}}[n-d]}\right)$ and

$$\boldsymbol{C}_{\boldsymbol{h}[n]|\bar{\boldsymbol{h}}_{\mathrm{Q}}[n-d]} = \boldsymbol{C}_{\boldsymbol{h}} - J_0^2(\omega d)\boldsymbol{C}_{\boldsymbol{h}}\boldsymbol{S}^{\mathrm{H}}\boldsymbol{W}^{\mathrm{H}}\boldsymbol{C}_{\bar{\boldsymbol{h}}_{\mathrm{Q}}}^{-1}\boldsymbol{W}\boldsymbol{S}\boldsymbol{C}_{\boldsymbol{h}}. \tag{2.16}$$

We see that only the statistic channel information remains for (2.15) if d is large. Then, the estimate in (2.14) equals $\boldsymbol{m}$ and the covariance matrix in (2.16) equals $\boldsymbol{C}_{\boldsymbol{h}}$.

2.2 Multiplicative Error Models

The error model (2.2) implies that the spatial channel characteristic is known, but the fading gain is unknown. The error is due to the experienced changes in the scattering, attenuation, and shadowing from obstacles in the vicinity of the receiver.

2.2.1 *Deterministic Channel and Shadow Fading*

For the ray-tracing model, (2.2) results from scenarios with either one strictly dominant LoS path from the transmitter to the receiver or multiple paths that are so close such that they can hardly be separated.[9] The channel error is then due to fluctuations of the shadowing and diffraction from obstacles. It is modeled by a log-dB-normal attenuation [175, Section 2.7], i.e., $\zeta = |1+\xi|^{-2}$ in Decibel (dB) is distributed as

[8] The case with many estimates is provided by [43].

[9] The separation is by the direction of departure at the multi-antenna transmitter, which can be estimated as the uplink direction of arrival using ESPRIT-like algorithms [186].

$$\ln(\zeta_{\mathrm{dB}}) \sim \mathcal{N}(m, \sigma_\zeta^2). \tag{2.17}$$

If $\zeta \in \mathbb{R}$ changes slowly, it can perfectly be estimated and provided to the transmitter. However, if the shadowing changes within the coherence time of the channel or the system delays are too large, the attenuation is a random parameter for the transmitter. Such a behavior is seen for mobile SatCom [59, 204], where the varying atmospheric effects cannot fully be captured due to the long delays for feedback. The parameters, i.e., the mean and variance for the shadow fading, are provided by empirical models for given frequency ranges [205]. The details and applications are provided in Chap. 6.

2.2.2 *Rank-One Channel Covariance Matrix*

A multiplicative channel error also arises from the complex Gaussian model (2.7) if the channel entries are fully correlated. Then, the covariance matrix is rank-one, i.e.,

$$\boldsymbol{C_h} = \sigma_{\boldsymbol{h}}^2 \boldsymbol{v}_1 \boldsymbol{v}_1^{\mathrm{H}}, \tag{2.18}$$

where $\boldsymbol{v}_1 \in \mathbb{C}^N$ is the unit-norm eigenvector to the single positive eigenvalue $\sigma_{\boldsymbol{h}}^2$.

With (2.18), the Rayleigh fading channel can be represented as

$$\boldsymbol{h}_{\mathrm{Rayleigh}} = w \sigma_{\boldsymbol{h}} \boldsymbol{v}_1, \tag{2.19}$$

where $w = 1 + \xi \sim \mathcal{N}_{\mathbb{C}}(0, 1)$ and $\bar{\boldsymbol{h}} = \sigma_{\boldsymbol{h}} \boldsymbol{v}_1$. For non-zero mean, $\boldsymbol{m}$ must additionally be aligned with $\boldsymbol{v}_1$ in order to rewrite (2.7) into

$$\boldsymbol{h} = \left(1 + \frac{\sigma_{\boldsymbol{h}}}{\|\boldsymbol{m}\|_2} w\right) \boldsymbol{m}. \tag{2.20}$$

Now, $\bar{\boldsymbol{h}} = \boldsymbol{m}$ and $\xi = \sigma_{\boldsymbol{h}} / \|\boldsymbol{m}\|_2 w \sim \mathcal{N}_{\mathbb{C}}(0, \sigma_{\boldsymbol{h}}^2 / \|\boldsymbol{m}\|_2^2)$ are the parameters for (2.2). For Rician fading (2.8), the rank-one covariance matrix $\boldsymbol{C}_z = \sigma_z^2 \boldsymbol{v}_1 \boldsymbol{v}_1^{\mathrm{H}}$ results in

$$\boldsymbol{h}_{\mathrm{Rician}} = \left(\sqrt{\frac{\kappa}{\kappa+1}} + \sqrt{\frac{\kappa}{\kappa+1}} \frac{\sigma_z}{\|\bar{z}\|_2} w\right) \bar{z}. \tag{2.21}$$

This model serves as a basis for the SatCom channel in [52, 206] and Sect. 6.1.

The MMSE estimation for (2.20) also results in the multiplicative model

$$\boldsymbol{h} = \left(1 + \frac{\sigma_{e_{\mathrm{MMSE}}}^2}{\|\bar{\boldsymbol{h}}_{\mathrm{MMSE}}\|_2} w\right) \bar{\boldsymbol{h}}_{\mathrm{MMSE}},$$

where the estimate and the error variance, respectively, read as [cf. (2.10) and (2.11)]

$$\bar{h}_{\mathrm{MMSE}} = \left(1 + \frac{\sigma_h^2 v_1^{\mathrm{H}} S^{\mathrm{H}} C_{\mathrm{tr}}^{-1}(y_{\mathrm{tr}} - Sm)}{\|m\|_2 \sigma_h^2 v_1^{\mathrm{H}} S^{\mathrm{H}} C_{\mathrm{tr}}^{-1} S v_1}\right) m,$$

$$\sigma_{e_{\mathrm{MMSE}}}^2 = \frac{\sigma_h^2}{1 + \sigma_h^2 v_1^{\mathrm{H}} S^{\mathrm{H}} C_{\mathrm{tr}}^{-1} v_1}.$$

Another model for $|1+\xi|^2$ would be Nakagami-m fading [207, 208]. The scalar Rayleigh distribution is then a special case for $m = 1$ (e.g., see [189]).[10] Since applications for vector channels are limited [28], it is beyond the scope of this work.

2.3 Multiplicative Approximations for Additive Fading

The accurate spatial channel knowledge of the above rank-one channel covariance matrix model substantially simplifies robust QoS and RB beamformer designs.[11] This motivates a rank-one approximation also for the additive error model (2.1):

$$h \approx (\hat{h}_1 + \hat{e}_1) v_1 = (1 + \hat{e}_1/\hat{h}_1)\hat{h}_1 v_1, \tag{2.22}$$

where $\hat{h}_1 \in \mathbb{C}$ and $\hat{e}_1 \sim \mathcal{N}_{\mathbb{C}}(0, \hat{\sigma}_1^2)$. For example, the unit-norm vector $v_1 \in \mathbb{C}^N$ and the parameters $\hat{h}_1 \in \mathbb{C}$ and $\hat{\sigma}_1^2 \in \mathbb{R}_+$ are chosen to minimize the MSE

$$(\hat{h}_1, \hat{\sigma}_1^2) = \underset{h, \hat{\sigma}, \hat{e}_1 \sim \mathcal{N}_{\mathbb{C}}(0, \hat{\sigma})}{\arg\min} \; \mathrm{E}[\|h - (h + \hat{e}_1) v_1\|_2^2]. \tag{2.23}$$

The solution is $\hat{h}_1 = v_1^{\mathrm{H}} \bar{h}$, $\hat{e}_1 = v_1^{\mathrm{H}} e$, and $\hat{\sigma}_1^2 = v_1^{\mathrm{H}} C_e v_1$, where v_1 is the dominant eigenvector of the channel's second order moment $\bar{h}\bar{h}^{\mathrm{H}} + C_e$.

Approximation (2.22) suits for scenarios where $C_e \approx \alpha \bar{h}\bar{h}^{\mathrm{H}}$, $\alpha \in \mathbb{R}_+$, e.g., for SatCom [210]. Otherwise, it is subject to errors in the null-space of v_1^{H}. Parameterizing this space by a subunitary matrix $V_{N-1} \in \mathbb{C}^{N \times N-1}$, the true error model reads as

$$h = (\hat{h}_1 + \hat{e}_1) v_1 + V_{N-1} \hat{e}_{N-1}, \tag{2.24}$$

where $\hat{e}_{N-1} \sim \mathcal{N}_{\mathbb{C}}(V_{N-1}^{\mathrm{H}} \bar{h}, V_{N-1}^{\mathrm{H}} C_e V_{N-1})$. The mismatch for neglecting the error $V_{N-1}\hat{e}_{N-1}$ remains small with high probability if $\mathrm{E}[\|\hat{e}_{N-1}\|_2^2]$ is small.[12]

[10] For example, Nakagami-m fading fits indoor-mobile [189] and land-mobile SatCom [209].

[11] Chapters 3 and 5 provide the details for the simplified optimizations.

[12] Markov's inequality [211, Proposition 5.4] implies $\Pr(\|\hat{e}_{N-1}\|_2^2 \geq \delta) \leq \epsilon$ if $\mathrm{E}[\|\hat{e}_{N-1}\|_2^2] < \epsilon\delta$.

The errors $\hat{e}_1$ and $\hat{\boldsymbol{e}}_{N-1}$ are generally correlated, i.e., $\mathrm{E}[\hat{\boldsymbol{e}}_{N-1}\hat{e}_1{}^*] = \boldsymbol{V}_{N-1}^{\mathrm{H}}\boldsymbol{C}_{\boldsymbol{e}}\boldsymbol{v}_1$, but become uncorrelated for a diagonal covariance matrix, e.g., $\boldsymbol{C}_{\boldsymbol{e}} = \sigma_e^2\mathbf{I}$. If uncorrelated errors are requested for the general case, the approximation can be based on thepre-whitened channel $\boldsymbol{C}_{\boldsymbol{e}}^{-1/2}\boldsymbol{h}$ instead of $\boldsymbol{h}$. Then, the error model becomes

$$\boldsymbol{C}_{\boldsymbol{e}}^{-1/2}\boldsymbol{h} = (\hat{h}_1 + \hat{e}_1)\boldsymbol{v}_1 + \boldsymbol{V}_{N-1}\hat{\boldsymbol{e}}_{N-1} \tag{2.25}$$

with the channel parameters $\hat{h}_1 = \boldsymbol{v}_1^{\mathrm{H}}\boldsymbol{C}_{\boldsymbol{e}}^{-1/2}\bar{\boldsymbol{h}}$, $\hat{e}_1 = \boldsymbol{v}_1^{\mathrm{H}}\boldsymbol{C}_{\boldsymbol{e}}^{-1/2}\boldsymbol{e}$, and $\hat{\sigma}_1^2 = 1$, where the main direction is $\boldsymbol{v}_1 = \|\boldsymbol{C}_{\boldsymbol{e}}^{-1/2}\bar{\boldsymbol{h}}\|_2^{-1}\boldsymbol{C}_{\boldsymbol{e}}^{-1/2}\bar{\boldsymbol{h}}$ and $\hat{\boldsymbol{e}}_{N-1} \sim \mathcal{N}_{\mathbb{C}}(\mathbf{0}, \mathbf{I}_{N-1})$.

Alternatively to these approximations, the estimation may be reduced to a one-dimensional space. The reduced ML estimation for (2.22) reads as (cf. [212])[13]

$$\hat{h}_{1\mathrm{RML}} = \arg\min_{\hat{h}_1} \|\boldsymbol{C}_{\mathrm{tr}}^{-1/2}(\boldsymbol{y}_{\mathrm{tr}} - \boldsymbol{S}\boldsymbol{v}_1\hat{h}_1)\|_2^2.$$

Then, the estimate is $\bar{\boldsymbol{h}}_{\mathrm{RML}} = \boldsymbol{v}_1\hat{h}_{1\mathrm{RML}}$, where $\hat{h}_{1\mathrm{RML}} = (\boldsymbol{v}_1^{\mathrm{H}}\boldsymbol{S}^{\mathrm{H}}\boldsymbol{C}_{\mathrm{tr}}^{-1}\boldsymbol{S}\boldsymbol{v}_1)^{-1}\boldsymbol{v}_1^{\mathrm{H}}$ $\boldsymbol{S}^{\mathrm{H}}\boldsymbol{C}_{\mathrm{tr}}^{-1}\boldsymbol{y}_{\mathrm{tr}}$ and $\hat{e}_1 = (\boldsymbol{v}_1^{\mathrm{H}}\boldsymbol{S}^{\mathrm{H}}\boldsymbol{C}_{\mathrm{tr}}^{-1}\boldsymbol{S}\boldsymbol{v}_1)^{-1}\boldsymbol{v}_1^{\mathrm{H}}\boldsymbol{S}^{\mathrm{H}}\boldsymbol{C}_{\mathrm{tr}}^{-1}\boldsymbol{n}_{\mathrm{tr}}$ with $\sigma_{\hat{e}_1}^2 = 1/(\boldsymbol{v}_1^{\mathrm{H}}\boldsymbol{C}_{\boldsymbol{e}\mathrm{ML}}^{-1}\boldsymbol{v}_1)$ are the channel parameters in the reduced domain. For scaled identity $\boldsymbol{C}_{\boldsymbol{e}\mathrm{ML}}$, the estimation error $\hat{e}_1$ is again uncorrelated of the unwanted modeling error.

To avoid that the disregarded error becomes significant, the channel's subspace may be tracked, e.g., with the subspace tracking algorithm from [214] or by estimating it in the long-term via minimizing the MSE between the pre-whitened ML estimate and its reduced version [185, 212]. The latter problem reads as

$$\boldsymbol{v}_1 = \underset{\|\boldsymbol{v}\|_2=1}{\arg\min}\ \mathrm{E}\left[\|\boldsymbol{C}_{\boldsymbol{e}\mathrm{ML}}^{-1/2}(\bar{\boldsymbol{h}}_{\mathrm{ML}} - \boldsymbol{v}\hat{h}_{1\mathrm{RML}})\|_2^2\right].$$

The solution is the unit-norm right-hand eigenvector of the largest eigenvalue λ for

$$\boldsymbol{C}_{\boldsymbol{e}\mathrm{ML}}^{-1}(\boldsymbol{m}\boldsymbol{m}^{\mathrm{H}} + \boldsymbol{C}_{\boldsymbol{h}})\boldsymbol{C}_{\boldsymbol{e}\mathrm{ML}}^{-1}\boldsymbol{v}_1 = \lambda\boldsymbol{C}_{\boldsymbol{e}\mathrm{ML}}^{-1}\boldsymbol{v}_1.$$

Similar to the reduced ML criterion, there is a reduced MMSE channel estimation based on $\bar{\boldsymbol{h}}_{\mathrm{RMMSE}} = \boldsymbol{v}_1\hat{h}_{1\mathrm{RMMSE}}$ [43], where $\hat{h}_{1\mathrm{RMMSE}}$ minimizes

$$\hat{h}_{1\mathrm{RMMSE}} = \arg\min_{\hat{h}_1} \mathrm{E}[\|\boldsymbol{h} - \boldsymbol{v}_1\hat{h}_1\|_2^2|\boldsymbol{y}_{\mathrm{tr}}].$$

Its solution $\hat{h}_{1\mathrm{RMMSE}} = \boldsymbol{v}_1^{\mathrm{H}}\bar{\boldsymbol{h}}_{\mathrm{MMSE}}$ is a post-processing of the MMSE estimate (2.10) and the error in the reduced space is $\hat{e}_{1\mathrm{RMMSE}} \sim \mathcal{N}_{\mathbb{C}}(0, \boldsymbol{v}_1^{\mathrm{H}}\boldsymbol{C}_{\boldsymbol{e}\mathrm{MMSE}}\boldsymbol{v}_1)$. Minimizing the remaining long-term MSE with respect to $\boldsymbol{v}_1$, i.e.,

[13] Alternatively, the observations may already be projected onto the channel's subspace for the estimation, e.g., $\hat{y}_{\mathrm{tr}} = \boldsymbol{v}_1^{\mathrm{H}}\boldsymbol{S}^{\mathrm{H}}\boldsymbol{y}_{\mathrm{tr}}$. Then, the reduced ML estimation is performed with $\hat{y}_{\mathrm{tr}}$ [213].

$$\boldsymbol{v}_1 = \operatorname*{arg\,min}_{\|\boldsymbol{v}\|_2=1} \mathrm{E}[\|\boldsymbol{h} - \boldsymbol{v}\boldsymbol{v}\bar{\boldsymbol{h}}_{\mathrm{MMSE}}\|_2^2],$$

the solution is the eigenvector of $\boldsymbol{m}\boldsymbol{m}^{\mathrm{H}} + \boldsymbol{C}_{\boldsymbol{h}}\boldsymbol{S}^{\mathrm{H}}(\boldsymbol{S}\boldsymbol{C}_{\boldsymbol{h}}\boldsymbol{S}^{\mathrm{H}} + \boldsymbol{C}_{\mathrm{tr}})^{-1}\boldsymbol{S}\boldsymbol{C}_{\boldsymbol{h}}$ that corresponds to the dominant eigenvalue

$$\lambda = \min\{\lambda \in \mathbb{R}_+ : \lambda\mathbf{I} - (\boldsymbol{m}\boldsymbol{m}^{\mathrm{H}} + \boldsymbol{C}_{\boldsymbol{h}}\boldsymbol{S}^{\mathrm{H}}(\boldsymbol{S}\boldsymbol{C}_{\boldsymbol{h}}\boldsymbol{S}^{\mathrm{H}} + \boldsymbol{C}_{\mathrm{tr}})^{-1}\boldsymbol{S}\boldsymbol{C}_{\boldsymbol{h}}) \succeq \mathbf{0}\}.$$

Chapter 3
Precoder Design for Ergodic Rates with Multiplicative Fading

While the QoS and RB optimizations with instantaneous rate constraints feature well-established solutions, solving these problems with ergodic rates is demanding. The QoS optimization with ergodic rates $R_k(\boldsymbol{t}) = \mathrm{E}[r_k(\boldsymbol{t}, \boldsymbol{h}_k)]$ reads as [cf. (1.9)]

$$\min_{p \geq 0,\, \boldsymbol{t}} p \quad \text{s.t.} \quad p^{-1}\boldsymbol{t} \in \mathcal{P}, \quad \boldsymbol{R}(\boldsymbol{t}) \in \mathcal{R}, \tag{3.1}$$

where $\boldsymbol{R}(\boldsymbol{t}) = [R_1(\boldsymbol{t}), \ldots, R_K(\boldsymbol{t})]^{\mathrm{T}}$ and $\mathcal{R}$ is the per-user target region (1.10). The corresponding ergodic RB optimization is given by [cf. (1.12)]

$$\max_{\rho,\, \boldsymbol{t}} \rho \quad \text{s.t.} \quad \boldsymbol{t} \in \mathcal{P}, \quad \rho^{-1}\boldsymbol{R}(\boldsymbol{t}) \in \mathcal{R}. \tag{3.2}$$

These problems inherently share many properties of the optimizations with instantaneous rates. First, at least one of the power constraints in $\mathcal{P}$ is tight at the optimum of (3.1) and (3.2). Second, (3.2) has a solution $\boldsymbol{t}$ that balances the ergodic rates with respect to their targets, i.e., $R_k(\boldsymbol{t})/\rho_k = \rho$ for all $k = 1, \ldots, K$. Moreover, the ergodic QoS and RB optimizations are inverse problems as well.[1] Hence, the optimum of (3.2) may also be found via a series of QoS optimizations if the solution for (3.1) were known.

The difficulty in solving (3.1) and (3.2) results from the involved numerical integrations of closed-form expressions for $R_k(\boldsymbol{t})$. These terms are no functions of some SINR (cf. Sect. 3.1). Therefore, there is a general lack of uplink–downlink duality [215] for ergodic rates. Moreover, direct convex reformulations are missing for ergodic rate requirements. Non-convex global optimization strategies, e.g., via a BB search, are in turn very demanding [140]. Direct optimizations with ergodic

[1]These properties follow directly by replacing $r_k(\boldsymbol{t})$ with $R_k(\boldsymbol{t})$ within Sect. A.1.

A. Gründinger, *Statistical Robust Beamforming for Broadcast Channels and Applications in Satellite Communication*, Foundations in Signal Processing, Communications and Networking 22, https://doi.org/10.1007/978-3-030-29578-3_3

rate constraints are therefore computationally intractable for a hundred of users and antennas, even though these optimizations are still solvable for perfect CSI.

A common approximation of the ergodic rate is based on the *average SINR* [71, 118, 216], where the expectation with respect to the random channel is separately in the nominator and denominator of (1.4).[2] Even though this approximation simplifies the problems, its relation to the ergodic rate remains unknown. Generally, the average SINR neither lower nor upper bounds the ergodic rate [217, Section 6]. The lower rate bound from [128, 218] considers also imperfect receiver CSI and has been studied for RB under additive Gaussian error knowledge by Joham et al. [219]. Due to self-interference, this bound is only tight if the channel knowledge is close to perfect. Alternatively, Chap. 4 addresses average MMSE bounds for the ergodic rate.

The (approximate) multiplicative error model (2.2) achieves a compromise between robustness and efficient solutions for the ergodic QoS and RB optimizations. This chapter extends the pre-published results and beamformer designs in this context [139–143]. Closed-form expressions of the ergodic rates are shown in Sect. 3.1. Since minimal requirements on these expressions feature no direct convex reformulation and dual uplink representation, tight rate bounds using either ZF beamforming or Jensen's inequality are studied in Sect. 3.2.[3] Based on these bounds, Sect. 3.3 addresses feasibility for the QoS optimization in setups with less transmit antennas than users. In particular, we prove that the polytopal feasibility region (1.31) from the perfect CSI scenario also holds for (3.1). The constructive proof suggests an adaptive beamformer design with a tight rate bound followed by a power allocation to achieve any point within the feasible region (see Sect. 3.4).

We further improve the performance with a SCS strategy (cf. [150]) that exploits locally tight constraint approximations with an SINR structure (see Sect. 3.5). It finds, a local solution for (3.1) via executing a series of SINR constrained QoS problems [142]. A similar SCS finds a solution of the ergodic RB problem [143].

We compare the attained optimum of the SCS also with the result of a global search based on a BB algorithm [140]. The BB method is an exhaustive iterative partitioning procedure and has exponential complexity in the number of ergodic rate requirements. Thus, it is only useful for benchmarks. The SCS performance matches the BB result and resides between the results of the ergodic rate lower and upper bounds. This suggests that the SCS procedure finds the global optimum for the vector BC optimizations at hand with polynomial complexity. The less complex ZF strategy falls behind these results in a SatCom related simulation setup, i.e., with a small angular spread of the receivers from the transmitters view (cf. Sect. 3.7).

---

[2]Thus, the average SINR uses the channels second moments $\boldsymbol{R}_k = \mathrm{E}[\boldsymbol{h}_k \boldsymbol{h}_k^{\mathrm{H}}]$ instead of $\boldsymbol{h}_k \boldsymbol{h}_k^{\mathrm{H}}$.

[3]Applying ZF beamforming for only a part of the receivers reduces the computational complexity. This is especially relevant if the number of antennas N and users K are large and $N \geq K$.

3.1 Closed-Form Rate Expressions

Inserting the multiplicative channel error model (2.2) into $r_k(\boldsymbol{t}, \boldsymbol{h}_k)$ from (1.3), the ergodic rates (1.35) are equivalently recast as[4]

$$R_k(\boldsymbol{t}) = \mathrm{E}\left[\log_2\left(1 + \frac{|\bar{\boldsymbol{h}}_k^{\mathrm{H}}\boldsymbol{t}_k|^2}{\zeta_k + \sum_{i\neq k}|\bar{\boldsymbol{h}}_k^{\mathrm{H}}\boldsymbol{t}_i|^2}\right)\right], \tag{3.3}$$

where $\zeta_k = |1 + \xi_k|^{-2}$ is the random attenuation—the noise enhancement—due to the erroneous channel knowledge. A further reformulation of (3.3) results in

$$R_k(\boldsymbol{t}) = B_k\Big(\sum_{i=1}^{K}|\bar{\boldsymbol{h}}_k^{\mathrm{H}}\boldsymbol{t}_i|^2\Big) - B_k\Big(\sum_{i\neq k}|\bar{\boldsymbol{h}}_k^{\mathrm{H}}\boldsymbol{t}_i|^2\Big) \tag{3.4}$$

where the function $B_k : \mathbb{R}_+ \to \mathbb{R}_+, x \mapsto B_k(x)$ is a substitute for[5]

$$B_k(x) = \mathrm{E}[b_k(\zeta_k^{-1}, x)] = \mathrm{E}[\log_2(1 + \zeta_k^{-1}x)]. \tag{3.5}$$

The exact representation of (3.5) depends on the distribution of ζ_k^{-1} (or ζ_k). If its evaluation requires a numerical integration, e.g., for a log-normal distributed attenuation ζ_k in dB (cf. Chap. 2), the values may also be deduced from a table, which is based on x and the distribution parameters of ζ_k.

For $\xi_k \sim \mathcal{N}_{\mathbb{C}}(0, \sigma_{\xi_k}^2)$, e.g., with Rician fading, the random variable $\zeta_k^{-1} = |1 + \xi_k|^2$ is chi-square distributed, i.e., $2\sigma_{\xi_k}^{-2}\zeta_k^{-1} \sim \mathcal{X}_2^2(2\sigma_{\xi_k}^{-2})$. The expected logarithm (3.5) then becomes (cf. [26, Equations (18) and (19)])

$$B_k(x) = \mathrm{e}^{-\sigma_{\xi_k}^{-2}} \int_0^\infty \log_2\left(1 + x\sigma_{\xi_k}^2 t\right)\mathrm{e}^{-t}{}_0F_1\left(;1;\sigma_{\xi_k}^{-2}t\right)\mathrm{d}t, \tag{3.6}$$

with the *confluent hypergeometric limit function* (e.g., see [220, 16.3])[6]

---

[4]To simplify the notation, the noise variance is without loss of generality set to $\sigma_k^2 = 1$.

[5]Instead of B_k in (3.5), we can alternatively use $A_k : \mathbb{R}_+ \to \mathbb{R}_+, x \mapsto A_k(x)$ with

$$A_k(x) = \mathrm{E}[a_k(\zeta_k, x)] = \mathrm{E}[\log_2(\zeta_k + x)] = B_k(x) + \mathrm{E}[\log_2(\zeta_k)].$$

While (3.5) leads to known closed-form expressions for Gaussian ξ_k [see (3.6) and (3.7)], A_k could be used if $\zeta_k = (1 + \xi_k)^{-2}$ were a positive quadratic form Gaussian random variables.

[6]This formula for the function ${}_0F_1(;b;z) = \Gamma(b)z^{\frac{1}{2}-\frac{1}{2}b}I_{b-1}(2\sqrt{z})$ [220, 16.3] follows with the series expansion of the modified Bessel function $I_{b-1}(z)$ [203, 13.3.1].

$$ {}_0F_1(;1;z) = z^{-\frac{1}{2}} I_0(2\sqrt{z}) = \frac{1}{\sqrt{z}} \sum_{m=0}^{\infty} \frac{z^m}{(m!)^2}. $$

Results for the numerical integration in (3.6) may be tabulated for varying $\sigma_{\xi_k}^2$ and x.

For rank-one Rayleigh fading (2.19), where $\xi_k \sim \mathcal{N}_{\mathbb{C}}(-1, 1)$ and $2\zeta_k^{-1} \sim \mathcal{X}_2^2$,[7] a closed-form expression for (3.5) reads as[221, Section IV]

$$ B_k(x) = \log_2(\mathrm{e})\mathrm{e}^{\frac{1}{x}}\,\mathrm{E}_1\left(\frac{1}{x}\right), \tag{3.7} $$

where $\mathrm{E}_1(x) = \int_x^\infty \frac{\mathrm{e}^{-t}}{t}\,\mathrm{d}t$ is the exponential integral function [203, Section 5.1].

The properties of (3.5) resemble those of a logarithmic function. In particular, B_k is obviously a concave and (sub-)logarithmically increasing function. However, the subtraction (and addition) of $B_k(x)$ and $B_k(y)$, $x \geq y \geq 0$, are different from the rules for logarithmic functions, especially $B_k(x) - B_k(y) \neq B_k((x-y)/(c_k+y))$, $c_k \in \mathbb{R}_+$. Therefore, (3.4) cannot be formulated as a function of a deterministic SINR. Existence of such a mapping was crucial for the QoS and RB solutions in Sect. 1.3. The conic constraints suited for disciplined convex programming solvers are inaccessible for ergodic rates. Moreover, the missing dual problem formulation prevents from employing the efficient uplink design algorithms from [67, 89, 90]. For applying these beamformer optimization methods, we present tight ergodic rate bounds in the next section.

3.2 Lower and Upper Rate Bounds

Pre-fixed precoding may be employed to bound the achievable rates and simplify (3.1) and (3.2) (see Sect. 3.4). Important examples are MRT, ZF precoding, and RZF precoding. These beamforming strategies can result in unwanted losses for per-antenna power restrictions [110]. To keep the beamformers adaptive, we suggest a (partial) ZF strategy and derive ergodic rate bounds with a logarithmic structure. These bounds enable ergodic QoS and RB solutions with the known frameworks and are the basis for the derivation of the QoS feasibility region in Sect. 3.3.

3.2.1 Generalized Zero-Forcing Lower Bound

The ergodic rate (3.4) is a difference of two integral functions in the beamformers. Only one integral function remains when we restrict to zero interference. In this way, we can simplify up to N ergodic rate constraints for a regular matrix $\bar{\boldsymbol{H}}$.

---

[7]With $\xi_k \sim \mathcal{N}_{\mathbb{C}}(-1, 1)$, the random variable is $\zeta_k^{-1} = |1+\xi_k|^2 = |w_k|^2$, where $w_k \sim \mathcal{N}_{\mathbb{C}}(0, 1)$.

By restricting to zero interference at receivers $k \in \mathcal{I} \subseteq \{1, \ldots, K\}$, i.e., $|\bar{\boldsymbol{h}}_k^{\mathrm{H}} \boldsymbol{t}_i|^2 = 0$ for $i \neq k$, we impose the additional subspace requirements (cf. [110])

$$\boldsymbol{t}_k = \boldsymbol{W}_k \boldsymbol{\tau}_k, \quad \bar{\boldsymbol{h}}_i^{\mathrm{H}} \boldsymbol{W}_k = \boldsymbol{0}, \quad i \in \mathcal{I}, \quad k \neq i, \quad k = 1, \ldots, K, \tag{3.8}$$

where $\boldsymbol{W}_k$ defines the subunitary basis for the beamformer's remaining subspace, i.e., $\boldsymbol{\tau}_k \in \mathbb{C}^{N-|\mathcal{I}|}$ and $\boldsymbol{W}_k = \boldsymbol{W}' \in \mathbb{C}^{N \times N-|\mathcal{I}|}$ with $\boldsymbol{W}'^{\mathrm{H}} \boldsymbol{W}' = \mathbf{I}_{N-|\mathcal{I}|}$ for indices $k \notin \mathcal{I}$, and $\boldsymbol{\tau}_k \in \mathbb{C}^{N-|\mathcal{I}|+1}$ and $\boldsymbol{W}_k \in \mathbb{C}^{N \times N-|\mathcal{I}|+1}$ with $\boldsymbol{W}_k^{\mathrm{H}} \boldsymbol{W}_k = \mathbf{I}_{N-|\mathcal{I}|+1}$ for $k \in \mathcal{I}$. Therewith, receivers $k \in \mathcal{I}$ experience no interference, but the beamformers can still cause interference for receivers $i \notin \mathcal{I}$ if $|\mathcal{I}| < K$.[8] The resulting ZF rates read as

$$R_k^{(\mathrm{ZF})}(\boldsymbol{W}_k \boldsymbol{\tau}_k) = B_k \left(|\bar{\boldsymbol{h}}_k^{\mathrm{H}} \boldsymbol{W}_k \boldsymbol{\tau}_k|^2 \right), \qquad k \in \mathcal{I}. \tag{3.9}$$

Here, $R_k^{(\mathrm{ZF})}$ is a concave function of the received Signal-to-noise ratio (SNR) $|\bar{\boldsymbol{h}}_k^{\mathrm{H}} \boldsymbol{W}_k \boldsymbol{\tau}_k|^2$. This enables a convex reformulation of the ergodic rate constraint $R_k^{(\mathrm{ZF})}(\boldsymbol{W}_k \boldsymbol{\tau}_k) \geq \rho \rho_k$ in $\boldsymbol{\tau}_k$, i.e.,[9]

$$\mathrm{Re}\left(\bar{\boldsymbol{h}}_k^{\mathrm{H}} \boldsymbol{W}_k \boldsymbol{\tau}_k\right) \geq \sqrt{B_k^{-1}(\rho \rho_k)}, \qquad k \in \mathcal{I}.$$

The power constraint set (1.33) remains an intersection of convex SOCs, i.e.,

$$\mathrm{vec}([\boldsymbol{W}_1 \boldsymbol{\tau}_1, \ldots, \boldsymbol{W}_K \boldsymbol{\tau}_K]) \in \mathcal{P}, \tag{3.10}$$

since (3.8) is linear in $\boldsymbol{\tau}_k$.[10] Hence, the beamformer solution for (3.1) and (3.2) is still found via conic optimization or, alternatively, using duality (see Sect. 3.2.3).

The partial ZF beamforming approach reduces the search from NK complex variables within $\boldsymbol{t}$ to the $(N-|\mathcal{I}|+1)K-(K-|\mathcal{I}|)$ entries of $\boldsymbol{\tau} = [\boldsymbol{\tau}_1^{\mathrm{T}}, \ldots, \boldsymbol{\tau}_K^{\mathrm{T}}]^{\mathrm{T}}$. In case of ZF beamforming for all receivers, i.e., $|\mathcal{I}| = K \leq N$, the number of complex decision variables reduces to $(N-K+1)K$. Then, the interconnection between the reduced beamformers $\boldsymbol{\tau}_k$ within the optimizations (3.1) and (3.2) is only due to the power constraint (3.10).[11] For $|\mathcal{I}| = K = N$, the generalized ZF beamformer optimization further simplifies to a power allocation. Then, the optimization is only over the K real scalars $|\tau_k|^2$, $k = 1, \ldots, K$, that represent the power allocations for transmission to the receivers, and the unit-norm beamforming vectors $\boldsymbol{w}_k \in \mathbb{C}^N$ are the scaled columns of $\bar{\boldsymbol{H}}(\bar{\boldsymbol{H}}^{\mathrm{H}} \bar{\boldsymbol{H}})^{-1}$.

[8] If ZF is realized only for a subset of the channels, the transmitter either needs perfect knowledge of the remaining channels [139] or applies the bounds in Sect. 3.2.2 to simplify the other constraints.

[9] The balancing target is $\rho = 1$ for the QoS optimization (3.1).

[10] The sum power constraint becomes $\sum_{k=1}^K \|\boldsymbol{t}_k\|_2^2 = \sum_{k=1}^K \|\boldsymbol{\tau}_k\|_2^2 \leq P$, for example, [cf. (1.11)].

[11] This generalized ZF beamformer design has also been considered by [110], but for perfect CSI.

The drawback of the ZF restriction is that it markedly deforms the QoS feasible region if $K > N$ (see Sect. 3.3). This results in performance losses also for finite transmit power. Therefore, the choice for $\mathcal{I}$ is generally a trade-off between low complexity and power efficient transmission.

3.2.2 *Bounds for Adaptive Beamforming*

Bounds on the ergodic rate (3.3), e.g., based on Jensen's inequality [102, p. 77], provide an alternative approximation that enables the perfect CSI beamforming methods for (3.1) and (3.2). For multiplicative channel errors (2.2), the lower and upper bounds under study have the following structure (cf. [141]):

$$R_k^{(\mathrm{B})}(\boldsymbol{t}) = \log_2\left(1 + \mathrm{SINR}_k^{(\mathrm{B})}(\boldsymbol{t})\right) + \mu_k^{(\mathrm{B})}, \tag{3.11}$$

with offset $\mu_k^{(\mathrm{B})} \in \mathbb{R}$ and noise enhancement $\nu_k^{(\mathrm{B})} \in \mathbb{R}_{++}$ within the SINR bound

$$\mathrm{SINR}_k^{(\mathrm{B})}(\boldsymbol{t}) = \frac{|\bar{\boldsymbol{h}}_k^{\mathrm{H}} \boldsymbol{t}_k|^2}{\nu_k^{(\mathrm{B})} + \sum_{i \neq k} |\bar{\boldsymbol{h}}_k^{\mathrm{H}} \boldsymbol{t}_i|^2}. \tag{3.12}$$

The values of $\mu_k^{(\mathrm{B})}$ and $\nu_k^{(\mathrm{B})}$ for the studied lower and upper bounds are shown in Table 3.1. The bound $R_k^{(\mathrm{B})}(\boldsymbol{t})$ can be negative if $\mu_k^{(\mathrm{B})} < 0$ and the SINR (3.12) is too small. Then, a non-negative version of the lower bound reads as

$$R_{k,+}^{(\mathrm{B})}(\boldsymbol{t}) = \max\left\{0, R_k^{(\mathrm{B})}(\boldsymbol{t})\right\}. \tag{3.13}$$

The offset and noise variance values for the first lower bound (LB1) and the first upper bound (UB1) are derived in Sect. A.3. These bounds require known and finite values for the expectations $\mathrm{E}[\zeta_k]$ and $\mathrm{E}[\log_2(\zeta_k)]$, e.g., when ζ_k (in dB) models a log-normal distributed attenuation. Then, the lower bound LB1 is tight for the ergodic rate. In particular, $R_k^{(\mathrm{LB1})}(\boldsymbol{t}) = R_k(\boldsymbol{t})$ when either the useful signal power

Table 3.1 Effective noise variance and offset for bounds on the ergodic rate

Bound (B)	Noise $\nu_k^{(\mathrm{B})}$	Offset $\mu_k^{(\mathrm{B})}$
LB1	$\mathrm{E}[\zeta_k]$	0
UB1	$\mathrm{E}[\zeta_k]$	$\mathrm{E}\left[\log_2(\mathrm{E}[\zeta_k]/\zeta_k)\right]$
LB2	$1/\mathrm{E}[\zeta_k^{-1}]$	$-\mathrm{E}\left[\log_2\left(\mathrm{E}[\zeta_k^{-1}]/\zeta_k^{-1}\right)\right]$
UB2	$1/\mathrm{E}[\zeta_k^{-1}]$	0
ALB	$\mathrm{e}^{\mathrm{E}[\ln(\zeta_k)]}$	$-d_k$
AUB	$\mathrm{e}^{\mathrm{E}[\ln(\zeta_k)]}$	d_k

$d_k = \max_{x \geq 0} \mathrm{E}[\log_2(\zeta_k + x)] - \log_2\left(\mathrm{e}^{\mathrm{E}[\ln(\zeta_k)]} + x\right)$

$$\mathtt{S}_k = |\bar{\boldsymbol{h}}_k^{\mathrm{H}} \boldsymbol{t}_k|^2 \tag{3.14}$$

is zero, i.e., $\mathtt{S}_k = 0$, or the useful signal power and the interference power

$$\mathtt{I}_k = \sum_{i \neq k} |\bar{\boldsymbol{h}}_k^{\mathrm{H}} \boldsymbol{t}_i|^2 \tag{3.15}$$

are both large, i.e., $\mathtt{S}_k \to \infty$ and $\mathtt{I}_k \to \infty$. The maximum approximation error $\mu_k^{(\mathrm{UB1})}$ is only reached if $\mathtt{I}_k = 0$ and $\mathtt{S}_k \to \infty$. The opposite behavior shows the upper bound UB1, which offset to LB1 is $\mu_k^{(\mathrm{UB1})}$.

The expected value $\mathrm{E}[\zeta_k]$ does not exist if the Probability density function (PDF) of $z_k = \zeta_k^{-1}$ has non-zero support at zero. For example, $\mathrm{E}[\zeta_k]$ is infinity if $z_k \sim \mathcal{E}$, i.e., z_k is exponentially distributed, which corresponds to rank-one Rayleigh fading (2.19). Then, the bounds LB2 and UB2 are available instead.[12] Now, UB2 is tight for the ergodic rate, that is, $R_k^{(\mathrm{UB2})}(\boldsymbol{t}) = R_k(\boldsymbol{t})$ when either $\mathtt{S}_k = 0$ or both $\mathtt{S}_k \to \infty$ and $\mathtt{I}_k \to \infty$. In contrast, the lower bound LB2 only meets the ergodic rate if $\mathtt{S}_k \to \infty$ and $\mathtt{I}_k = 0$ hold jointly.

Closed-form expressions for the offset $\mu_k^{(\mathrm{LB2})}$ and the noise $\nu_k^{(\mathrm{LB2})}$ are available when ζ_k^{-1} stems from the multiplicative channel model (2.18). For exponentially distributed $z_k = \zeta_k^{-1}$ with $\mathrm{E}[z_k] = 1$, i.e., Rayleigh fading, the offset is

$$\mu_k^{(\mathrm{LB2})} = \mathrm{E}[\log_2(z_k)] = -\log_2(\mathrm{e})\gamma,$$

where $\gamma \approx 0.577216$ is the Euler–Mascheroni constant [203, p. 229]. Alternatively, if $2z_k \sim \mathcal{X}_2^2(2\sigma_{\xi_k}^2)$ because $\zeta_k^{-1} = |1+\xi_k|^2$ with $\xi_k \sim \mathcal{N}_{\mathbb{C}}(0, \sigma_{\xi_k}^2)$ [cf. (2.20)], the offset is given by[13]

$$\mu_k^{(\mathrm{LB2})} = \mathrm{E}[\log_2(z_k)] - \log_2(\mathrm{E}[z_k]) = \log_2(\mathrm{e})\,\mathrm{E}_1(\sigma_{\xi_k}^{-2}) - \log_2(\sigma_{\xi_k}^2).$$

Table 3.1 also provides the parameters of the alternative ergodic rate bounds ALB and AUB, which are derived in Sect. A.3. The offset for this pair of lower and upper

---

[12] See Sect. A.3 for the derivation of these bounds based on Jensen's inequality and B_k.

[13] According to Lapidoth and Moser [222, Lemma 10.1] (see also [223, Theorem 3]), the expected value for the logarithm of a non-central chi-square-distributed random variable $y = \sum_{i=1}^{N} |x_i + m_i|^2$ of degree $2N$, with $m_i \in \mathbb{C}$ and i.i.d. $x_i \sim \mathcal{N}_{\mathbb{C}}(0, 1)$, is given by

$$\mathrm{E}[\ln(y)] = \begin{cases} \ln(s^2) + \mathrm{E}_1(s^2) + \sum_{j=1}^{N-1} (-1)^j \left[\mathrm{e}^{-s^2} (j-1)! - \frac{(N-1)!}{j(N-1-j)!} \right] \left(\frac{1}{s^2} \right)^j, & s^2 > 0 \\ \psi(N), & s^2 = 0, \end{cases}$$

where $s^2 = \sum_{j=1}^{N} |m_i|^2$, $\mathrm{E}_1(\cdot)$ is the exponential integral [203, Chapter 5], and $\psi(N)$ is the Digamma function [203, Section 6.3], e.g., $\mathrm{E}[\ln(y)] = \ln(s^2) + \mathrm{E}_1(s^2)$ for $N = 1$ [224, Appendix B].

bound is symmetric around the logarithmic term of (3.11). The distance between these bounds can be smaller than for LB2 and UB2 for other distributions of ζ_k^{-1}. For Rayleigh fading, i.e., $2\zeta_k^{-1} \sim \mathcal{X}_2^2$, the numerically computed offset is (cf. [143])

$$d_k \approx \log_2(\mathrm{e})0.1709.$$

3.2.3 QoS and Balancing Optimization with Rate Bounds

Replacing the ergodic rates in (3.1) by either of the bounds from Sects. 3.2.1 and 3.2.2, the QoS optimization with explicit rate constraints reads as[14]

$$\min_{p\geq 0,\, \boldsymbol{t}} p \quad \text{s.t.} \quad p^{-1}\boldsymbol{t} \in \mathcal{P}, \quad R_k^{(\mathrm{B})}(\boldsymbol{t}) \geq \rho_k, \quad k=1,\ldots,K. \tag{3.16}$$

With the bounds from Sect. 3.2.2, the structure of the rate constraints is similar to that for perfect CSI [see (1.10)], but with $\nu_k^{(\mathrm{B}),-1/2}\bar{\boldsymbol{h}}_k$ replacing the channels $\boldsymbol{h}_k$.

However, a non-zero $\mu_k^{(\mathrm{B})}$ changes the demand on the logarithmic term of (3.11). The targets are more demanding for negative offsets, i.e., for LB2 and ALB, and less demanding for positive offsets, i.e., for UB1 and AUB. The positive offset from UB1 or AUB may even be larger than some target rates, e.g., $\rho_i \leq \mu_i^{(\mathrm{B})}$ for $i \in \mathcal{A}^{\mathrm{c}} \subseteq \{1,\ldots,K\}$. Then, the corresponding constraints are always satisfied, also for $\boldsymbol{t}_i = \boldsymbol{0}$, and can be neglected for the QoS optimization. For the remaining problem, all solution approaches from Sect. 1.3 apply if the targets $\rho_k \geq \max\{\mu_k^{(\mathrm{B})}, 0\}$ are feasible.

These solutions remain valid for the partial ZF strategy from Sect. 3.2.1 with constraints $R_k^{(\mathrm{ZF})}(\boldsymbol{W}_k\boldsymbol{\tau}_k) \geq \rho_k$ for $k \in \mathcal{I}$ and $R_k^{(\mathrm{B})}(\boldsymbol{W}\boldsymbol{\tau}) \geq \rho_k$ for $k \in \mathcal{I}^{\mathrm{c}}$.[15] Then, the QoS optimization is with respect to the reduced precoder $\boldsymbol{\tau} = \mathrm{vec}([\boldsymbol{\tau}_1,\ldots,\boldsymbol{\tau}_K])$, i.e.,

$$\min_{p\geq 0,\, \boldsymbol{t}} p \quad \text{s.t.} \quad p^{-1}\boldsymbol{W}\boldsymbol{\tau} \in \mathcal{P}, \quad \mathrm{SINR}_k^{(\mathrm{B})}(\boldsymbol{W}\boldsymbol{\tau}) \geq \vartheta_k, \quad k=1,\ldots,K, \tag{3.17}$$

where $\boldsymbol{W} = \mathrm{bdiag}(\boldsymbol{W}_1,\ldots,\boldsymbol{W}_K)$ defines the subspace for the extended beamformers. The SINRs with ZF beamformers now read as [cf. (3.9)]

$$\mathrm{SINR}_k^{(\mathrm{B})}(\boldsymbol{W}\boldsymbol{\tau}) = \begin{cases} \dfrac{|\bar{\boldsymbol{h}}_k^{\mathrm{H}}\boldsymbol{W}_i\boldsymbol{\tau}_i|^2}{\nu_k^{(\mathrm{B})}+\sum_{i\neq k}|\bar{\boldsymbol{h}}_k^{\mathrm{H}}\boldsymbol{W}_i\boldsymbol{\tau}_i|^2}, & k\in\mathcal{I}^{\mathrm{c}}, \\ |\bar{\boldsymbol{h}}_k^{\mathrm{H}}\boldsymbol{W}_k\boldsymbol{\tau}_k|^2 & k\in\mathcal{I}, \end{cases} \tag{3.18}$$

[14]Using $R_k^{(\mathrm{B})}$ instead of $R_{k,+}^{(\mathrm{B})}$ does not change the result as $\rho_k > 0$ requires $R_k^{(\mathrm{B})}(\boldsymbol{t}) > 0$.

[15]The set $\mathcal{I}^{\mathrm{c}}$ stands for the complement subset to $\mathcal{I}$, i.e., $\mathcal{I}^{\mathrm{c}} = \{1,\ldots,K\}\setminus\mathcal{I}$.

and the targets are $\vartheta_k = B_k^{-1}(\rho_k)$ for $k \in \mathcal{I}$, while $\vartheta_k = 2^{\rho_k - \mu_k^{(\mathrm{B})}} - 1$ for $k \in \mathcal{I}^{\mathrm{c}}$. The ZF strategy reduces the complexity of the convex SOC formulation compared to only employing the bounds, because it replaces the SINR requirements with $|\mathcal{I}|$ linear inequality and only $K - |\mathcal{I}|$ instead of K SOC constraints, i.e., (cf. Sect. 3.2.1)

$$\min_{\boldsymbol{\tau},p} \ p \quad \text{s.t.} \quad \begin{cases} \mathrm{Re}(\bar{\boldsymbol{h}}_k^{\mathrm{H}} \boldsymbol{W}_k \boldsymbol{\tau}_k) - \sqrt{\vartheta_k} \in \mathbb{R}_+, & k \in \mathcal{I}, \\ \left(\mathrm{Re}(\bar{\boldsymbol{h}}_k^{\mathrm{H}} \boldsymbol{W}_k \boldsymbol{\tau}_k), \sqrt{\vartheta_k} \begin{bmatrix} \boldsymbol{T}_{\setminus\{k\}}^{\mathrm{H}} \bar{\boldsymbol{h}}_k \\ \nu_k^{(\mathrm{B}),1/2} \end{bmatrix} \right) \in \mathcal{L}^K, & k \in \mathcal{I}^{\mathrm{c}}, \end{cases} \tag{3.19}$$
$$\left(p\sqrt{P_\ell}, \boldsymbol{A}_\ell^{1/2} \boldsymbol{W} \boldsymbol{\tau} \right) \in \mathcal{L}^{NK}, \quad \ell = 1, \ldots, L,$$

where we substituted $\boldsymbol{T}_{\setminus\{k\}} = [\boldsymbol{W}_1 \boldsymbol{\tau}_1, \ldots, \boldsymbol{W}_{k-1} \boldsymbol{\tau}_{k-1}, \boldsymbol{W}_{k+1} \boldsymbol{\tau}_{k+1}, \ldots, \boldsymbol{W}_K \boldsymbol{\tau}_K]$.

Due to the partial ZF condition, also the dual formulation to (3.17) changes to

$$\max_{\boldsymbol{\mu} \geq \mathbf{0}} \min_{\boldsymbol{\lambda} \geq \mathbf{0}, \boldsymbol{u}} \sum_{i=1}^{K} \lambda_i \quad \text{s.t.} \quad \sum_{\ell=1}^{L} P_\ell \mu_\ell \leq 1, \quad \mathrm{SINR}_k^{(\mathrm{B,ul})} \geq \vartheta_k, \quad k = 1, \ldots, K, \tag{3.20}$$

where $\boldsymbol{u} = [\boldsymbol{u}_1^{\mathrm{T}}, \ldots, \boldsymbol{u}_K^{\mathrm{T}}]^{\mathrm{T}}$ comprises the uplink equalizers and $\boldsymbol{\lambda} = [\lambda_1, \ldots, \lambda_K]^{\mathrm{T}}$ the power allocation, i.e., the Lagrangian multipliers for the downlink SINR constraints. The outer optimization over the Lagrangian multipliers for the downlink power constraints $\boldsymbol{\mu} = [\mu_1, \ldots, \mu_K]^{\mathrm{T}}$ jointly minimizes the SINRs [cf. (1.34)]

$$\mathrm{SINR}_k^{(\mathrm{B,ul})} = \frac{\nu_k^{(\mathrm{B})-1} |\bar{\boldsymbol{h}}_k^{\mathrm{H}} \boldsymbol{W}_k \boldsymbol{u}_k|^2 \lambda_k}{\sum_{i \in \mathcal{I}^{\mathrm{c}} \setminus \{k\}} \nu_i^{(\mathrm{B})-1} |\bar{\boldsymbol{h}}_i^{\mathrm{H}} \boldsymbol{W}_k \boldsymbol{u}_k|^2 \lambda_i + \sum_{\ell=1}^{L} \mu_\ell \| \boldsymbol{A}_{k,\ell}^{1/2} \boldsymbol{W}_k \boldsymbol{u}_k \|_2^2}. \tag{3.21}$$

Due to the ZF requirements (3.8), the uplink filters $\boldsymbol{W}_k \boldsymbol{u}_k$, $k = 1, \ldots, K$ are orthogonal to the interfering channels to the downlink receivers $k \in \mathcal{I}$.

Similarly to (3.16), the RB problem with ergodic rate bounds reads as [cf. (3.2)]

$$\max_{\rho, \boldsymbol{t}} \ \rho \quad \text{s.t.} \quad \boldsymbol{t} \in \mathcal{P}, \quad R_{k,+}^{(\mathrm{B})}(\boldsymbol{t}) \geq \rho \rho_k, \quad k = 1, \ldots, K. \tag{3.22}$$

Depending on the employed bounds, the solution for this problem differs from the perfect CSI counterpart. The optimum of (3.22) will be zero if $\mu_i^{(\mathrm{B})} < 0$, $i = 1, \ldots, K$, and $\boldsymbol{t} \in \mathcal{P}$ is too restrictive. Then, there is no precoder $\boldsymbol{t}$ that achieves $R_k^{(\mathrm{B})}(\boldsymbol{t}) > 0$ for all $k = 1, \ldots, K$.[16] In the contrary, the optimum of (3.22) is strictly positive if $\mu_k^{(\mathrm{B})} > 0$ for all $k = 1, \ldots, K$, even for $\boldsymbol{t} = \mathbf{0}$. For low transmit power and with different bounds for the receivers, the ratios $R_k^{(\mathrm{B})}(\boldsymbol{t})/\rho_k$ can also be unbalanced at the optimum of (3.22), e.g., the optimum ρ may satisfy $\rho \rho_k \geq \mu_k^{(\mathrm{B})} =$

[16] The optimum of (3.22) could even become negative if $R_{k,+}^{(\mathrm{B})}(\boldsymbol{t})$ was replaced by $R_k^{(\mathrm{B})}(\boldsymbol{t})$.

0 for $k \in \mathcal{I}$ and $\rho\rho_i < \mu_i^{(\mathrm{B})} > 0$ for $k \in \mathcal{I}^{\mathrm{c}}$. Then, the optimal beamformers are $\boldsymbol{t}_i = \boldsymbol{0}$ for $i \in \mathcal{I}^{\mathrm{c}}$ and the remaining beamformers can be designed with the active constraints $k \in \mathcal{I}$.

These situations have to be detected by the beamformer design strategy. For example, solving (3.22) via a bisection over ρ, the number of active constraints is inherently known for each candidate ρ'. Then, it suffices to test whether a $\boldsymbol{t} \in \mathcal{P}$ exists that satisfies the active rate constraints $k \in \mathcal{I}$. A convex interior-point solver may be employed for the test (e.g., see [79]). Alternatively, we test feasibility via solving the above SINR constrained power minimization with the targets

$$\vartheta_k(\rho') = [2^{\rho'\rho_k - \mu_k^{(\mathrm{B})}} - 1]^+, \quad k = 1, \ldots, K. \tag{3.23}$$

Alternatively, the iterative balancing algorithms from Sect. 1.3 may be applied for the Lagrangian dual SINR balancing formulation

$$\begin{aligned} \min_{\boldsymbol{\mu} \geq \boldsymbol{0}} \max_{\rho, \boldsymbol{u}, \boldsymbol{\lambda} \geq \boldsymbol{0}} \quad & \rho \text{ s.t. } \sum_{i=1}^{K} \lambda_i \leq 1, \ \sum_{\ell=1}^{L} P_\ell \mu_\ell \leq 1, \\ & \mathrm{SINR}_k^{(\mathrm{B,ul})} \geq \vartheta_k(\rho), \\ & k = 1, \ldots, K, \end{aligned} \tag{3.24}$$

with the dual uplink SINR formulation

$$\mathrm{SINR}_k^{(\mathrm{B,ul})} = \frac{\nu_k^{(\mathrm{B})-1} |\bar{\boldsymbol{h}}_k^{\mathrm{H}} \boldsymbol{u}_k|^2 \lambda_k}{\sum_{i \neq k} \nu_i^{(\mathrm{B})-1} |\bar{\boldsymbol{h}}_i^{\mathrm{H}} \boldsymbol{u}_k|^2 \lambda_i + \sum_{\ell=1}^{L} \mu_\ell \|\boldsymbol{A}_{k,\ell}^{1/2} \boldsymbol{u}_k\|_2^2}. \tag{3.25}$$

To solve (3.24), the algorithms require some sequential target adaptation to account for the continuous but non-linear target function (3.23). Furthermore, the Lagrangian multiplier λ_k to the kth downlink SINR constraint must be zero if $\rho\rho_k - \mu_k^{(\mathrm{B})} < 0$. We ensure this by the next presented algorithms: a two-step iterative process similar to [225, Section III] for the inner uplink balancing problem and an outer projected gradient search similar to [165, Section III.D].

Remark 3.1 For additional ZF constraints, the balancing optimization must be based on the dual SINR formulation (3.21). The optimization and derivation steps remain the same, but with the effective channels $\boldsymbol{W}_k^{\mathrm{H}} \bar{\boldsymbol{h}}_i$ instead of $\bar{\boldsymbol{h}}_i$ for the uplink SINR. The downlink beamformers are then given by (3.8).

3.2.3.1 Iterative Inner Rate Balancing Optimization

The iterative solution for the inner maximization of (3.24) exploits the interference function representation of the uplink SINR requirements (1.19) and the global convergence of the fixed point update (1.18), which is even valid under a normalization in each iteration [226, Section II]. Let the per-user interference be given by[17]

$$I_k(\boldsymbol{\lambda}; \boldsymbol{\mu}) = \lambda_k / \operatorname{SINR}_k^{(\mathrm{B,ul})}, \quad k = 1, \ldots, K. \tag{3.26}$$

The so obtained interference function $\boldsymbol{I}(\boldsymbol{\lambda}; \boldsymbol{\mu})$ is standard (cf. Definition 1.1) even when inserting the SINR maximizing MMSE filters

$$\boldsymbol{u}_k = \boldsymbol{Y}_k^{-1}(\boldsymbol{\lambda}, \boldsymbol{\mu})\bar{\boldsymbol{h}}_k, \quad k = 1, \ldots, K, \tag{3.27}$$

with $\boldsymbol{Y}_k(\boldsymbol{\lambda}, \boldsymbol{\mu}) = \sum_{i=1}^K \nu_i^{(\mathrm{B})-1}\bar{\boldsymbol{h}}_i\bar{\boldsymbol{h}}_i^{\mathrm{H}}\lambda_i + \sum_{\ell=1}^L \mu_\ell \boldsymbol{A}_{k,\ell}$. Then, (3.26) equals

$$I_k(\boldsymbol{\lambda}; \boldsymbol{\mu}) = \frac{\nu_k^{(\mathrm{B})}}{\bar{\boldsymbol{h}}_k^{\mathrm{H}}\boldsymbol{Y}_k^{-1}(\boldsymbol{\lambda}, \boldsymbol{\mu})\bar{\boldsymbol{h}}_k} - \lambda_k, \quad k = 1, \ldots, K. \tag{3.28}$$

With (3.28), the SINR constraints and the power limitation alternatively read as

$$\boldsymbol{\lambda} \geq \boldsymbol{\Gamma}(\rho)\boldsymbol{I}(\boldsymbol{\lambda}; \boldsymbol{\mu}), \tag{3.29a}$$

$$1 \geq \mathbf{1}^{\mathrm{T}}\boldsymbol{\lambda}, \tag{3.29b}$$

respectively, where $\boldsymbol{\Gamma}(\rho) = \operatorname{diag}(\vartheta_1(\rho), \ldots, \vartheta_K(\rho)) \in \mathbb{R}^{K \times K}$ comprises the SINR targets from (3.23). Both constraints hold with equality at the optimum if there is an achievable upper bound for balancing level ρ and a non-zero lower bound for power allocation $\boldsymbol{\lambda}$ (see Proposition A.1).

A normalized fixed point search similar to the power iteration from [226] has shown to jointly converge to the solution power allocation and RB factor [225, Section III]. The iteration is a variation of the globally convergent update rule (1.18):

$$\boldsymbol{\lambda}^{(n+1)} = \boldsymbol{\Gamma}(\rho^{(n+1)})\boldsymbol{I}(\boldsymbol{\lambda}^{(n)}; \boldsymbol{\mu}). \tag{3.30}$$

The normalization by the choice of $\rho^{(n+1)}$ is to ensure (3.29b), i.e., $\mathbf{1}^{\mathrm{T}}\boldsymbol{\lambda}^{(n+1)} = 1$. Directly inserting (3.30), the update of the balancing level reads as

$$\rho^{(n+1)} = \max\left\{\rho \in \mathbb{R}_+ : \mathbf{1}^{\mathrm{T}}\boldsymbol{\Gamma}(\rho^{(n+1)})\boldsymbol{I}(\boldsymbol{\lambda}^{(n)}; \boldsymbol{\mu}) - 1 = 0\right\}. \tag{3.31}$$

This root is found via a line search, e.g., a bisection or a Newton-type method [227, Section 1.4]. Iteratively performing these power update and normalization steps, the sequence $\rho^{(n)}$ shows a similar convergence behavior as the iteration from [226]. It is neither strictly converging from below nor from above, but merely alternates

[17] To simplify exposition, the target function is excluded from the interference function.

between iterations above and below the convergence point. Moreover, the two-step iteration keeps the complexity for an implementation and the computations per iteration low compared to direct downlink solutions.

Remark 3.2 The SINR balancing power allocation via repeated eigenvector and uplink filter computations [89] can be employed for an alternative solution. Since this method was introduced for linear target functions ϑ, a direct application for RB fails. An additional adaptation of the balancing level is required for finding $(\rho, \boldsymbol{\lambda})$, which increases the number of eigenvector computations.

3.2.3.2 Outer Worst-Case Noise Optimization

The inner iteration is sufficient for solving (3.22) for a single sum power requirement, because the dual variable $\mu_1 = 1/P_1$ in this case. For multiple power constraints, the remaining outer optimization searches the worst-case $\boldsymbol{\mu} \in \mathbb{R}_+^L$, which minimizes the inner problem's optimum $\rho(\boldsymbol{\mu})$:

$$\min_{\boldsymbol{\mu} \geq \mathbf{0}} \rho(\boldsymbol{\mu}) \quad \text{s.t.} \quad \sum_{\ell=1}^{L} P_\ell \mu_\ell \leq 1. \tag{3.32}$$

For the solution, we might employ a projected subgradient algorithm similar to [63, 120] or the fixed point search from [118]. The next detailed Projected gradient (PG) algorithm speeds up convergence in comparison to these methods.

For the PG algorithm, a gradient descent step reads as

$$\boldsymbol{\mu}^{(j+1)} = \boldsymbol{\Pi}_{\mathcal{C}}\big(\boldsymbol{\mu}^{(j)} - a_j \boldsymbol{\delta}\big).$$

The orthogonal projection $\boldsymbol{\Pi}_{\mathcal{C}} : \mathbb{R}_+^L \rightarrow \mathcal{C}$ into the simplex $\mathcal{C} = \{\boldsymbol{\mu} \in \mathbb{R}_+^L : \sum_{\ell=1}^{L} P_\ell \mu_\ell \leq 1\}$ ensures that an update along the descent direction $\boldsymbol{\mu}^{(j)} - a_j \boldsymbol{\delta}$, with step size $a_j \in \mathbb{R}_+$, results in a feasible point. This projection reads as (cf. [228])

$$\boldsymbol{\Pi}_{\mathcal{C}}(\boldsymbol{x}) = [\boldsymbol{x} - \gamma \boldsymbol{p}]^+, \tag{3.33}$$

where $\boldsymbol{p} = [P_1, \ldots, P_L]^{\mathrm{T}}$, $[\cdot]^+$ performs an elementwise projection onto the non-negative orthant[18] and γ is chosen to satisfy $\sum_{\ell=1}^{L} P_\ell [x_\ell - \gamma P_\ell]^+ = 1$. Here, the step size is selected via Armijo's rule that diminishes a_j until ρ decreases [227].

Since a closed-form expression for $\rho(\boldsymbol{\mu})$ is missing, we exploit equality of (3.29a) and (3.29b) for the gradient calculation. Assuming $\vartheta_k(\rho) > 0$ for $\mathcal{I} = \{1, \ldots, |\mathcal{I}|\}$ and $\vartheta_i(\rho) = 0$ for $i \in \mathcal{I}^{\mathrm{c}} = \{|\mathcal{I}| + 1, \ldots, K\}$ to this end,[19] we only consider the remaining equalities

[18] The projection onto the non-negative orthant itself reads as $[z]^+ = \max(0, z)$ for $z \in \mathbb{R}$.

[19] This indexing is achieved via a reordering, such that $\rho_1 \rho - \mu_1^{(\mathrm{B})} \geq \ldots \geq \rho_K \rho - \mu_K^{(\mathrm{B})}$ for $\rho \in \mathbb{R}_+$.

$$\sum_{i=1}^{|\mathcal{I}|} \lambda_i = 1, \quad \lambda_k = \vartheta_k(\rho) I_k(\boldsymbol{\lambda}; \boldsymbol{\mu}), \quad k \in \mathcal{I}. \tag{3.34}$$

Based on the derivatives of (3.34) with respect to the entries of $\boldsymbol{\mu}$, the entries of the gradient $\boldsymbol{\delta} = [\delta_1, \ldots, \delta_L]^{\mathrm{T}}$ can be shown to read as

$$\delta_m = \frac{\partial \rho}{\partial \mu_m} = -\frac{\mathbf{1}^{\mathrm{T}}(\mathbf{I} - \boldsymbol{\Gamma}\boldsymbol{\Psi})^{-1}\boldsymbol{\Gamma}\boldsymbol{n}_m}{\mathbf{1}^{\mathrm{T}}(\mathbf{I} - \boldsymbol{\Gamma}\boldsymbol{\Psi})^{-1}\boldsymbol{D}\boldsymbol{I}(\boldsymbol{\lambda}; \boldsymbol{\mu})}, \quad m = 1, \ldots, L, \tag{3.35}$$

where $\boldsymbol{D} = \mathrm{diag}(d_1, \ldots, d_{|\mathcal{I}|})$ comprises the derivatives of the targets, i.e., $d_i = \frac{\partial \vartheta_i(\rho)}{\partial \rho}$, and $\boldsymbol{\Gamma} = \mathrm{diag}(\vartheta_1(\rho), \ldots, \vartheta_{|\mathcal{I}|}(\rho))$. Furthermore, $\boldsymbol{\Psi}$ and $\boldsymbol{n}_m$ have the entries

$$[\boldsymbol{\Psi}]_{k,i} = \frac{\partial I_k(\boldsymbol{\lambda}; \boldsymbol{\mu})}{\partial \lambda_i}, \qquad [\boldsymbol{n}_m]_k = \frac{\partial I_k(\boldsymbol{\lambda}; \boldsymbol{\mu})}{\partial \mu_m}, \qquad k, i = 1, \ldots, |\mathcal{I}|. \tag{3.36}$$

The target derivatives are $\frac{\partial \vartheta_k(\rho)}{\partial \rho} = \vartheta_k(\rho)\rho_k$ for $\vartheta_k(\rho) = 2^{\rho\rho_k - \mu_k^{(\mathrm{B})}}$ and the derivatives of the interference function $I_k(\boldsymbol{\lambda}; \boldsymbol{\mu})$ for $\boldsymbol{\Psi}$ and $\boldsymbol{n}_m$ read as

$$\frac{\partial I_k(\boldsymbol{\lambda}; \boldsymbol{\mu})}{\partial \lambda_i} = \begin{cases} 0 & i = k \\ \nu_k^{(\mathrm{B})} \frac{|\bar{\boldsymbol{h}}_k^{\mathrm{H}} \boldsymbol{Y}^{-1} \bar{\boldsymbol{h}}_i|^2 \nu_i^{(\mathrm{B}),-1}}{(\bar{\boldsymbol{h}}_k^{\mathrm{H}} \boldsymbol{Y}^{-1} \bar{\boldsymbol{h}}_k)^2} = \frac{\lambda_k}{\vartheta_k(\rho)} \frac{|\bar{\boldsymbol{h}}_k^{\mathrm{H}} \boldsymbol{Y}^{-1} \bar{\boldsymbol{h}}_i|^2 \nu_i^{(\mathrm{B}),-1}}{\bar{\boldsymbol{h}}_k^{\mathrm{H}} \boldsymbol{Y}^{-1} \bar{\boldsymbol{h}}_k} & i \neq k \end{cases} \tag{3.37}$$

$$\frac{\partial I_k(\boldsymbol{\lambda}; \boldsymbol{\mu})}{\partial \mu_m} = \nu_k^{(\mathrm{B})} \frac{\bar{\boldsymbol{h}}_k^{\mathrm{H}} \boldsymbol{Y}^{-1} \boldsymbol{A}_m \boldsymbol{Y}^{-1} \bar{\boldsymbol{h}}_k}{(\bar{\boldsymbol{h}}_k^{\mathrm{H}} \boldsymbol{Y}^{-1} \bar{\boldsymbol{h}}_k)^2} = \frac{\lambda_k}{\vartheta_k(\rho)} \frac{\bar{\boldsymbol{h}}_k^{\mathrm{H}} \boldsymbol{Y}^{-1} \boldsymbol{A}_m \boldsymbol{Y}^{-1} \bar{\boldsymbol{h}}_k}{\bar{\boldsymbol{h}}_k^{\mathrm{H}} \boldsymbol{Y}^{-1} \bar{\boldsymbol{h}}_k}, \tag{3.38}$$

respectively, where the second equation follows with (3.34).

Existence of the inverse matrix $(\mathbf{I} - \boldsymbol{\Gamma}\boldsymbol{\Psi})^{-1}$ in (3.35) is shown with the matrix reformulation $\mathbf{I} - \boldsymbol{\Gamma}\boldsymbol{\Psi} = \mathbf{I} - \boldsymbol{\Lambda}\boldsymbol{\Psi}'$, where $\boldsymbol{\Lambda} = \mathrm{diag}(\lambda_1, \ldots, \lambda_{|\mathcal{I}|})$ and

$$[\boldsymbol{\Psi}']_{k,i} = \begin{cases} 0 & i = k, \\ \frac{|\bar{\boldsymbol{h}}_k^{\mathrm{H}} \boldsymbol{Y}^{-1} \bar{\boldsymbol{h}}_i|^2 \nu_i^{(\mathrm{B}),-1}}{\bar{\boldsymbol{h}}_k^{\mathrm{H}} \boldsymbol{Y}^{-1} \bar{\boldsymbol{h}}_k} & i \neq k. \end{cases}$$

The matrix $\mathbf{I} - \boldsymbol{\Lambda}\boldsymbol{\Psi}'$ is a Z-matrix [229] with non-positive off-diagonal entries and ones at the diagonal position. In particular, $\mathbf{I} - \boldsymbol{\Lambda}\boldsymbol{\Psi}'$ is a non-singular M-matrix according to [229, Theorem 6.2.3, (M36)][20] because $\boldsymbol{\Lambda}^{-1}(\mathbf{I} - \boldsymbol{\Lambda}\boldsymbol{\Psi})\boldsymbol{\Lambda}$ is diagonally dominant, that is, the sums of the row entries are non-negative. Therefore, the inverse $(\mathbf{I} - \boldsymbol{\Gamma}\boldsymbol{\Psi})^{-1}$ exists and all its entries are positive. Hence, also the entries of the gradient (3.35) are non-negative.

[20]Berman and Plemmons [229, Chapter 6] provide an extensive discussion about non-singular M-matrices including 50 equivalent statements for their definition [229, Theorem 6.2.3].

3.2.3.3 Uplink–Downlink Transformation

With the gradient (3.35), the PG method ensures convergence to a solution of (3.32). The primal beamformers $\boldsymbol{t}_k$ are then reconstructed from the dual solution $\boldsymbol{\mu}$, $\boldsymbol{\lambda}$, and $\boldsymbol{u}$ as has been detailed by Yu and Lan [74]. In particular, the downlink beamformers are $\boldsymbol{t}_k = \sqrt{\theta_k}\boldsymbol{u}_k$ for $k = 1, \ldots, |\mathcal{I}|$ and $\boldsymbol{t}_i = \boldsymbol{0}$ for $i \in \mathcal{I}^{\mathrm{c}}$. The scaling factors $\theta_k > 0$ are the solution of a $|\mathcal{I}|$-dimensional linear equation system for $\mathrm{SINR}_k = \mathrm{SINR}_k^{(\mathrm{ul})}$, $k \in \mathcal{I}$ (e.g., see [5, Section II] and [83, Section II.B]).

3.3 Quality-of-Service Feasibility Region

The QoS problem (3.16) with given rate target vector $\boldsymbol{\rho} = [\rho_1, \ldots, \rho_K]^{\mathrm{T}}$ can be infeasible when $K > N$, even though the transmit power is unbounded. While Hunger et al. [65] provided the perfect CSI feasibility region from Lemma 1.1 and (1.31), the region of feasible QoS targets has remained unknown for ergodic rate requirements, also for the rank-one channel model. Directly applying the proof of [65, Theorem III.1] fails due to the lack of uplink–downlink duality for ergodic rates. The main result in this section is that the QoS feasible regions for (1.9) and (3.1), with perfect CSI and ergodic rates, respectively, coincide for the multiplicative channel error model (2.2) (see Theorem 3.1). To show this, we first provide the QoS feasible region for (3.16), which employs the ergodic rate bounds (3.11).

Lemma 3.1 *Let the channels in $\bar{\boldsymbol{H}} = [\bar{\boldsymbol{h}}_1, \ldots, \bar{\boldsymbol{h}}_K] \in \mathbb{C}^{N\times K}$ be regular according to Definition 1.2. A target $\boldsymbol{\rho} \in \mathbb{R}_+^K$ is feasible for* (3.16) *if $\boldsymbol{\rho} \in \mathcal{F}_{\mathrm{rate}}^{(\mathrm{B})}$, where*

$$\mathcal{F}_{\mathrm{rate}}^{(\mathrm{B})} = \left\{\boldsymbol{\rho} \in \mathbb{R}_+^K : \sum_{k=1}^{K} 2^{-\max\left\{\rho_k - \mu_k^{(\mathrm{B})}, 0\right\}} \geq K - N\right\}. \tag{3.39}$$

The proof in Sect. A.4 exploits Lemma 1.1 and relies on a reformulation of the rate constraints within (3.16) as equivalent uplink MMSE requirements.

The region $\mathcal{F}_{\mathrm{rate}}^{(\mathrm{B})}$ can be a sub- or a superset or equal to the perfect CSI feasible rate region $\mathcal{F}_{\mathrm{rate}}$ (1.31), depending on the employed ergodic rate bounds (cf. Table 3.1):

1. $\mathcal{F}_{\mathrm{rate}}^{(\mathrm{B})} \supset \mathcal{F}_{\mathrm{rate}}$ if $\mu_k^{(\mathrm{B})} > 0$, i.e., for the upper bounds UB1 and AUB;
2. $\mathcal{F}_{\mathrm{rate}}^{(\mathrm{B})} = \mathcal{F}_{\mathrm{rate}}$ if $\mu_k^{(\mathrm{B})} = 0$, i.e., for the two bounds LB1 and UB2;
3. $\mathcal{F}_{\mathrm{rate}}^{(\mathrm{B})} \subset \mathcal{F}_{\mathrm{rate}}$ if $\mu_k^{(\mathrm{B})} < 0$, i.e., for the lower bounds LB2 and ALB.

For case 3, the feasible region $\mathcal{F}_{\mathrm{rate}}^{(\mathrm{B})}$ can even be empty if $K \gg N$. For example, let $\mu^{(\mathrm{B})} = \mu_k^{(\mathrm{B})}$ for all $k = 1, \ldots, K$. Then, the sum bound within (3.39) is violated also for $\boldsymbol{\rho} = \boldsymbol{0}$ if $K > N(1 - 2^{\mu^{(\mathrm{B})}})^{-1}$. This is in contrast to the QoS feasible region with ergodic rates, which cannot be empty. Furthermore, also case 1 does not reflect the properties of the QoS feasible region with ergodic rates. Intuitively, this region must be a subset or equal to the perfect CSI rate region. Equality is suggested by case 2, where $\mathcal{F}_{\text{rate}}^{(\text{B})} = \mathcal{F}_{\text{rate}}$ holds for the ergodic rate bounds LB1 and UB2, respectively. In fact, the feasible regions for (1.9) and (3.1) coincide.

Theorem 3.1 *Let the channels in* $\bar{\boldsymbol{H}} = [\bar{\boldsymbol{h}}_1, \ldots, \bar{\boldsymbol{h}}_K] \in \mathbb{C}^{N \times K}$ *be regular. A rate target* $\boldsymbol{\rho}$ *is feasible for* (3.1) *if* $\boldsymbol{\rho} \in \mathcal{F}_{\text{rate}} = \left\{\boldsymbol{\rho} \in \mathbb{R}_+^K : \sum_{k=1}^K 2^{-\rho_k} \geq K - N\right\}$.

The proof in the remainder of this section is constructive in the sense that it provides a feasible beamforming scheme for (3.1) in all three cases: (1) $\mathrm{E}[\zeta_k]$ and $\mathrm{E}[\zeta_k^{-1}]$ are both finite, (2) $\mathrm{E}[\zeta_k]$ exists and $\mathrm{E}[\zeta_k^{-1}]$ is infinite, and (3) $\mathrm{E}[\zeta_k^{-1}]$ exists but $\mathrm{E}[\zeta_k]$ is infinite. Cases (2) and (3) additionally exploit the following lemmas, which are based on the properties of A_k and a_k and B_k and b_k, respectively.

Lemma 3.2 *The ergodic rate* R_k *and its lower bound* $R_k^{(\text{LB1})}$ *differ at most by the value of the function* $\delta_k^{(\text{LB1})}: \mathbb{C}^{NK} \to \mathbb{R}_+$*:*

$$\begin{aligned} R_k^{(\text{LB1})}(\boldsymbol{t}) \leq R_k(\boldsymbol{t}) \leq R_k^{(\text{LB1})}(\boldsymbol{t}) + \delta_k^{(\text{LB1})}(\boldsymbol{t}), \\ \delta_k^{(\text{LB1})}(\boldsymbol{t}) = A_k(\mathtt{I}_k(\boldsymbol{t})) - a_k(\mathrm{E}[\zeta_k], \mathtt{I}_k(\boldsymbol{t})), \end{aligned} \tag{3.40}$$

where $\mathtt{I}_k(\boldsymbol{t})$ *is the interference* (3.15). *If* $\mathtt{I}_k(\boldsymbol{t}) > 0$, *then* $\delta_k^{(\text{LB1})}(\alpha \boldsymbol{t})$ *decreases monotonically in the scaling factor* $\alpha > 1$, $\delta_k^{(\text{LB1})}(\mathbf{0}) = \mu_k^{(\text{UB1})}$, *and* $\lim_{\alpha \to \infty} \delta_k^{(\text{LB1})}(\alpha \boldsymbol{t}) = 0$.

Lemma 3.3 *Similarly,* R_k *and* $R_k^{(\text{UB2})}$ *differ at most by* $\delta_k^{(\text{UB2})}: \mathbb{C}^{NK} \to \mathbb{R}_+$*:*

$$\begin{aligned} R_k^{(\text{UB2})}(\boldsymbol{t}) - \delta_k^{(\text{UB2})}(\boldsymbol{t}) \leq R_k(\boldsymbol{t}) \leq R_k^{(\text{UB2})}(\boldsymbol{t}), \\ \delta_k^{(\text{UB2})}(\boldsymbol{t}) = B_k(\mathtt{I}_k(\boldsymbol{t})) - b_k(\mathrm{E}[\zeta_k^{-1}], \mathtt{I}_k(\boldsymbol{t})) - \mu_k^{(\text{LB2})}. \end{aligned} \tag{3.41}$$

Here, $\delta_k^{(\text{UB2})}(\alpha \boldsymbol{t})$ *decreases monotonically with increasing* α *and has the limit values* $\delta_k^{(\text{UB2})}(\mathbf{0}) = -\mu_k^{(\text{LB2})}$ *and* $\lim_{\alpha \to \infty} \delta_k^{(\text{UB2})}(\alpha \boldsymbol{t}) = 0$.

Proof (Proof of Theorem 3.1) Let $\mathcal{F}_{\boldsymbol{R}}$ denote the feasible region for the QoS optimization (3.1), that is, any target vector $\boldsymbol{\rho} \in \mathcal{F}_{\boldsymbol{R}}$ is attainable. To prove that equality hold for $\mathcal{F}_{\boldsymbol{R}} = \mathcal{F}_{\text{rate}}$, we distinguish three scenarios based on the existence of $\mathrm{E}[\zeta_k]$, $\mathrm{E}[\zeta_k^{-1}]$, or both.

1. Let $\mathrm{E}[\zeta_k]$ and $\mathrm{E}[\zeta_k^{-1}]$ be positive and finite for all $k = 1, \ldots, K$. Then,

$$R_k^{(\text{LB1})}(\boldsymbol{t}) \leq R_k(\boldsymbol{t}) \leq R_k^{(\text{UB2})}(\boldsymbol{t})$$

holds for all $p^{-1}\boldsymbol{t} \in \mathcal{P}$ with $p \in (0, \infty)$. The second inequality implies the relation $\mathcal{F}_{\boldsymbol{R}} \subseteq \mathcal{F}_{\text{rate}}^{(\text{UB2})} = \mathcal{F}_{\text{rate}}$, i.e., there is no $\boldsymbol{\rho} \in \mathcal{F}_{\boldsymbol{R}}$ outside of $\mathcal{F}_{\text{rate}}$. The first inequality in turn implies $\mathcal{F}_{\text{rate}}^{(\text{LB1})} = \mathcal{F}_{\text{rate}} \subseteq \mathcal{F}_{\boldsymbol{R}}$, i.e., all $\boldsymbol{\rho} \in \mathcal{F}_{\text{rate}}$ are attainable.

Hence, we conclude that

$$\mathcal{F}_{\text{rate}} = \mathcal{F}_{\text{rate}}^{(\text{LB1})} \subseteq \mathcal{F}_{\boldsymbol{R}} \subseteq \mathcal{F}_{\text{rate}}^{(\text{UB2})} = \mathcal{F}_{\text{rate}}.$$

2. Next, assume that only $\mathrm{E}[\zeta_k]$ exists, while $\mathrm{E}[\zeta_k^{-1}]$ is infinite. In this case, the first inequality of (3.40) again imposes $\mathcal{F}_{\text{rate}} = \mathcal{F}_{\text{rate}}^{(\text{LB1})} \subseteq \mathcal{F}_{\boldsymbol{R}}$. To proof the converse, i.e., $\mathcal{F}_{\boldsymbol{R}} \subseteq \mathcal{F}_{\text{rate}}$, we assume the contrary. Let $\boldsymbol{\rho} \notin \mathcal{F}_{\text{rate}}$ be attainable with ergodic rates, i.e., $\boldsymbol{\rho} \in \text{int}(\mathcal{F}_{\boldsymbol{R}})$. Then, there is a $\boldsymbol{t}$ such that $\rho_k \leq R_k(\boldsymbol{t})$ for all $k = 1, \ldots, K$. Since $R_k(\alpha \boldsymbol{t})$ strictly increases with $\alpha \in \mathbb{R}_+$, Lemma 3.2 imposes the series of inequalities

$$\rho_k \leq R_k(\boldsymbol{t}) \leq \lim_{\alpha \to \infty} R_k(\alpha \boldsymbol{t}) \leq \lim_{\alpha \to \infty} R_k^{(\text{LB1})}(\alpha \boldsymbol{t}) + \delta_k^{(\text{LB1})}(\alpha \boldsymbol{t}) = \lim_{\alpha \to \infty} R_k^{(\text{LB1})}(\alpha \boldsymbol{t})$$

for all $k \in \mathcal{I} = \{k \in \mathbb{N} : \mathtt{I}_k(\boldsymbol{t}) > 0, 1 \leq k \leq K\}$ and

$$\rho_k \leq R_k(\boldsymbol{t}) \leq \lim_{\alpha \to \infty} R_k^{(\text{LB1})}(\alpha \boldsymbol{t})$$

for $k \notin \mathcal{I}$, where $\mathtt{I}_k(\boldsymbol{t}) = 0$ and $R_k^{(\text{LB1})}(\alpha \boldsymbol{t})$ grows unbounded with α. The fact that $\lim_{\alpha \to \infty} [R_1^{(\text{LB1})}(\alpha \boldsymbol{t}), \ldots, R_K^{(\text{LB1})}(\alpha \boldsymbol{t})]^{\mathrm{T}} \in \mathcal{F}_{\text{rate}}^{(\text{LB1})} = \mathcal{F}_{\text{rate}}$ finally provides the intended contradiction.
3. Finally, let $\mathrm{E}[\zeta_k^{-1}]$ exist, while $\mathrm{E}[\zeta_k]$ is infinite, e.g., because $2\zeta_k^{-1} \sim \mathcal{X}_2^2$. For this case, the proof is based on (3.41). The second inequality imposes $\mathcal{F}_{\boldsymbol{R}} \subseteq \mathcal{F}_{\text{rate}}^{(\text{UB2})} = \mathcal{F}_{\text{rate}}$, i.e., any $\boldsymbol{\rho} \notin \mathcal{F}_{\text{rate}}$ is unattainable even with infinite transmit power. To show that $\mathcal{F}_{\text{rate}} \subseteq \mathcal{F}_{\boldsymbol{R}}$, we demonstrate that all $\boldsymbol{\rho} \in \mathcal{F}_{\text{rate}}$ can be reached. To this end, we consider an arbitrary but fixed $\boldsymbol{\rho} \in \text{int}(\mathcal{F}_{\text{rate}})$. As this target vector resides in the interior, there exists a $\boldsymbol{t}$ with finite norm satisfying $R_k^{(\text{UB2})}(\boldsymbol{t}) \geq \rho_k$ for all $k = 1, \ldots, K$.[21]

   To show that $\boldsymbol{\rho}$ is also attainable with ergodic rates (3.3), we find a $\boldsymbol{t}'$ such that

$$R_k(\boldsymbol{t}') \geq R_k^{(\text{UB2})}(\boldsymbol{t}), \quad \forall k = 1, \ldots, K. \tag{3.42}$$

We construct this precoder via an appropriate scaling $\boldsymbol{t}' = \alpha \boldsymbol{t}$ with $\alpha \geq 1$, for example. For (3.42) to hold, we require a large enough α that satisfies

$$R_k(\alpha \boldsymbol{t}) \geq R_k^{(\text{UB2})}(\alpha \boldsymbol{t}) - \delta_k(\alpha \boldsymbol{t}) \geq R_k^{(\text{UB2})}(\boldsymbol{t}), \quad \forall k = 1, \ldots, K. \tag{3.43}$$

The first inequality in this series stems from Lemma 3.3. For the second inequality in (3.43), we distinguish two cases with $\mathtt{I}_k(\boldsymbol{t}) > 0$ and $\mathtt{I}_k(\boldsymbol{t}) = 0$. If the interference is strictly positive, i.e., $\mathtt{I}_k(\boldsymbol{t}) > 0$, then $R_k^{(\text{UB2})}(\alpha \boldsymbol{t})$ increases with α, i.e.,

[21] This precoder can be found numerically, as we have suggested in the previous section.

$$\lim_{\alpha\to\infty} R_k^{(\mathrm{UB2})}(\alpha \boldsymbol{t}) = \log_2\left(1 + \frac{|\bar{\boldsymbol{h}}_k^{\mathrm{H}}\boldsymbol{t}_k|^2}{\sum_{i\neq k}|\bar{\boldsymbol{h}}_k^{\mathrm{H}}\boldsymbol{t}_i|^2}\right) \geq R_k^{(\mathrm{UB2})}(\boldsymbol{t}).$$

Moreover, $\delta_k^{(\mathrm{UB2})}(\alpha \boldsymbol{t})$ decreases with α and $\lim_{\alpha\to\infty}\delta_k^{(\mathrm{UB2})}(\alpha \boldsymbol{t}) = 0$ (see Lemma 3.3). Thus, there is a finite $\alpha > 1$ that fulfills (3.42). Otherwise, if $\mathtt{I}_k(\boldsymbol{t}) = 0$ exactly, then the ergodic rate bound

$$R_k^{(\mathrm{UB2})}(\alpha \boldsymbol{t}) = \log_2(1 + \alpha\,\mathrm{E}[\zeta_k^{-1}]|\bar{\boldsymbol{h}}_k^{\mathrm{H}}\boldsymbol{t}_k|^2)$$

increases unboundedly with α, while $\delta_k^{(\mathrm{UB2})}(\alpha \boldsymbol{t}) \leq \delta_k^{(\mathrm{UB2})}(\boldsymbol{0}) = -\mu_k^{(\mathrm{LB2})}$. Hence, there is also an α that satisfies (3.43). This completes the converse proof. □

Note that the proof required fully adaptive beamforming. Insisting on ZF beamforming to a subset of users $i \in \mathcal{I}$ (see Sect. 3.2.1) deforms this feasible QoS region if $K > N$. The degrees of freedom for the beamformers $\boldsymbol{t}_k, k \notin \mathcal{I}$ are reduced by one with each ZF constraint, which results into the feasibility region

$$\mathcal{F}_{\mathrm{rate}}^{(\mathrm{ZF})} = \left\{\boldsymbol{\rho} \in \mathbb{R}_+^K : \sum_{k\notin\mathcal{I}} 2^{-\rho_k} \geq K - |\mathcal{I}| - (N - |\mathcal{I}|)\right\}. \tag{3.44}$$

Here, the summation is over $k \notin \mathcal{I}$, while there are no bounds for targets ρ_i, $i \in \mathcal{I}$.

3.4 Post-Processing Power Allocation

When the beamformers are pre-fixed, e.g., by either of the beamforming solutions with the ergodic rate bounds from Sect. 3.2.3, power allocation can serve as a post-processing for the QoS optimization (3.1) and balancing problem (3.2). We even find a QoS feasible $\boldsymbol{t}$ for (3.1) via a rescaling of the solution $\boldsymbol{t}'$ for the QoS optimization with the ergodic rate bound UB2 (see proof of Theorem 3.1). The next power allocation improves upon the performance of the precoder rescaling.

Given a fixed precoder $\boldsymbol{t}'$, we search an improved $\boldsymbol{t}$ for (3.1) or (3.2) of the form

$$\boldsymbol{t}_k = \sqrt{p_k}\boldsymbol{t}'_k, \quad k = 1, \ldots, K.$$

For this rescaling of the beamformers with $p_k \in \mathbb{R}_+$, $k = 1, \ldots, K$, a second QoS or RB optimization, respectively, is performed (cf. [140, 141]) using the actual ergodic rates (3.3). To simplify expositions for these optimizations, let [cf. (3.3)]

$$\tilde{R}_k(\boldsymbol{p}) = R_k(\mathrm{vec}(\boldsymbol{T}'\,\mathrm{diag}(p_1^{1/2}, \ldots, p_K^{1/2})), \tag{3.45}$$

where $\boldsymbol{T}' = [\boldsymbol{t}'_1, \ldots, \boldsymbol{t}'_K]$. The set of jointly admissible power vectors $\boldsymbol{p} = [p_1, \ldots, p_K]^{\mathrm{T}}$ is denoted by

$$\mathcal{T} = \{\boldsymbol{p} \in \mathbb{R}_+^K : \tilde{\boldsymbol{A}}\boldsymbol{p} \le \mathbf{1}\}, \tag{3.46}$$

where $[\tilde{\boldsymbol{A}}]_{\ell,i} = P_\ell^{-1}\|\boldsymbol{A}_{i,\ell}^{1/2}\boldsymbol{t}_i'\|_2^2 \ge 0$, and $\mathrm{null}\{\tilde{\boldsymbol{A}}\} \cap \mathbb{R}_+^K = \{\mathbf{0}\}$.[22] This set $\mathcal{T}$ is a convex compact polytope with non-empty interior. In case of a single sum power constraint, i.e., $L = 1$ and $\boldsymbol{A}_{i,1} = \mathbf{I}$ for all $i = 1, \ldots, K$, only the half-space constraint $\tilde{\boldsymbol{a}}^{\mathrm{T}}\boldsymbol{p} \le 1$ bounds $\mathcal{T}$ from above, where $\tilde{\boldsymbol{a}}^{\mathrm{T}} = P^{-1}[\|\boldsymbol{t}_1'\|_2^2, \ldots, \|\boldsymbol{t}_K'\|_2^2] > \mathbf{0}$.

3.4.1 Post-Processing for QoS Optimization

With (3.45) and (3.46), the QoS power allocation problem reads as

$$\min_{P>0,\boldsymbol{p}\ge\mathbf{0}} \ P \quad \text{s.t. } P^{-1}\boldsymbol{p} \in \mathcal{T}, \quad \tilde{R}_k(\boldsymbol{p}) \ge \rho_k, \quad k = 1, \ldots, K. \tag{3.47}$$

Here, we squared the objective compared to (3.1) for a simplified notation. While this problem could be recast into a Linear program (LP) for instantaneous rate constraints, such a reformulation is unavailable for ergodic rate requirements.

Problem (3.47) has the equivalent min–max formulation

$$\min_{\boldsymbol{p}\ge\mathbf{0}} \max_{\ell} \ \mathbf{e}_\ell^{\mathrm{T}}\tilde{\boldsymbol{A}}\boldsymbol{p} \quad \text{s.t.} \quad \boldsymbol{p} \ge \boldsymbol{I}(\boldsymbol{p};\boldsymbol{\rho}), \tag{3.48}$$

where $\boldsymbol{I}(\cdot;\boldsymbol{\rho}) : \mathbb{R}_+^K \to \mathbb{R}_{++}^K$ with $\boldsymbol{I}(\boldsymbol{p};\boldsymbol{\rho}) = [I_1(\boldsymbol{p};\rho_1), \ldots, I_K(\boldsymbol{p};\rho_K)]^{\mathrm{T}}$ has the function entries[23]

$$I_k(\boldsymbol{p};\rho_k) = \frac{p_k\rho_k}{\tilde{R}_k(\boldsymbol{p})}, \quad k = 1, \ldots, K. \tag{3.49}$$

The objective is the maximum of a set of linear functions, which is convex, but (3.49) is generally non-convex.[24] For the solutions of (3.48), we exploit that $\boldsymbol{I}(\cdot;\boldsymbol{\rho})$ is a standard interference function according to Definition 1.1.

Lemma 3.4 *The function $\boldsymbol{I}(\cdot;\boldsymbol{\rho}) : \mathbb{R}_+^K \to \mathbb{R}_{++}^K$, $\boldsymbol{I}(\boldsymbol{p};\boldsymbol{\rho}) = [I_1(\boldsymbol{p};\rho_1), \ldots, I_K(\boldsymbol{p};\rho_K)]^{\mathrm{T}}$ with entries $I_k(\boldsymbol{p};\rho_K)$ from* (3.49) *is a positive sublinearly increasing monotone map.*

---

[22] These conditions ensure that none of the powers $p_k, k = 1, \ldots, K$ increases unboundedly.

[23] Several other interference functions may be found for ergodic rates. The advantage of this version is that it only requires an evaluation of the involved $\tilde{R}_k$ (3.4) for evaluating $I_k(\boldsymbol{p}), k = 1, \ldots, K$.

[24] Similar to the case with instantaneous rates, (3.49) appears to be concave. Its bounds $p_k/\tilde{R}_k^{(\mathrm{B})}(\boldsymbol{p})$ are concave. Moreover, $p_k/\tilde{r}_k(\boldsymbol{p})$ is concave and $\tilde{R}_k$ has similar properties as $\tilde{r}_k$.

Proof (Proof of Lemma 3.4) Positivity of $I_k(\boldsymbol{p}; \rho_k)$ is by definition of $\tilde{R}_k(\boldsymbol{p})$ and

$$\lim_{p_k \to 0} p_k / \tilde{R}_k([p_k, 0, \ldots, 0]^{\mathrm{T}}) = \frac{1}{\log_2(\mathrm{e}) |\bar{\boldsymbol{h}}_k^{\mathrm{H}} \boldsymbol{t}'_k|^2 \, \mathrm{E}[\zeta_k^{-1}]}.$$

Monotonicity of $I_k(\boldsymbol{p}; \rho_k)$ in p_i, $i \neq k$ follows because $\tilde{R}_k(\boldsymbol{p})$ decreases when increasing the interference. Moreover, since $\tilde{R}_k(\boldsymbol{p})$ increases sublinearly in p_k, also $p_k / \tilde{R}_k(\boldsymbol{p})$ increases sublinearly in $p_k > 0$. Finally, sublinearity of $I_k(\boldsymbol{p}; \rho_k)$ follows as $\tilde{R}_k(\boldsymbol{p}) < \tilde{R}_k(\alpha \boldsymbol{p})$ for $\alpha > 1$, thus, $I_k(\alpha \boldsymbol{p}) < \alpha I_k(\boldsymbol{p})$. □

Due to this property, the solution with minimum sum power is characterized by

$$\boldsymbol{p}^\star = \boldsymbol{I}(\boldsymbol{p}^\star; \boldsymbol{\rho}), \tag{3.50}$$

which is the *minimum point* of the feasible set, i.e., the set of all $\boldsymbol{p} \in \mathbb{R}_+^K$ that satisfies $\boldsymbol{p} \geq \boldsymbol{I}(\boldsymbol{p}^\star; \boldsymbol{\rho})$ [230, Theorem 5.57]. The corresponding optimum is[25]

$$P^\star = \max_\ell \mathbf{e}_\ell^{\mathrm{T}} \tilde{\boldsymbol{A}} \boldsymbol{p}^\star. \tag{3.51}$$

This solution is the convergence point of the power iteration [cf. (1.18)][26]

$$\boldsymbol{p}^{(n+1)} = \boldsymbol{I}(\boldsymbol{p}^{(n)}; \boldsymbol{\rho}), \tag{3.52}$$

which only requires an evaluation of $\tilde{R}_k(\boldsymbol{p}^{(n)})$, $k = 1, \ldots, K$ (3.4) for each update. While (3.48) can have multiple solutions in general, $\boldsymbol{p}^\star$ is unique when $\tilde{\boldsymbol{A}} > \boldsymbol{0}$ [230, Corollary 5.58] or $\boldsymbol{p}^\star$ is a strictly minimal point of the feasible set (see Sect. A.5).

3.4.2 Post-Processing for Ergodic Rate Balancing

The power allocation problem for ergodic RB with (3.45) and (3.46) is

$$\max_{\rho, \boldsymbol{p} \geq \boldsymbol{0}} \ \rho \quad \text{s.t.} \ \boldsymbol{p} \in \mathcal{T}, \quad \tilde{R}_k(\boldsymbol{p}) \geq \rho \rho_k, \quad k = 1, \ldots, K. \tag{3.53}$$

The optimum of this problem and (3.47) is closely connected via an inversion similar to Sect. 1.2. We solve (3.53) via the normalized fixed point iteration from [226].

To this end, we write the interference function representation of (3.53) as

$$\max_{\boldsymbol{p} \geq \boldsymbol{0}} \min_k \frac{p_k}{I_k(\boldsymbol{p}; \rho_k)} \quad \text{s.t.} \ \tilde{\boldsymbol{A}} \boldsymbol{p} \leq \boldsymbol{1}, \qquad k = 1, \ldots, K. \tag{3.54}$$

---

25 It is impossible to further reduce the objective because $\tilde{\boldsymbol{A}} \geq \boldsymbol{0}$ and $\boldsymbol{p}' \ngeq \boldsymbol{I}(\boldsymbol{p}'; \boldsymbol{\rho})$ if $\exists i : p'_i < p^\star_i$.

26 Newton methods would also require the derivative of $\tilde{R}_k$ with respect to $\boldsymbol{p}^{(n)}$ in each iteration.

This formulation is similar to the balancing problem in [231] and [230, Section 5.6.3], due to the compact convex polytope for the power constraint set, but it includes a non-affine standard interference function similar to [226].

A solution for (3.54) is described by a tuple $(z^\star, \boldsymbol{p}^\star) \in \mathbb{R}_+ \times \mathbb{R}_+^K$ that satisfies

$$z^\star \boldsymbol{p}^\star = \boldsymbol{I}(\boldsymbol{p}^\star; \boldsymbol{\rho}). \tag{3.55}$$

This equation characterizes eigenvalue eigenvector pairs $(z, \boldsymbol{p})$ for the strictly positive standard interference function $\boldsymbol{I}(\cdot; \boldsymbol{\rho})$. Such pairs are positive, inverse monotonic, i.e., $z' > z$ implies $\boldsymbol{p}' < \boldsymbol{p}$ if $z'\boldsymbol{p}' = \boldsymbol{I}(\boldsymbol{p}'; \boldsymbol{\rho})$, and $\boldsymbol{p}$ is unique for a given z (cf. [226, Lemma 3.1]). Using these properties, Nuzman showed that there is exactly one solution for (3.55) if $\boldsymbol{p}$ is norm-bounded, e.g., $\|\tilde{\boldsymbol{A}}\boldsymbol{p}\|_\infty = 1$ with $\tilde{\boldsymbol{A}} \geq \boldsymbol{0}$ [226, Theorem 3.2]. This unique eigenvalue eigenvector pair is found in the limit of the normalized iteration [226, Section III].

$$\boldsymbol{p}^{(n+1)} = \frac{\boldsymbol{I}(\boldsymbol{p}^{(n)}); \boldsymbol{\rho}}{\|\tilde{\boldsymbol{A}}\boldsymbol{I}(\boldsymbol{p}^{(n)}; \boldsymbol{\rho})\|_\infty}, \qquad z^{(n)} = \frac{1}{\rho^{(n)}} = \|\tilde{\boldsymbol{A}}\boldsymbol{I}(\boldsymbol{p}^{(n)}; \boldsymbol{\rho})\|_\infty. \tag{3.56}$$

The sequence $\rho^{(n)}$ is generally neither increasing nor decreasing in n. Moreover, $\rho^{(n)}$ is usually unattainable for $\boldsymbol{p}^{(n)}$ if n is finite, i.e., $\boldsymbol{p}^{(n)} \ngeq \rho^{(n)}\boldsymbol{I}(\boldsymbol{p}^{(n)}; \boldsymbol{\rho})$. Therefore, we further reduce the balancing factor ρ to satisfy $\rho = \min_k p_k/I_k(\boldsymbol{p}; \rho_k)$, when (3.56) reaches the termination criteria $\|\boldsymbol{p}^{(n+1)} - \boldsymbol{p}^{(n)}\|_2 < \epsilon$.

Vučić and Schubert [92] found a variation of (3.56) for monotone maps that are *linear* with respect to a scaling of the power vector (Sect. 1.3). They achieve this linearity property by extending the power allocation with the noise component and the interference function with the power constraint similar to [11, Chapters 4 and 5].

3.5 Sequential Approximation Strategy

Using derivatives for the ergodic rate (3.3), we propose a local design of the beamformers via a sequential QoS power minimization (3.1) and balancing optimization (3.2), respectively.[27] Based on locally tight approximations, we reformulate the ergodic rate constraints into SINR-like requirements [142, 143]. Then, we solve the SINR constrained problems using fixed point iterations in the dual uplink.

The difference of the two sub-logarithmic functions B_k (3.4) prevents the direct reformulation of the ergodic rate requirement into an SINR constraint. With the inverse function B_k^{-1}, i.e., $B_k^{-1}(B_k(x)) = x$ for $x \geq 0$,[28] a constraint for the ratio between the useful signal power and the interference reads as

$$\frac{|\bar{\boldsymbol{h}}_k^{\mathrm{H}}\boldsymbol{t}_k|^2}{g_k(\mathrm{I}_k(\boldsymbol{t}); \rho\rho_k)} \geq 1, \tag{3.57}$$

[27] For example, see [150] for the theory behind the sequential inner approximation strategy.

[28] We evaluate $B_k^{-1}(y)$ with $y > 0$ numerically, using fixed point methods.

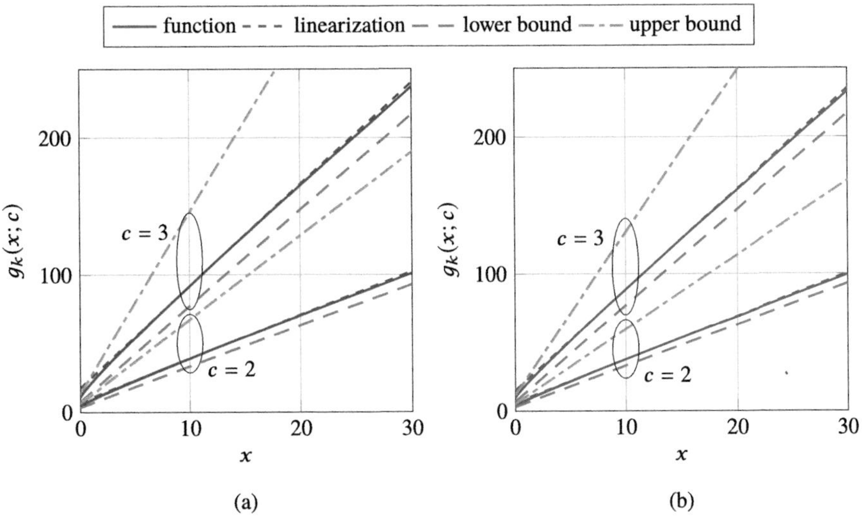

Fig. 3.1 Plot of $g_k(x; c)$, its lower bound $g_k^{(\mathrm{UB2})}(x; c)$, upper bound $g_k^{(\mathrm{LB2})}(x; c)$, and a linearization at $\tilde{x} = 8$. (**a**) Rayleigh fading $2\zeta_k^{-1} \sim \mathcal{X}_2^2$. (**b**) Rician fading $2\zeta_k^{-1} \sim \mathcal{X}_2^2(4)$

where the continuous function $g_k(\cdot; c) : \mathbb{R}_+ \to \mathbb{R}_+$ with parameter c reads as

$$g_k(x; c) = B_k^{-1}\big(c + B_k(x)\big) - x. \tag{3.58}$$

This function is non-linear increasing with $x > 0$.[29] In particular, it is apparently to be concave increasing and close to linear for the rank-one Rayleigh and Rician fading examples with $2\zeta_k^{-1} \sim \mathcal{X}_2^2$ and $2\sigma_{\xi_k}^{-2}\zeta_k^{-1} \sim \mathcal{X}_2^2(2\sigma_{\xi_k}^{-2})$, respectively, from Fig. 3.1.[30] An upper and a lower bound for (3.58), based on LB2 and UB2, respectively, read as

$$g_k(x; c) \geq g_k^{(\mathrm{UB2})}(x; c) = (2^c - 1)\big(x + \nu_k^{(\mathrm{UB2})}\big) \tag{3.59}$$

$$g_k(x; c) \leq g_k^{(\mathrm{LB2})}(x; c) = (2^{c-\mu_k^{(\mathrm{LB2})}} - 1)\big(x + \nu_k^{(\mathrm{UB2})}\big). \tag{3.60}$$

The functions' non-linearity also prevents from a conic reformulation of the constraints via the disciplined convex programming ruleset [102, 232]. Therefore, approximations are necessary for applying standard solvers for conic convex problems. A first order Taylor expansion at $\tilde{x} \geq 0$ reads as

$$g_k(x; c) = g_k(\tilde{x}; c) + g_k'(\tilde{x}; c)(x - \tilde{x}) + O\big((x - \tilde{x})^2\big) \tag{3.61}$$

[29] The increase in c is approximately exponential due to the close relationship to the SINR.

[30] A rigorous mathematical proof for this observation is still missing.

and the first order derivatives of g_k and B_k are given by

$$g_k'(x; c) = \frac{B_k'(x)}{B_k'\big(B_k^{-1}(c + B_k(x))\big)} - 1 \tag{3.62}$$

$$B_k'(x) = \mathrm{E}\Big[\log_2(\mathrm{e})\frac{1}{\zeta_k + x}\Big]. \tag{3.63}$$

For the closed-form examples of $B_k(x)$, i.e., (3.6) and (3.7), (3.63) reads as[31]

$$B_k'(x) = \mathrm{e}^{-\sigma_{\xi_k}^{-2}} \int_0^\infty \log_2(\mathrm{e}) \frac{\sigma_{\xi_k}^2 t}{1 + x\sigma_{\xi_k}^2 t} \mathrm{e}^{-t}\, {}_0F_1\big(;1;\sigma_{\xi_k}^{-2} t\big)\,\mathrm{d}t \tag{3.64}$$

$$B_k'(x) = \log_2(\mathrm{e})\frac{1}{x}\Big(1 - \frac{1}{x}B_k(x)\Big), \tag{3.65}$$

respectively. Skipping the error term in (3.61), we approximate (3.57) as

$$\frac{|\bar{\boldsymbol{h}}_k^{\mathrm{H}} \boldsymbol{t}_k|^2}{\frac{\beta_k(\rho)}{\alpha_k(\rho)} + \mathtt{I}_k(\boldsymbol{t})} \geq \alpha_k(\rho), \tag{3.66a}$$

where $\alpha_k : \mathbb{R}_+ \to \mathbb{R}_+$ and $\beta_k : \mathbb{R}_+ \to \mathbb{R}_+$ are defined as

$$\alpha_k(\rho) = g_k'(\mathtt{I}_k(\boldsymbol{t}); \rho\rho_k) \tag{3.66b}$$

$$\beta_k(\rho) = g_k(\mathtt{I}_k(\boldsymbol{t}); \rho\rho_k) - \alpha_k(\rho)\mathtt{I}_k(\boldsymbol{t}). \tag{3.66c}$$

Both depend on the balancing level ρ and the interference at $\boldsymbol{t} \in \mathcal{P}$. The target $\alpha_k(\rho)$ approaches $2^{\rho\rho_k} - 1$ if $\mathtt{I}_k(\boldsymbol{t}) \to \infty$ because the ergodic rate bound UB2 is tight in this case. Similarly, $\alpha_k(\rho)$ approaches $2^{\rho\rho_k - \mu_k^{(\mathrm{LB2})}} - 1$ if $\mathtt{I}_k(\boldsymbol{t}) = 0$ (see Fig. 3.1).

3.5.1 *Sequential Quality-of-Service Optimization*

Given (3.66a), we approximate (3.1) with the SINR constrained QoS optimization[32]

$$\min_{p \geq 0,\, \boldsymbol{t}} \; p \quad \text{s.t.} \quad p^{-1}\boldsymbol{t} \in \mathcal{P}, \quad \text{s.t.} \quad \frac{|\bar{\boldsymbol{h}}_k^{\mathrm{H}} \boldsymbol{t}_k|^2}{\frac{\beta_k}{\alpha_k} + \sum_{i \neq k} |\bar{\boldsymbol{h}}_k^{\mathrm{H}} \boldsymbol{t}_i|^2} \geq \alpha_k, \quad k = 1, \ldots, K. \tag{3.67}$$

[31] Here, we exploited $\frac{\mathrm{d}}{\mathrm{d}z}\mathrm{e}^z\, \mathrm{E}_1(z) = \mathrm{e}^z\, \mathrm{E}_1(z) - \frac{1}{z}$ [203, Equation 5.1.2.7] for Rayleigh fading (3.65).

[32] We shortened notations by substituting $\alpha_k := \alpha_k(1)$ and $\beta_k := \beta_k(1)$.

Algorithm 1 Sequential QoS power minimization with ergodic rate constraints

Require: $\bar{\boldsymbol{h}}_k, k = 1, \ldots, K, \boldsymbol{t}^{(0)}, \epsilon \leftarrow 10^{-4}$
1: **repeat**
2: $\quad n \leftarrow n + 1$
3: $\quad \alpha_k = g_k'(\mathtt{I}_k(\boldsymbol{t}^{(n-1)}); \rho_k)$ {constraint approximation (3.66a)}
$\quad \beta_k = g_k(\mathtt{I}_k(\boldsymbol{t}^{(n-1)}); \rho_k) - \alpha_k \mathtt{I}_k(\boldsymbol{t}^{(n-1)})$
4: $\quad (\boldsymbol{t}^{(n)}, p^{(n)}) = \mathrm{QoSA}(\{\alpha_k\}, \{\beta_k\})$ {approximate QoS optimization (3.67)}
5: **until** $\|\boldsymbol{t}^{(n)} - \boldsymbol{t}^{(n-1)}\|_2 < \epsilon$
6: **return** $(\boldsymbol{t}^{(n)}, p^{(n)})$

We can solve either via a reformulation into a convex problem or by using SINR uplink–downlink duality if the problem is feasible. For $\alpha_k > 0$ and regular $\bar{\boldsymbol{H}}$, (3.67) is feasible if the sum of the MMSE targets $\varepsilon_k = 1/(1 + \alpha_k)$ is smaller than $K - N$.

Note that the solution $(\boldsymbol{t}^\star, p^\star)$ of (3.67) only approximates that of (3.1). For concave g_k, $p^\star$ upper bounds the optimum of (3.1) and $\boldsymbol{t}^\star$ is feasible but suboptimal for the initial problem formulation (3.1). Only if $\mathtt{I}_k(\boldsymbol{t}^\star) = \mathtt{I}_k(\boldsymbol{t}')$ holds for $k = 1, \ldots, K$, where $\boldsymbol{t}'$ denotes the precoder from the approximation point, the tuple $(\boldsymbol{t}^\star, p^\star)$ is a local solution for (3.1). Then, the approximation (3.66) is tight.

This motivates the Sequential convex search (SCS) in Algorithm 1. In iteration n, problem (3.1) is approximated with the SINR constrained QoS minimization (3.67). We denote its solution by $(\boldsymbol{t}^{(n)}, p^{(n)}) = \mathrm{QoSA}(\{\alpha_k^{(n-1)}\}, \{\beta_k^{(n-1)}\})$. The approximation parameters $\alpha_k^{(n-1)}$ and $\beta_k^{(n-1)}$ from (3.66) are computed with $\mathtt{I}_k(\boldsymbol{t}^{(n-1)})$, $k = 1, \ldots, K$. Here, $\boldsymbol{t}^{(n-1)}$ is the solution of the previous iteration. The algorithm terminates if the distance between two consecutive elements of the series $(\boldsymbol{t}^{(n)})_n$ falls below the threshold $\epsilon > 0$.

Convergence to a local solution of (3.1) is ensured if, first, g_k is concave and continuously differentiable for $k = 1, \ldots, K$, and second, if the initial $\boldsymbol{t}^{(0)}$ results in feasible SINR targets $\alpha_k^{(0)}$, $k = 1, \ldots, K$.[33] Then, the constraints of (3.67) form a *locally tight inner approximation* for the rate constraints of (3.1) as (cf. [150]):

$$g_k(\mathtt{I}_k(\boldsymbol{t}^{(n)}); \rho_k) \leq \alpha_k^{(n-1)} \mathtt{I}_k(\boldsymbol{t}^{(n)}) + \beta_k^{(n-1)} \tag{3.68}$$

$$g_k(\mathtt{I}_k(\boldsymbol{t}^{(n)}); \rho_k) = \alpha_k^{(n)} \mathtt{I}_k(\boldsymbol{t}^{(n)}) + \beta_k^{(n)} \tag{3.69}$$

$$g_k'(\mathtt{I}_k(\boldsymbol{t}^{(n)}); \rho_k) = \alpha_k^{(n)}. \tag{3.70}$$

These properties ensure that $\boldsymbol{t}^{(n)}$ is a feasible precoder for the QoS optimization (3.67) in iteration $n+1$ and that the objective decreases in each iteration, that is, $p^{(n+1)} \leq p^{(n)}$. Hence, $(p^{(n)})_n$ and even $(\boldsymbol{t}^{(n)})_n$ converge to a local solution for (3.1) according to [150, Theorem 1 and Corollary 1]. Figure 3.2 depicts convergence in the objective for a rank-one Rayleigh fading scenario with $K = N = 4$.

[33] A feasible starting point $\boldsymbol{t}^{(0)}$ is found by rescaling the solution of (3.16) based on UB2 (or LB1).

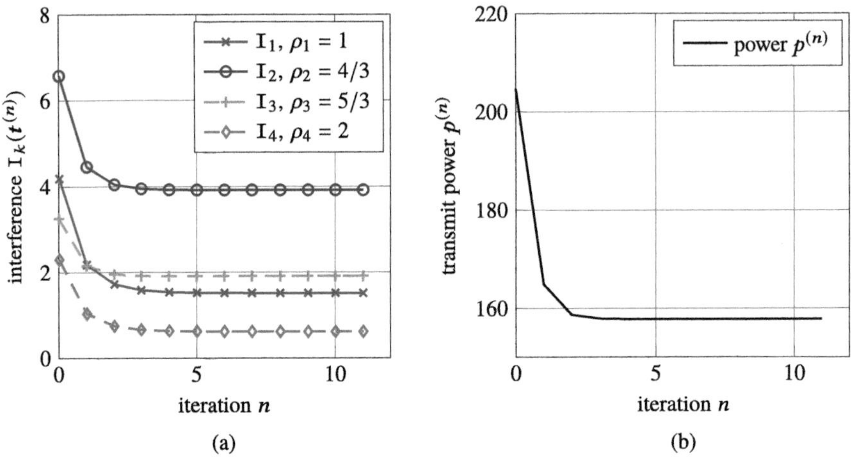

Fig. 3.2 Minimal power and interference vs. number of iterations until convergence for Rayleigh fading with $K = N = 4$ and the above targets. (**a**) Interference. (**b**) Transmit power

3.5.2 *Sequential Ergodic Rate Balancing*

Solving the ergodic RB problem (3.2) via repeatedly applying sequential convex QoS optimizations is computationally demanding. Instead, we propose a direct sequential search for (3.2) that provides a local solution. Now, the search iterates between the approximation (3.66) and solving the resulting rate maximization

$$\max_{\rho,\, \boldsymbol{t}} \rho \quad \text{s.t.} \quad \boldsymbol{t} \in \mathcal{P}, \quad \frac{|\bar{\boldsymbol{h}}_k^{\mathrm{H}} \boldsymbol{t}_k|^2}{\frac{\beta_k(\rho)}{\alpha_k(\rho)} + \mathrm{I}_k(\boldsymbol{t})} \geq \alpha_k(\rho), \quad k = 1, \ldots, K. \tag{3.71}$$

The SINR requirements herein differ from the constraints for instantaneous CSI (1.4) in that the noise power $\beta_k(\rho)/\alpha_k(\rho)$ now depends on the balancing target ρ. Nevertheless, (3.71) enables an algorithmic solution in the dual uplink.[34]

Here, we recast the dual uplink balancing problem as

$$\min_{\boldsymbol{\mu} \geq \mathbf{0}} \max_{\rho, \boldsymbol{u}, \boldsymbol{\lambda} \geq \mathbf{0}} \rho \ \text{ s.t. } \sum_{i=1}^{K} \frac{\beta_i(\rho)}{\alpha_i(\rho)} \lambda_i \leq 1, \ \sum_{\ell=1}^{L} P_\ell \mu_\ell \leq 1, \ \ \mathrm{SINR}_k^{(\mathrm{ul})} \geq \alpha_k(\rho),$$
$$k = 1, \ldots, K, \tag{3.72}$$

[34]Despite the effective noise power $\beta_k(\rho)/\alpha_k(\rho)$ depends on the balancing value ρ, the duality transformation remains the same as with instantaneous rates, e.g., with the dual uplink formulations from [67, 104] for a sum power constraint and from [74, 118, 121] for general power constraints.

where the uplink SINR reads as [cf. (1.34)][35]

$$\mathrm{SINR}_k^{(\mathrm{ul})} = \frac{|\bar{\boldsymbol{h}}_k^{\mathrm{H}} \boldsymbol{u}_k|^2 \lambda_k}{\sum_{i \neq k} |\bar{\boldsymbol{h}}_i^{\mathrm{H}} \boldsymbol{u}_k|^2 \lambda_i + \sum_{\ell=1}^{L} \mu_\ell \|\boldsymbol{A}_{k,\ell}^{1/2} \boldsymbol{u}_k\|_2^2}. \tag{3.73}$$

The vector $\boldsymbol{u} = \mathrm{vec}([\boldsymbol{u}_1, \ldots, \boldsymbol{u}_K])$ comprises the uplink filters and $\boldsymbol{\lambda} = [\lambda_1, \ldots, \lambda_K]^{\mathrm{T}}$ the uplink powers, i.e., the Lagrangian multipliers to the downlink SINR constraints, for the inner maximization. The outer optimization over $\boldsymbol{\mu} = [\mu_1, \ldots, \mu_K]^{\mathrm{T}}$, which comprises the Lagrangian multipliers for the downlink power constraints, searches the uplink noise covariance matrices that jointly minimize the SINRs.

For the exemplary single sum power limitation (see [143]), i.e., $L = 1$ and $\boldsymbol{A}_{k,1} = \mathbf{I}$ for $k = 1, \ldots, K$, the solution of the outer minimization is $\mu = 1/P_1$. Additionally substituting $\bar{\boldsymbol{\lambda}} = P_1 \boldsymbol{\lambda}$, the optimization has a similar structure as (1.15), namely

$$\max_{\rho, \boldsymbol{u}, \bar{\boldsymbol{\lambda}} \geq \mathbf{0}} \ \rho \quad \text{s.t.} \quad \sum_{i=1}^{K} \frac{\beta_i(\rho)}{\alpha_i(\rho)} \bar{\lambda}_i \leq P_1, \quad \mathrm{SINR}_k^{(\mathrm{ul})} \geq \alpha_k(\rho), \quad k = 1, \ldots, K, \tag{3.74}$$

where the uplink SINR reads as [cf. 1.34]

$$\mathrm{SINR}_k^{(\mathrm{ul})} = \frac{|\bar{\boldsymbol{h}}_k^{\mathrm{H}} \boldsymbol{u}_k|^2 \bar{\lambda}_k}{\sum_{i \neq k} |\bar{\boldsymbol{h}}_i^{\mathrm{H}} \boldsymbol{u}_k|^2 \bar{\lambda}_i + \|\boldsymbol{u}_k\|_2^2}. \tag{3.75}$$

The transformation between the dual uplink filters and the downlink beamformers is via a K-dimensional linear equation system based on $\mathrm{SINR}_k = \mathrm{SINR}_k^{(\mathrm{ul})}$ and $\boldsymbol{t}_k = \sqrt{\theta_k} \boldsymbol{u}_k$, $k = 1, \ldots, K$ (e.g., see[5, Section II] and [83, Section II.B]).

The inner maximization of (3.72) over $\boldsymbol{u}$ and $\boldsymbol{\lambda}$ is solved with the MMSE filters

$$\boldsymbol{u}_k = \Big(\sum_{i=1}^{K} \bar{\boldsymbol{h}}_i \bar{\boldsymbol{h}}_i^{\mathrm{H}} \lambda_i + \sum_{\ell=1}^{L} \mu_\ell \boldsymbol{A}_{k,\ell} \Big)^{-1} \bar{\boldsymbol{h}}_k \tag{3.76}$$

and the convergence point $\boldsymbol{\lambda}^\star$ of the two-step iterative process defined by (3.30) and (3.31). Here, the diagonal (SINR) target matrix comprises the slopes of the linear approximations for g_k, $k = 1, \ldots, K$, i.e., $\boldsymbol{\Gamma}(\rho) = \mathrm{diag}(\alpha_1(\rho), \ldots, \alpha_K(\rho))$. Convergence is again ensured because the inner solution $\boldsymbol{\lambda}$ of (3.74) monotonically increases with ρ. Moreover, the complexity for the computations in each iteration is low. If we tabulate $\alpha_k(\rho)$ and $\beta_k(\rho)$ and insert the MMSE filters (3.76), computing the inverse of $\boldsymbol{Y}_k(\boldsymbol{\lambda}^{(n)}, \boldsymbol{\mu}) = \sum_{i \neq k} \bar{\boldsymbol{h}}_i \bar{\boldsymbol{h}}_i^{\mathrm{H}} \lambda_i^{(n)} + \sum_{\ell=l1}^{L} \mu_\ell \boldsymbol{A}_{k,\ell}$ within

[35] This is a simplified notation in comparison to the formulation of [143] (cf. Sect. 3.2.3). It is obtained by a substitution of the usual multipliers $\bar{\lambda}_i$ with $\lambda_i = \bar{\lambda}_i \alpha_i(\rho)/\beta_i(\rho)$, $i = 1, \ldots, K$.

$$I_k(\boldsymbol{\lambda}^{(n)};\boldsymbol{\mu}) = \lambda_k^{(n)}\left(\frac{1}{\bar{\boldsymbol{h}}_k^{\mathrm{H}}\boldsymbol{Y}_k^{-1}(\boldsymbol{\lambda}^{(n)},\boldsymbol{\mu})\bar{\boldsymbol{h}}_k} - 1\right), \quad k = 1,\ldots,K,$$

dominates the computational complexity of the power update.[36]

The PG update procedure from Sect. 3.2.3 again finds a solution for the outer minimization in (3.72). However, the gradient computation differs because the power constraint is now a function of ρ. The equalities for its computation read as[37]

$$\sum_{i=1}^{K}\frac{\beta_i(\rho)}{\alpha_i(\rho)}\lambda_i = 1, \quad \lambda_k = \alpha_k(\rho)I_k(\boldsymbol{\lambda};\boldsymbol{\mu}), \quad k = 1,\ldots,K. \tag{3.77}$$

The derivative for these equations with respect to μ_m result in the equation system

$$\begin{bmatrix} \phi(\rho) & \boldsymbol{\gamma}^{\mathrm{T}}(\rho) \\ -\boldsymbol{D}(\rho)\boldsymbol{I}(\boldsymbol{\lambda};\boldsymbol{\mu}) & \mathbf{I}-\boldsymbol{\Gamma}(\rho)\boldsymbol{\Psi} \end{bmatrix}\begin{bmatrix} \frac{\partial\rho}{\mu_m} \\ \frac{\partial\boldsymbol{\lambda}}{\mu_m} \end{bmatrix} = \begin{bmatrix} 0 \\ \boldsymbol{\Gamma}(\rho)\boldsymbol{n}_m \end{bmatrix}, \tag{3.78}$$

where $\phi(\rho) = \sum_{i=1}^{K}\lambda_i\frac{\partial}{\partial\rho}\frac{\beta_i(\rho)}{\alpha_i(\rho)}$ and $\boldsymbol{\gamma}(\rho) = [\beta_1(\rho)/\alpha_1(\rho),\ldots,\beta_K(\rho)/\alpha_K(\rho)]^{\mathrm{T}}$, while the diagonal matrix $\boldsymbol{D}(\rho) = \mathrm{diag}(d_1(\rho),\ldots,d_K(\rho))$ comprises again the derivatives of the (SINR) targets, i.e., $d_i(\rho) = \frac{\partial\alpha_i(\rho)}{\partial\rho}$. The entries of $\boldsymbol{\Psi}$ and $\boldsymbol{n}_m$ are that of (3.36).[38] The derivative of ρ with respect to μ_m follows from (3.78) via block matrix inversion and Schur's complement relative to the matrix $\mathbf{I}-\boldsymbol{\Gamma}(\rho)\boldsymbol{\Psi}$:

$$\begin{aligned}\delta_m = \frac{\partial\rho}{\partial\mu_m} &= -\phi^{-1}\boldsymbol{\gamma}^{\mathrm{T}}(\rho)\big(\mathbf{I}-\boldsymbol{\Gamma}(\rho)\boldsymbol{\Psi} - \phi^{-1}\boldsymbol{D}\boldsymbol{I}(\boldsymbol{\lambda};\boldsymbol{\mu})\boldsymbol{\gamma}^{\mathrm{T}}(\rho)\big)^{-1}\boldsymbol{\Gamma}(\rho)\boldsymbol{n}_m \\ &= \frac{\boldsymbol{\gamma}^{\mathrm{T}}(\rho)(\mathbf{I}-\boldsymbol{\Gamma}(\rho)\boldsymbol{\Psi})^{-1}\boldsymbol{\Gamma}(\rho)\boldsymbol{n}_m}{-\phi} \\ &\quad \left(1+\frac{\boldsymbol{\gamma}^{\mathrm{T}}(\rho)(\mathbf{I}-\boldsymbol{\Gamma}(\rho)\boldsymbol{\Psi})^{-1}\boldsymbol{D}(\rho)\boldsymbol{I}(\boldsymbol{\lambda};\boldsymbol{\mu})}{\phi - \boldsymbol{\gamma}^{\mathrm{T}}(\rho)(\mathbf{I}-\boldsymbol{\Gamma}(\rho)\boldsymbol{\Psi})^{-1}\boldsymbol{D}(\rho)\boldsymbol{I}(\boldsymbol{\lambda};\boldsymbol{\mu})}\right),\end{aligned} \tag{3.79}$$

where the second form follows via the matrix inversion lemma.

Alternatively, the outer search for $\boldsymbol{\mu} \in \mathbb{R}_+^L$ may be solved via a projected subgradient method (e.g., see [120]). The update of $\boldsymbol{\mu}$ then requires a diminishing step size rule to ensure convergence [227],[39] but it avoids computing the derivative of $\alpha_k(\rho)$ and $\beta_k(\rho)$. The ℓth component of the subgradient $\boldsymbol{\delta} = [\delta_1,\ldots,\delta_L]^{\mathrm{T}}$ is

---

[36] Here, the matrix $\boldsymbol{Y}_k(\boldsymbol{\lambda}^{(n)},\boldsymbol{\mu})$ is independent of $\rho^{(n+1)}$ in contrast to the version in [143].

[37] All approximate interference constraints are active since $\alpha_k(\rho)$ is strictly positive if $\rho \geq 0$.

[38] Here, the effective noise variance is $v_k^{(\mathrm{B})} = 1$.

[39] The projected subgradient method does not result in a decreasing objective within each iteration.

$$\delta_\ell = -P_\ell + \sum_{i=1}^{K} \boldsymbol{t}_i^{\mathrm{H}} \boldsymbol{A}_{i,\ell} \boldsymbol{t}_i,$$

where the beamformers $\boldsymbol{t}_i$ are computed via the uplink-to-downlink transformation from Sect. 3.2.3, based on the dual variables from the previous iteration (3.72).

3.6 Branch and Bound Method

A global solution via a BB algorithm (e.g., see [151]) serves as a benchmark for the above beamformer designs. It finds a solution of the non-convex downlink formulations for (3.1) and (3.2) up to a pre-defined accuracy.[40] The BB formulation is similar to [16] and exploits the partly convex monotone structure of the ergodic rates in the useful signal power (3.14) and the interference (3.15).

In particular, we consider the generalized problem formulation (cf. [140]):

$$\min_{(a,\boldsymbol{t})\in\mathcal{A},\boldsymbol{b}\in\mathbb{R}_+^K} \quad a \quad \text{s.t.} \quad \mathtt{I}(\boldsymbol{t}) \leq \boldsymbol{b}, \quad \boldsymbol{f}(\boldsymbol{t},\boldsymbol{b}) \leq \boldsymbol{0}, \tag{3.80}$$

where $\mathcal{A} = \mathbb{R}_+ \times \mathbb{C}^{NK}$ is a convex set. Furthermore, the convex constraint $\mathtt{I}(\boldsymbol{t}) \leq \boldsymbol{b}$ bounds the square root values of the interferences by $\boldsymbol{b} \in \mathbb{R}_+^K$, that is,

$$\mathbf{e}_k^{\mathrm{T}} \mathtt{I}(\boldsymbol{t}) = \sqrt{\mathtt{I}_k(\boldsymbol{t})} = \|\bar{\boldsymbol{h}}_k^{\mathrm{H}} \boldsymbol{T}_{\backslash\{k\}}\|_2, \quad k = 1, \ldots, K, \tag{3.81}$$

with $\mathtt{I}_k(\boldsymbol{t})$ from (3.15). Finally, the function $\boldsymbol{f}(\boldsymbol{t},\boldsymbol{b})$ is entry-wise convex in $\boldsymbol{t}$ for fixed $\boldsymbol{b}$ and monotonically increasing in $\boldsymbol{b}$ for fixed $\boldsymbol{t}$. The basis for this function is a reformulation of the ergodic rate requirements $R_k(\boldsymbol{t}) \geq \rho\rho_k$, $k = 1, \ldots, K$, with $\rho = 1$ for the QoS optimization (3.1). To obtain a convex form in $\boldsymbol{t}$, we bound the useful signal power at user k in terms of the experienced interference, i.e., [cf. (3.57)]

$$|\bar{\boldsymbol{h}}_k^{\mathrm{H}} \boldsymbol{t}_k|^2 \geq g_k(\mathtt{I}_k(\boldsymbol{t}); \rho). \tag{3.82}$$

Since $g_k(\mathtt{I}_k(\boldsymbol{t}); \rho)$ increases with $\mathtt{I}_k(\boldsymbol{t})$ and $\mathcal{P}$ is independent w.r.t. a phase rotation of the beamformers, we require $\bar{\boldsymbol{h}}_k^{\mathrm{H}} \boldsymbol{t}_k$ to be real and recast (3.82) as

$$\bar{\boldsymbol{h}}_k^{\mathrm{H}} \boldsymbol{t}_k \geq \sqrt{g_k(\mathtt{I}_k(\boldsymbol{t}); \rho)}, \quad k = 1, \ldots, K. \tag{3.83}$$

[40] These problems are non-convex because the ergodic rates are neither convex nor concave functions in $\boldsymbol{t}$ and a convex reformulation of $R_k(\boldsymbol{t}) \geq \rho\rho_k$ is missing.

Finally, we define the kth entry of the slack variable $\boldsymbol{b} \in \mathbb{R}_+^K$ as $b_k = \sqrt{\mathtt{I}_k(\boldsymbol{t})}$, relax this equality to an inequality to obtain (3.81), and define the entries of $\boldsymbol{f}(\boldsymbol{t}, \boldsymbol{b})$ as

$$f_k(\boldsymbol{t}, b_k; \rho) = \sqrt{g_k(b_k^2; \rho)} - \bar{\boldsymbol{h}}_k^{\mathrm{H}} \boldsymbol{t}_k, \quad k = 1, \ldots, K. \tag{3.84}$$

While the number of variables increases by this notation, the resulting QoS optimization becomes a standard Second-order cone program (SOCP) for given $\boldsymbol{b}$ and the solution of the balancing optimization can again be found via a line search over ρ and a sequence of feasibility tests if $\boldsymbol{b}$ was given.

Given a convex solution of (3.80) for fixed $\boldsymbol{b}$, the search for $\boldsymbol{b}$ remains. This is the task of the BB procedure, which alternates between branching—splitting—the search space of $\boldsymbol{b}$ and bounding the optimum for each subset until the optimum is met with pre-defined accuracy (cf. [151]). Let the initial index set be $\mathcal{I} = \{1\}$ and the interference box be $\mathcal{B}_1 = [\boldsymbol{b}^{\min}, \boldsymbol{b}^{\max}]$, which contains the optimal interference vector $\boldsymbol{b}^{\min} \leq \boldsymbol{b}^\star \leq \boldsymbol{b}^{\max}$ and has labels for a lower bound $L(\mathcal{B}_1)$ and an upper bound $U(\mathcal{B}_1)$ of the optimum $a^\star$. Then, the following two steps are repeated [16]:

1. *Branching:* In each iteration, the box $\mathcal{B}_l$ with minimum lower bound, i.e.,

$$\mathcal{B}_l = \operatorname*{arg\,min}_{m \in \mathcal{I}} L(\mathcal{B}_m),$$

is divided into two disjoint sub-boxes $\mathcal{B}_j$, $j \notin \mathcal{I}$. Here, a standard bisection along the longest edge of $\mathcal{B}_l$ is applied, such that the two sub-boxes are of equal size.

2. *Bounding:* For each new sub-box $\mathcal{B}_j = [\boldsymbol{b}^{\min}, \boldsymbol{b}^{\max}]$, we calculate the bounds as

$$L(\mathcal{B}_j) = \min \left\{ a \in \mathbb{R}_+ : (a, \boldsymbol{t}) \in \mathcal{A}, \boldsymbol{f}(\boldsymbol{t}, \boldsymbol{b}^{\min}) \leq \boldsymbol{0}, \mathtt{I}(\boldsymbol{t}) \leq \boldsymbol{b}^{\max} \right\}, \tag{3.85a}$$

$$U(\mathcal{B}_j) = \min \left\{ a \in \mathbb{R}_+ : (a, \boldsymbol{t}) \in \mathcal{A}, \boldsymbol{f}(\boldsymbol{t}, \boldsymbol{b}^{\max}) \leq \boldsymbol{0}, \mathtt{I}(\boldsymbol{t}) \leq \boldsymbol{b}^{\max} \right\}. \tag{3.85b}$$

These bounds guarantee that $L(\mathcal{B}_j) \to U(\mathcal{B}_j)$ for $\boldsymbol{b}^{\min} \to \boldsymbol{b}^{\max}$. Finally, the index set $\mathcal{I}$ of the active boxes is updated according to the following rule:

$$\mathcal{I} = \left\{ m \in (\mathcal{I} \setminus \{l\}) \cup \{j\} \middle| \, L(\mathcal{B}_m) + \epsilon < \min_n U(\mathcal{B}_n) \right\}.$$

We remark that $\min_{l \in \mathcal{I}} L(\mathcal{B}_l) + \epsilon \geq \min_{l \in \mathcal{I}} U(\mathcal{B}_l)$ at the convergence point of this procedure. If the tolerance is $\epsilon > 0$, the algorithm converges in finitely many iterations. However, being essentially an exhaustive search strategy, the complexity of the BB method is exponential in the dimension of $\boldsymbol{b}$, i.e., K in this case, where the basis increases unbounded with decreasing ϵ [233, Theorem 4]. Therefore, this technique is only tractable for a small number of ergodic rate constraints.

To find the global solution of the QoS optimization (3.1) with the above BB method, i.e., to rewrite the QoS problem into the form of (3.80), we set $a = p$ and[41]

$$\mathcal{A} = \{(p, \boldsymbol{t}) \in \mathbb{R}_+ \times \mathbb{C}^{NK} : p^{-1}\boldsymbol{t} \in \mathcal{P}\},$$

where the constraint $p^{-1}\boldsymbol{t} \in \mathcal{P}$ is an intersection of SOCs as in (1.4.1). Furthermore, to find an initial box $\mathcal{B}_1$ containing $\boldsymbol{b}^\star$, we set $\boldsymbol{b}^{\min} = \boldsymbol{0}$. For $\boldsymbol{b}^{\max}$ of $\mathcal{B}_1$, we first find a feasible $\boldsymbol{t}$ with the ergodic rate bounds UB2 and the post-processing of Sect. 3.4. Given the so obtained $p^{\max}$, we set the entries of $\boldsymbol{b}^{\max}$ to the square root of[42]

$$\mathtt{I}_k^{\max} = (K-1)(p^{\max})^2\|\bar{\boldsymbol{h}}_k\|_2^2 > \mathtt{I}_k^\star.$$

For the ergodic RB problem (3.2), $a = -\rho$ and $\mathcal{A} = \mathbb{R}_+ \times \mathcal{P}$. Thus, the transmit power is fixed, but $\boldsymbol{f}(\boldsymbol{t}, \boldsymbol{b}; \rho)$ additionally depends on the balancing parameter ρ [cf. (3.84)]. As a consequence, (3.84) is non-convex even for given interference $\boldsymbol{b}$. However, we can again calculate the lower and upper bounds in the *Bounding* step of the BB algorithm, i.e., (3.85a) and (3.85b), with a series of SOCPs. For example, we use a bisection over ρ and check whether a feasible $\boldsymbol{t}$ exists via the corresponding SOCP power minimization.

3.7 Numerical Optimization Results for Ergodic Rates

The numerical evaluation of the QoS power minimization and the ergodic RB optimization is for the geometric model of a GEO-stationary satellite communication scenario. The satellite is directed to Munich (11° east and 48° north) and equipped with a rectangular antenna array of N elements and serves K users that are randomly placed within 1° east to 21° east and 40° north to 56° north. With the Free space loss (FSL) for the gains (cf. Sect. 6.1), the approximate Vandermonde channels $\bar{\boldsymbol{h}}_k$, $k = 1, \ldots, K$, and scattering only close to the ground, the system suites for the multiplicative channel error model. The channels are normalized such that $\|\bar{\boldsymbol{h}}_k\|_2 = 1$ if a user k were exactly located at 11° east and 48° north. The curvature of the earth and the angle to the satellite result in channels that are not unit-norm. Instantaneous channels are created at random according to $\boldsymbol{h}_k = (1 + \xi_k)\bar{\boldsymbol{h}}_k$, $k = 1, \ldots, K$, where $\zeta_k^{-1} = |1 + \xi_k|^2$ is Rayleigh or Rician distributed with $2\zeta_k^{-1} \sim \mathcal{X}_2^2$ or $2\zeta_k^{-1} \sim \mathcal{X}_2^2(2)$, respectively.[43]

---

[41]This BB problem formulation can be extended such that $(p, \boldsymbol{t})$ resides in a combination of the generalized transmit power constraint set $\mathcal{P}$ and the convex reformulations of instantaneous rate requirements if the transmitter is perfectly aware of some of the users' channels [140].

[42]For the bound, we assume that all $\boldsymbol{t}_i$, $i \neq k$ are collinear with $\bar{\boldsymbol{h}}_k$ and $\|\boldsymbol{t}_i\|_2^2 = p^{\max}$.

[43]In other words, the multiplicative factor is $1 + \xi_k \sim \mathcal{N}_\mathbb{C}(0, 1)$ or $1 + \xi_k \sim \mathcal{N}_\mathbb{C}(1/\sqrt{2}, 1/2)$.

To compare the performance to partial ZF from Sect. 3.2.1, the K users are splitted into two parts. For users $k = 1, \ldots, \lfloor K/2 \rfloor$, the transmitter has perfect knowledge of the states $\boldsymbol{h}_k$, while only knowledge of $\bar{\boldsymbol{h}}_k$ is available for users $k = \lfloor K/2 \rfloor + 1, \ldots, K$. The corresponding QoS optimization reads as [cf. (3.1)]

$$\min_{p \geq 0,\, \boldsymbol{t}} p \quad \text{s.t.} \quad p^{-1}\boldsymbol{t} \in \mathcal{P}, \quad \begin{cases} r_k(\boldsymbol{t}) \geq \rho_k, & k = 1, \ldots, \lfloor K/2 \rfloor, \\ R_k(\boldsymbol{t}) \geq \rho_k, & k = \lfloor K/2 \rfloor + 1, \ldots, K, \end{cases} \tag{3.86}$$

and the related RB optimization has the explicit form [cf. (3.2)]

$$\max_{\rho,\, \boldsymbol{t}} \rho \quad \text{s.t.} \quad \boldsymbol{t} \in \mathcal{P}, \quad \begin{cases} r_k(\boldsymbol{t}) \geq \rho\rho_k, & k = 1, \ldots, \lfloor K/2 \rfloor, \\ R_k(\boldsymbol{t}) \geq \rho\rho_k, & k = \lfloor K/2 \rfloor + 1, \ldots, K. \end{cases} \tag{3.87}$$

For SatCom, this represents a setup with fixed and mobile terminals in the same frequency band [139–143]. The ergodic rate approximations are then only applied for mobile users $k = \lfloor K/2 \rfloor + 1, \ldots, K$. To analyze the proposed approximations and reduce the computational complexity for the benchmarking BB algorithm, the simulation results are for a sum power constraint within $\mathcal{P}$.

3.7.1 Power Minimization Results

Since the BB algorithm from Sect. 3.6 is exponential in the number of ergodic constraints, simulations with benchmarks are for two small systems: a *fully loaded system* where $K = N = 4$ and an *overloaded system* with $K = 6$ users and $N = 4$ transmit antennas. The basic targets of the users are $\rho'_{2i-1} = 1$ and $\rho'_{2i} = 2$, $i = 1, \ldots, K/2$. These targets are scaled by ρ to plot the optimum of the QoS optimization versus increasing target values $\rho_k = \rho\rho_k{}'$, $k = 1, \ldots, K$ in Fig. 3.3.

Figure 3.3a, c show the curves for the *fully loaded system* and the *overloaded system*, respectively. These curves are an average over ten realizations of $\bar{\boldsymbol{h}}_k$, $k = 1, \ldots, K$ for random user placements. All finite rate requirements are feasible in the fully loaded system, while the feasible range for ρ is limited by $\rho \leq \rho_{\max} \approx 1.1284$ for the overloaded system. Both figures depict two upper and lower bounds for the QoS optimum. The lower bounds are obtained with $R_k^{(\mathrm{UB2})}$ and $R_k^{(\mathrm{AUB})}$, $k = \lfloor K/2 \rfloor, \ldots, K$, and the upper bounds are the results with LB2 and ALB. The curve of the BB algorithm lies between the upper and lower bounds and meets the global optimum with an accuracy of $\epsilon = 10^{-2}$.

The alternative rate bounds AUB and ALB are slightly tighter than the rate bounds UB2 and LB2, respectively, in Fig. 3.3a. For this setup, also the results of the partial ZF scheme are close to the upper bounds, even though the partial ZF requirements reduced the degrees of freedom for beamforming. The two beamformers $\boldsymbol{t}_i^{\mathrm{ZF}}$, $i = 1, 2$ are orthogonal to $\bar{\boldsymbol{h}}_k$, $k = 3, 4$ for this case.

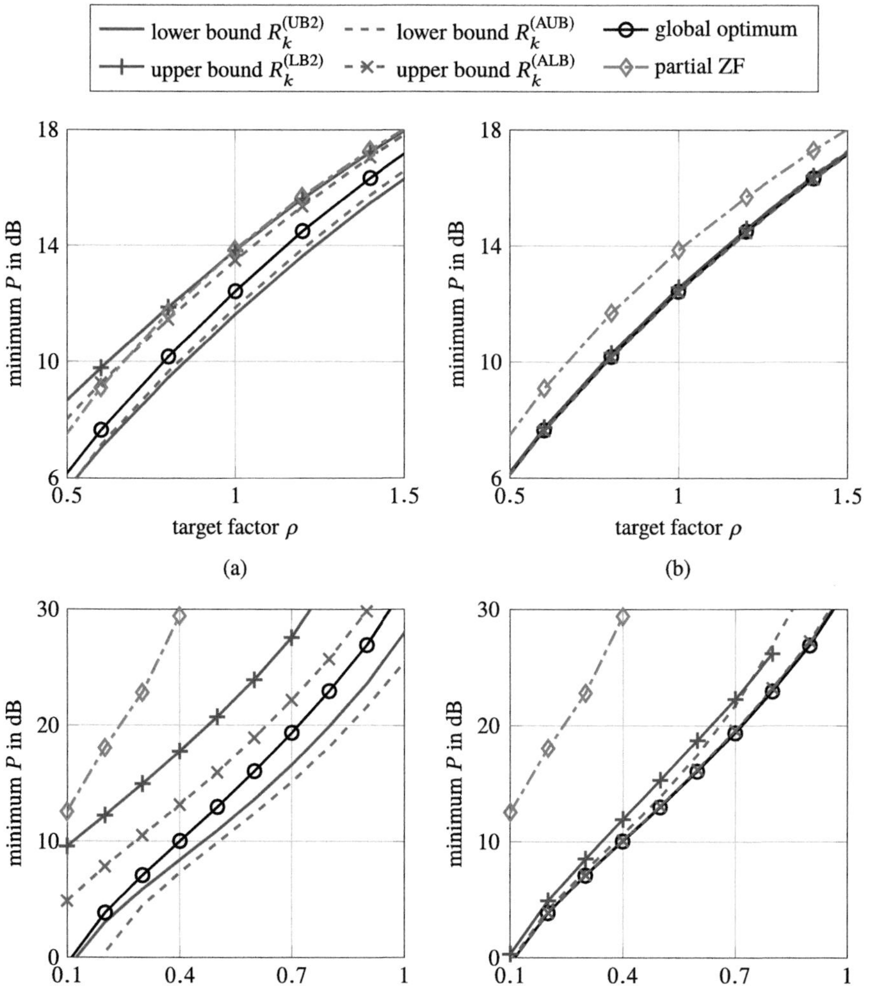

Fig. 3.3 Minimum sum transmit power vs. a common increase in the target level ρ, i.e., $\rho_k = \rho\rho_k{}'$, for the QoS optimization with ergodic rates for users $k = \lfloor K/2 \rfloor, \ldots, K$ using partial ZF, the ergodic rate bounds, or the globally optimal BB strategy. (**a**) $N = K = 4$ antennas and users. (**b**) $N = K = 4$ and post-processing. (**c**) $N = 4$ and $K = 6$. (**d**) $N = 4$, $K = 6$, and post-processing

After the post-processing power allocation, only minor differences are visible between the bounds and the global optimum. The power allocation almost compensates for the suboptimal beamforming. Therewith, the imperfect beamforming based on the rate bounds plus the post-processing achieves close to optimal performance with small increase of the computational complexity compared to rate balancing for perfect CSI. The partial ZF scheme does not improve by a post-processing since the beamformers are already optimized using the actual ergodic rate expressions.

For the overloaded setup (see Fig. 3.3c), the lower bound based on UB2 is tighter than that for AUB. The previous lower bound is even tight for small ρ. In contrast, the upper bound from LB2 is looser than that with ALB. In other words, the post-processing power allocation is only able to compensate for the suboptimal beamforming based on the ergodic rate approximations with the bounds UB2 and ALB. The suboptimal beamformer design based on the other two rate bounds clearly results in a loss of transmit power that increases with ρ to more than 3 dB.

The partial ZF beamforming strategy strictly reduces the available degrees of freedom for beamforming in the overloaded system. Only one degree of freedom remains for the beamformers $\boldsymbol{t}_k$, $k = 1, 2, 3$ with ergodic rate constraints for users $k = 4, 5, 6$. Therefore, these beamformers are collinear. This restriction deforms the achievable rate region, so that the common target factor is bounded by $\rho \leq \rho_{\max} = -\log_2(\sqrt{3} - 1) \approx 0.4500$. This is also visible within the RB optimization results.

3.7.2 Rate Balancing Results

Figure 3.4 depicts the ergodic RB results versus the total sum transmit power P in dB as an average over ten channel mean realizations $\bar{\boldsymbol{h}}_k$, $k = 1, \ldots, K$ for the same fully loaded and overloaded systems as before. We clearly see that the optimum ρ increases unbounded with P in dB for the fully loaded setup. In contrast, ρ saturates in the high power regime for the overloaded setup. Using partial ZF to overcome the difficulties with the ergodic rates, the saturation level is further decreased. This is a result from the shrinked feasible region (3.44) for partial ZF beamforming.

Performing the optimization with $R_k^{(\text{UB2})}$ or $R_k^{(\text{AUB})}$ instead of R_k, $k = \lfloor K/2 \rfloor, \ldots, K$ results in upper bounds for the global optimum. Here, we calculated the BB solution with an accuracy of $\epsilon = 10^{-2}$. The upper bound UB2 is closer to the global optimum than AUB. It becomes tight for $P \leq 0$ dB and $P \geq 30$ dB, i.e., when either the noise dominates over the interference or the interference becomes large (cf. Sect. 3.2.2).

Similarly, lower bounds for the RB optimization are obtained with $R_k^{(\text{LB2})}$ or $R_k^{(\text{ALB})}$ (see Fig. 3.4a, c). The lower bound ALB exceeds LB2 due to its smaller worst-case error and both bounds outperform the partial ZF strategy for $P \geq 10$ dB.

Surprisingly, there is almost no difference between the suboptimal beamforming with the ergodic rate bounds and the global optimum after post-processing (see Fig. 3.4b, d). A performance degradation in the high power regime is only visible for LB2 in Fig. 3.4d. When the RB optimum with the lower bounds is zero, i.e., for $P \leq 5$ dB, we used the MRT strategy for the default beamformers.

Increasing the system dimensions to $N = K = 100$ or $N = 100$ and $K = 120$ for the fully loaded or overloaded setups, respectively, the implications of the small setups remain (see Fig. 3.5). However, a higher transmit power P is required to

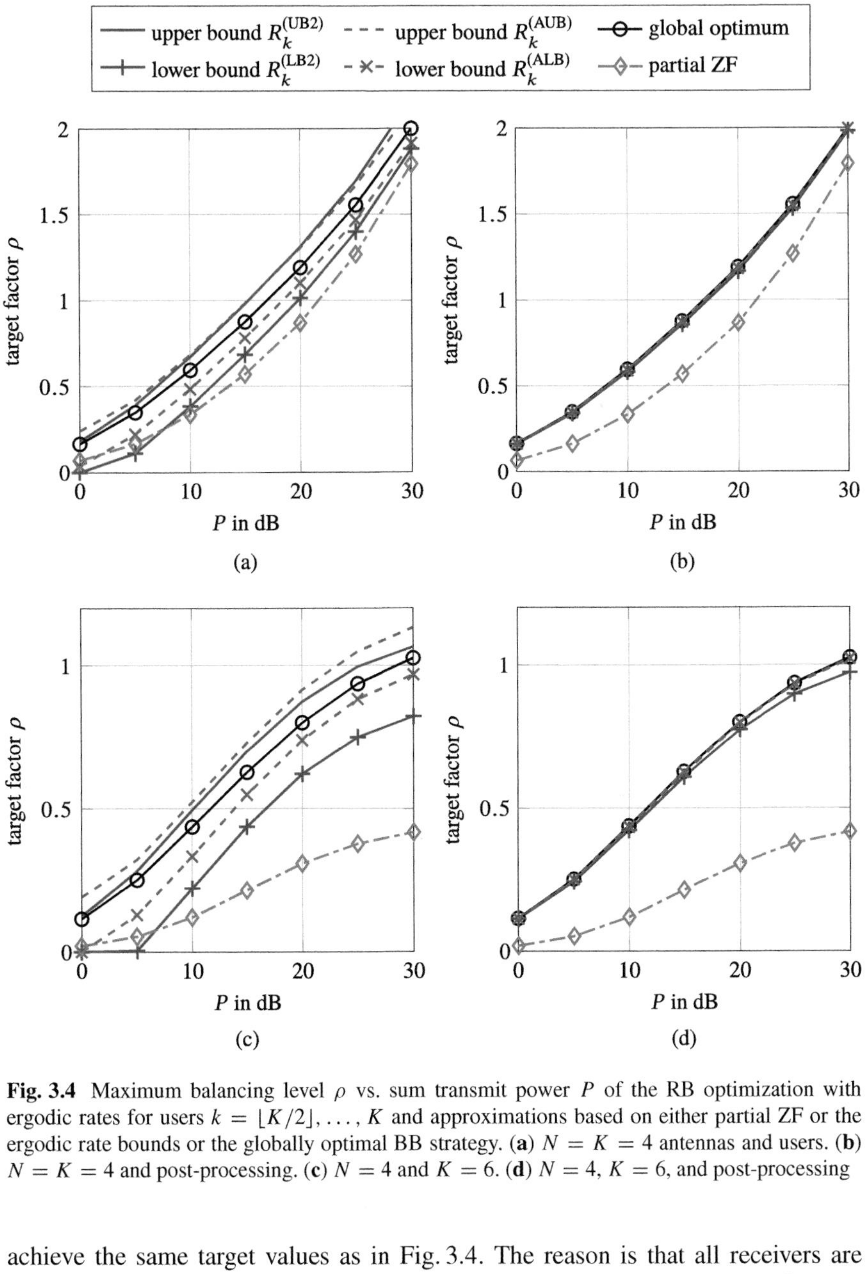

Fig. 3.4 Maximum balancing level ρ vs. sum transmit power P of the RB optimization with ergodic rates for users $k = \lfloor K/2 \rfloor, \ldots, K$ and approximations based on either partial ZF or the ergodic rate bounds or the globally optimal BB strategy. (**a**) $N = K = 4$ antennas and users. (**b**) $N = K = 4$ and post-processing. (**c**) $N = 4$ and $K = 6$. (**d**) $N = 4$, $K = 6$, and post-processing

achieve the same target values as in Fig. 3.4. The reason is that all receivers are placed in the same area as for the small user setups, which increases the interference. The lower bounds in Fig. 3.5a, c are zero when P is too small for satisfying $R_k^{(LB2)} > 0$ or $R_k^{(ALB)} > 0$ for all $k = \lfloor K/2 \rfloor, \ldots, K$. After post-processing, the balancing result is close to the UB2 bound and, therefore, also close to the global optimum.

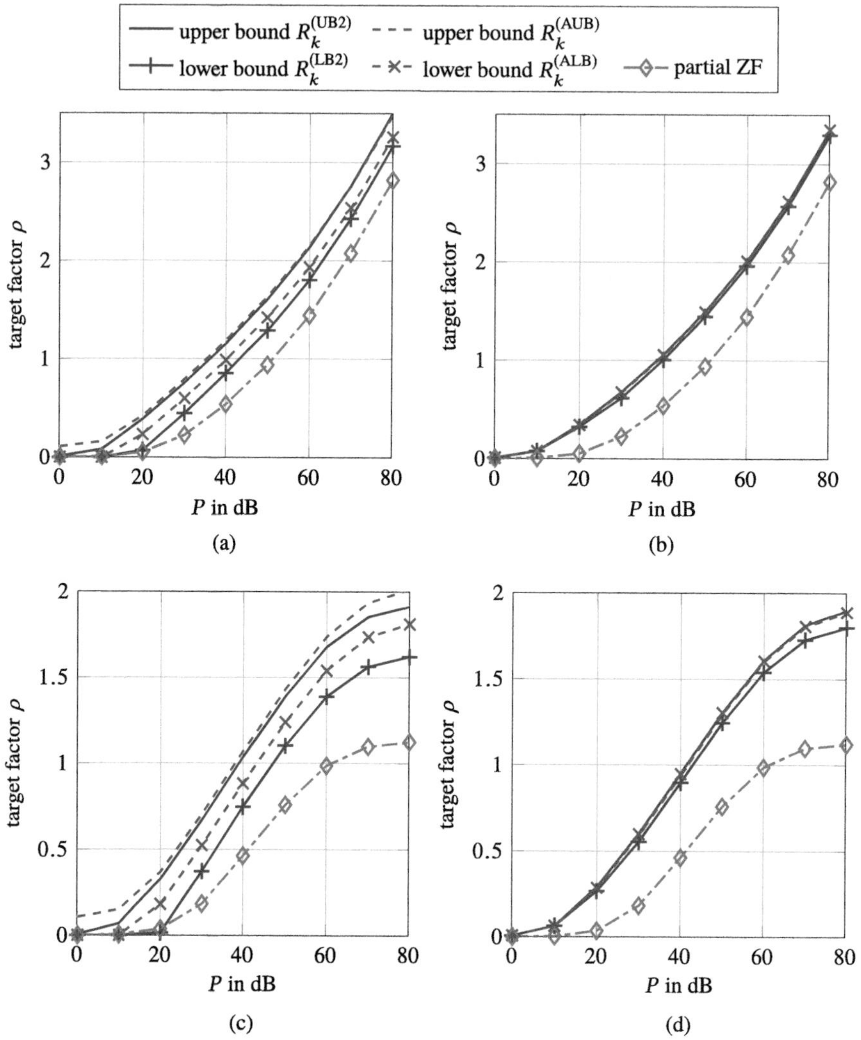

Fig. 3.5 Maximum balancing level ρ vs. sum transmit power P of the RB optimization with ergodic rates for users $k = \lfloor K/2 \rfloor, \ldots, K$ with $K = 100$ or $K = 120$ receivers and approximations based on either partial ZF or the ergodic rate bounds. (**a**) $N = K = 100$ antennas and users. (**b**) $N = K = 100$ and post-processing. (**c**) $N = 100$ antennas and $K = 120$ users. (**d**) $N = 100$, $K = 120$, and post-processing

3.7.3 Results for Sequential QoS Optimization and Balancing

To depict the performance of the SCS from Sect. 3.5, we consider the previous scenarios with $N = 4$. The QoS optimum and the resulting RB level, respectively,

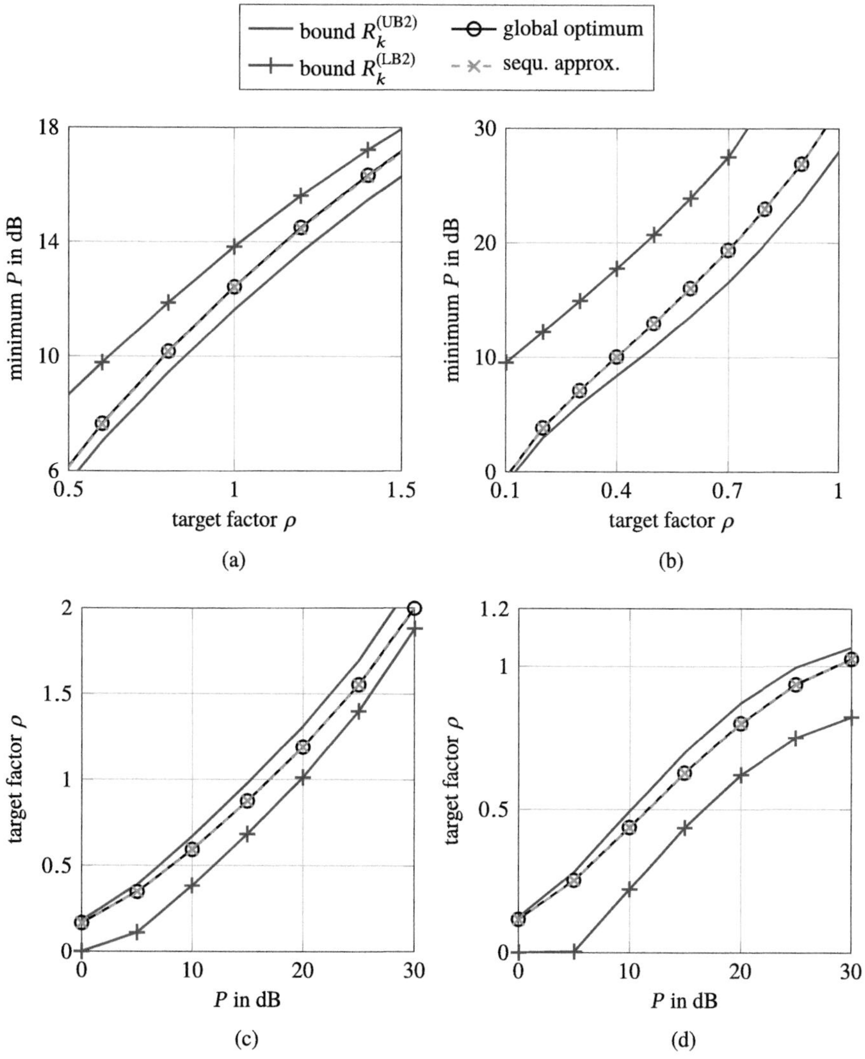

Fig. 3.6 QoS optimization and rate balancing results for the sequential convex approximation strategy for the ergodic rate users $k = \lfloor K/2 \rfloor, \ldots, K$ in comparison to the ergodic rate bounds LB2 and UB2 and the globally optimal BB strategy. (**a**) $N = K = 4$ antennas and users. (**b**) $N = 4$ and $K = 6$. (**c**) $N = K = 4$ and users. (**d**) $N = 4$ and $K = 6$

are shown in Fig. 3.6. The reference curves for this scheme are the lower and upper bounds based on $R_k^{(UB2)}$ and $R_k^{(LB2)}$ and the global optimum from the BB algorithm.

Figure 3.6a, b show that the SCS strategy meets the global optimum for the QoS optimization. In fact, the numerical simulation results are within the termination accuracy of the global optimization, i.e., smaller than $\epsilon = 10^{-2}$. Therewith, the SCS even outperforms the optimizations with ergodic rate bounds and a post-processing

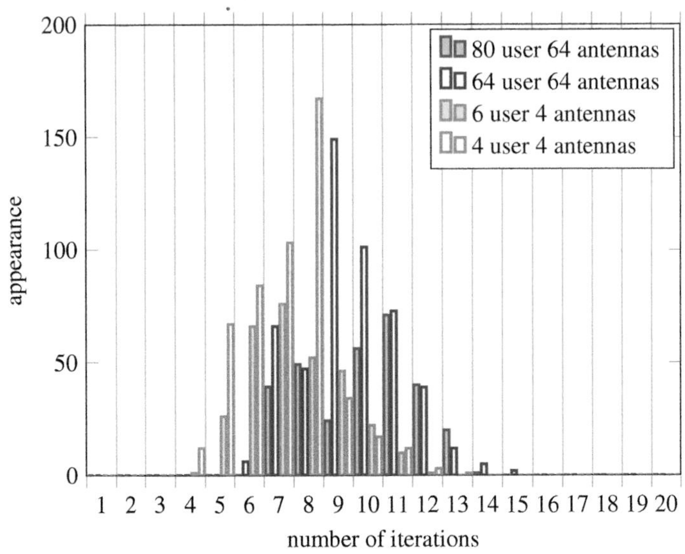

Fig. 3.7 Histogram of the iteration number for convergence of the sequential convex approximation method for the QoS optimization with ergodic rate constraints.

power allocation. Similarly, the SCS apparently reaches the global optimum of the RB problem with a sequence of balancing optimizations (see Fig. 3.6c, d).

The computational complexity of the SCS method is about M times that of a standard QoS or RB method, where M denotes the number of approximation steps until convergence. Fortunately, this number is small. To measure M and analyze the convergence speed of Algorithm 1, we performed simulations with ergodic rate constraints only, i.e., $R_k(\boldsymbol{t}) \geq \rho_k$ for $k = 1, \ldots, K$, and 100 channel realizations for either $N = 4$ and $K = 4$ or $K = 6$, or $N = 64$ and $K = 64$ or $K = 80$. As before, the individual targets are scaled with a common factor $\rho = 0.25, 0.5, 1, 2, 4$ in the fully loaded and $\rho = 0.25, 0.5, 1$ in the overloaded setup for the QoS optimization.

Figure 3.7 depicts a histogram of the iterations until convergence for these systems. We see that the iteration number slightly increases with the system dimensions, but remains below 16 for all scenarios, which indicates the fast convergence of the strategy. Hence, a local solution is found with less than 16 times the computational complexity of the standard QoS optimization.

A convergence analysis of the RB optimization with $N = K = 4$ and only ergodic rates is shown in Fig. 3.8. The 100 channel realizations for the averaged results are based on multiplicative Rician fading with mean $m_k = \mathrm{E}[1 + \xi_k] = \frac{1}{\sqrt{2}}$ and variance $\sigma_{\xi_k}^2 = \frac{1}{2}$, $k = 1, \ldots, K$. Figure 3.8a shows the plots of the balancing result ρ versus P in dB for the lower and upper bounds based on LB2 and UB2, a reference curve from a bisection over sequential QoS optimizations, and the direct sequential balancing strategy from Sect. 3.5.2. The sequential ergodic RB curve lies between the bounds and achieves the same performance as the reference curve.

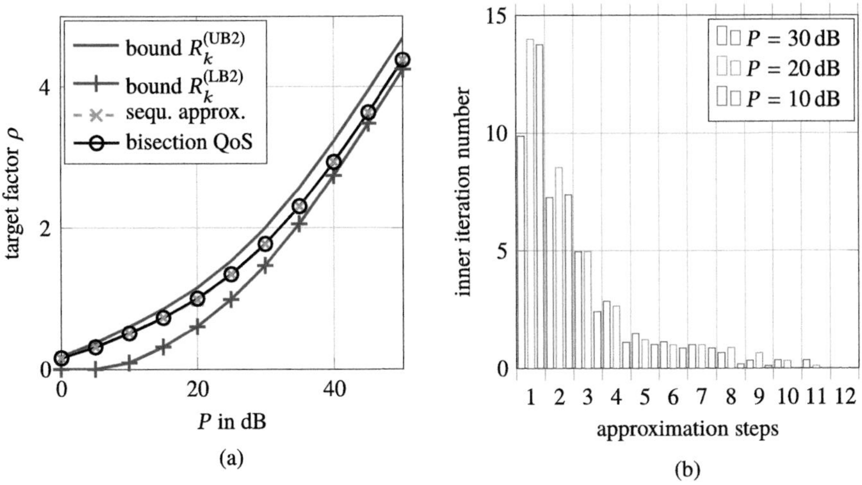

Fig. 3.8 The ergodic RB results of the sequential convex search with the average number of iterations until convergence for $N = K = 4$, 100 channel realizations, and multiplicative Rician fading. (**a**) Performance. (**b**) Iterations

We remark that the sequential RB strategy is less demanding than a single balancing optimization times the number of approximations. The number of inner fixed point iterations decreases. While the average number of inner RB iterations is about 14 for the first SINR approximation, it decreases to one for the ninth SINR approximation. The more accurate the approximation is, the less inner balancing iterations are required until convergence. Figure 3.8b indicates that about ten approximations are sufficient to achieve a local optimum with reasonable accuracy.

Chapter 4
Mean Square Error Transceiver Design for Additive Fading

For the additive channel model and imperfect CSI at the multi-antenna transmitter, rate based beamformer optimizations are difficult to solve directly. The closed-form expressions for ergodic rates involve numeric integrations and are hardly known for other scenarios than Rayleigh or Rician fading [25, 26, 132]. This prevents a reformulation of the QoS and RB optimizations into convex form (cf. Chap. 3).

Literature on multi-user communication focuses on the so-called average SINR and MSE metrics instead [43, 104, 116, 118, 121, 135, 136, 178, 234, 235]. To obtain the average SINR, the average is taken separately for the useful signal power and interference [118] and the average MSEs follow by assuming the same imperfect CSI at the transmitter and receivers' ends (e.g., [116, 136]). The beauty of these metrics are the closed-form expressions with only the channels' first and second order moments while the general SINR and MSE structure remain. Therefore, the QoS and balancing solutions from perfect CSI [67, 68, 154], using uplink–downlink duality, also apply for these metrics [104, 116, 118, 135, 236].

The disadvantage is the loose approximation that these metrics provide for the ergodic rate when the receivers' channel information is more accurate than the knowledge at the transmitter. The average SINR generally neither lower nor upper bounds the ergodic rate[1] and the above average MSE fits better to the achievable rate with imperfect receiver CSI [219] than to the scenario with perfect receiver CSI.

This chapter therefore focuses on QoS and balancing optimizations for average MSEs, but with the MMSE receive equalizers for perfect receiver CSI (1.6), i.e.,

$$\mathrm{AMSE}_k(\boldsymbol{t}) = \mathrm{E}[\mathrm{MMSE}_k(\boldsymbol{h}_k, \boldsymbol{t})]. \tag{4.1}$$

[1] In fact, the average SINR results in an upper rate bound for multiplicative fading (see Sect. 3.2).

A. Gründinger, *Statistical Robust Beamforming for Broadcast Channels and Applications in Satellite Communication*, Foundations in Signal Processing, Communications and Networking 22, https://doi.org/10.1007/978-3-030-29578-3_4

Section 4.1 shows that this metric provides a lower bound for the ergodic rate. Therefore, the generic QoS power minimization with MMSE requirements

$$\min_{p\geq 0,\, \boldsymbol{t}} p \quad \text{s.t.} \quad p^{-1}\boldsymbol{t} \in \mathcal{P}, \quad \mathbf{AMSE}(\boldsymbol{t}) \in \mathcal{M} \tag{4.2}$$

and the strongly connected MSE balancing minimization

$$\min_{\varepsilon,\, \boldsymbol{t}} \varepsilon \quad \text{s.t.} \quad \boldsymbol{t} \in \mathcal{P}, \quad \varepsilon^{-1}\,\mathbf{AMSE}(\boldsymbol{t}) \in \mathcal{M} \tag{4.3}$$

are conservative approximations for QoS and balancing optimization with ergodic rates, respectively, for an appropriate target MSE region $\mathcal{M}$. For example, if the target region is a hypercube with the bounds $\varepsilon_k \in (0, 1]$, $k = 1, \ldots, K$, i.e.,

$$\mathcal{M} = \left\{ \mathbf{MSE} \in [0, 1]^K : \mathrm{MSE}_k \leq \varepsilon_k,\; k = 1, \ldots, K \right\}, \tag{4.4}$$

the solution of (4.2) ensures reliable communication at $-\log_2(\varepsilon_k)$. The MSE balancing optimization (4.3) ensures the data rate $-\log_2(\varepsilon\varepsilon_k)$ if $\varepsilon\varepsilon_k < 1$ and deactivates communication to receivers with $\varepsilon\varepsilon_k \geq 1$ (cf. Sect. 4.5).

Important target regions besides (4.4) either bound the sum-MSE as

$$\mathcal{M} = \left\{ \mathbf{MSE} \in [0, 1]^K : \sum_{k=1}^{K} \mathrm{MSE}_k \leq \varepsilon \right\} \tag{4.5}$$

or a weighted sum-MSE. The previous region simplifies (4.3) to a sum-MSE minimization with imperfect CSI [68, 116, 136, 145, 157, 161, 162]. The latter weighted sum-MSE constraint may approximate the sum rate (cf. [155, 156, 237] and Sect. 4.6).

All these MSE regions are special cases of the convex polytope

$$\mathcal{M} = \left\{ \mathbf{MSE} \in [0, 1]^K : \sum_{k\in\mathcal{G}_j} w_k\,\mathrm{MSE}_k \leq \varepsilon_j,\; j = 1, \ldots, M \right\}, \tag{4.6}$$

with M disjoint groups of receivers $\mathcal{G}_i \subseteq \{1, \ldots, K\}$ and weights $w_k \in \mathbb{R}_+$, $k \in \mathcal{G}_i$. For the per-user requirements, $M = K$ and $\mathcal{G}_i = \{i\}$ hold, while $M = 1$ and $\mathcal{G}_1 = \{1, \ldots, K\}$ for the sum-MSE constraint. The results of this chapter are derived with (4.6) and the results for (4.4) and (4.5) are corollaries of these derivations.

Since closed-form expressions for the MMSEs (see Sect. 4.2) are too complicated for the precoder design, an ACS strategy is suggested for QoS and balancing optimization. It alternates between an update of the receive equalizers and the precoder [68]. The precoder update is transformed to a dual uplink filter design. Sections 4.3 and 4.5 show a concise duality derivation via Lagrangian multiplier theory for the QoS and min–max MSE optimization with general power constraints.

Related ACS strategies for sum-MSE minimization and imperfect CSI were proposed by Joham et al. [162] and Bogale and Vandendorpe [116]. They introduce duality for a single sum power constraint and typical power restrictions, respectively, but only for sum MSE minimization. These dualities are special cases of the duality in this chapter. The dual approach simplifies implementations and decreases the complexity compared to direct solutions in the broadcast domain. Kim et al. [238] propose an even less complex precoder update for a sum-MSE minimization with per-antenna power constraints. Therein, a Gauss–Seidel iteration sequentially updates those parts of the precoder that contribute to the same constraint.

While Sects. 4.1–4.5 focus on the target region (4.5) with fixed targets ε_k and weights w_k, $k = 1, \ldots, K$, related works also update these parameters within the ACS to approximate the ergodic rates. For example, the weighted MSE minimization of Christensen et al. [155] became a popular strategy for finding local solutions of the weighted sum rate maximization. This strategy provably converges [34] and was successfully applied to several other communication scenarios, e.g., interfering BCs [156, 237, 239] and relay channels [159].[2] A brief discussion on this locally tight bounding strategy for ergodic RB follows in Sect. 4.6.

4.1 Mean Square Error Based Rate Bounds

The motivation for optimizations (4.2) and (4.3) is the close connection between the MMSE and the mutual information. Guo et al. [152] showed that the MMSE between the received signal and its estimate is a derivative of the mutual information with respect to the transmit SNR.[3] This relation is independent of the input distribution and even holds for some non-Gaussian channels [240].

For zero-mean Gaussian inputs, several other relations between the mutual information and the MMSE have been identified. For example, the MMSE of the estimated and transmitted data signal is affine in the derivative of the rate with respect to the variance of the transmitted data signal i.e., MMSE $= \sigma_s^2(1 - \frac{\mathrm{d}}{\mathrm{d}\sigma_s^2} r)$ [157, 158, Lemma 1] and

$$\mathrm{E}[\mathrm{MMSE}] = \sigma_s^2\big(1 - \tfrac{\mathrm{d}}{\mathrm{d}\sigma_s^2}\,\mathrm{E}[r]\big). \tag{4.7}$$

We used this relation to derive a closed-form average MMSE expression in Sect. 4.2.

Moreover, the relation $r = -\log_2(\mathrm{MMSE})$ determines the mutual information and the MMSE by each other. This relation becomes $r \geq -B^{-1}\log_2(B^{-1}\mathrm{MMSE})$

---

[2] The references about the weighted MSE approximation for weighted sum-rate maximization are only a small subset of the available works. There exist many related publications citing [155].

[3] While the MMSE definition in [152] is between the received signal for (interference and) noiseless transmission and the actual received signal with conditional mean estimator, we define the (M)MSE between the intended transmit signal and the corresponding linear MMSE estimate instead.

for a B-dimensional transmit vector signal (multiple parallel streams) [153]. These formulas provided insights for the rate based uplink–downlink duality [154] and the QoS feasible region of the MIMO BC [115] using results for the MSE based duality and feasible region. With Jensen's inequality, this connection also results in an MMSE lower bound for the ergodic mutual information:

$$\mathrm{E}[r] \geq -B^{-1}\log_2(B^{-1}\,\mathrm{E}[\mathrm{MMSE}]) \geq -B^{-1}\log_2(B^{-1}\,\mathrm{E}[\mathrm{MSE}]), \tag{4.8}$$

where $B = 1$ for single-stream transmission. The latter inequality is valid for any suboptimal receive filter, e.g., based on imperfect receiver CSI [116, 136, 234].

The first order Taylor expansion of the rate at an MMSE point w^{-1}, i.e.,

$$r = -\log_2(\mathrm{MMSE}) = \log_2(w) + \log_2(\mathrm{e})(1 - w\mathrm{MMSE}) + o(|\mathrm{MMSE} - w^{-1}|^2) \tag{4.9}$$

with the non-negative error $o(|\mathrm{MMSE} - w^{-1}|^2)$, has an inherent weighted MSE interpretation. Christensen et al. [155] have exploited the multi-stream version of this relation for solving a weighted sum rate maximization in the MIMO BC via a Weighted sum mean square error (WSMSE) minimization. Besides the weighted sum rate maximization [34, 156, 159],[4] this relation is also useful for rate balancing [237]. Neglecting the error term, (4.9) lower bounds the ergodic rate as

$$\mathrm{E}[r] \geq \mathrm{E}[\log_2(w)] + \log_2(\mathrm{e})(1 - \mathrm{E}[w\mathrm{MMSE}]) \tag{4.10}$$

where the expectation is over the inner product of the weighting and the MMSE. Section 4.6 shows that the proposed duality framework is also suitable for the precoder update of the weighted MSE optimization approach with imperfect transmitter CSI and perfect transmitter CSI.

4.2 Closed-Form Mean Square Error Expressions

Most literature on MSEs with imperfect CSI focuses on the same local channel knowledge at the transmitter and the receivers, e.g., [116, 178, 234, 241].[5] The reason is the simple description of the average MMSE with the first and second order moments of the channels. In contrast, average MMSE expressions with imperfect CSI only at the transmitter but perfect MMSE filters involve numerical integrations. The average is over the ratios of quadratic forms in the Gaussian random channels

[4]Reference [34] already considers imperfect CSI. Therein, averaging is achieved by a stochastic update rule, which increases the involved samples for the sample average approximations in each iteration. Thereby, the corresponding ACS algorithm still almost surely converges to a KKT point.

[5]Each receiver has imperfect knowledge about its corresponding channel, e.g., user k knows the mean and covariance of $\boldsymbol{h}_k \sim \mathcal{N}_\mathbb{C}(\bar{\boldsymbol{h}}_k, \boldsymbol{C}_k)$, and the transmitter has this knowledge for all channels.

[see (4.12)], which evaluation requires either a numerical integration or an infinite series expansion [242–244].[6] Integral MSE expressions for Rician and Rayleigh fading are shown next.

Given the MMSE from (1.7), the kth user's average MMSE reads as

$$\mathrm{AMSE}_k = 1 - \mathrm{RQ}_k(\boldsymbol{t}_k, \boldsymbol{Q}), \tag{4.11}$$

where $\boldsymbol{Q} = \sum_{i=1}^{K} \boldsymbol{t}_i \boldsymbol{t}_i^{\mathrm{H}}$ denotes the transmit covariance matrix and

$$\mathrm{RQ}_k(\boldsymbol{x}, \boldsymbol{Z}) = \mathrm{E}\left[\frac{|\boldsymbol{h}_k^{\mathrm{H}} \boldsymbol{x}|^2}{\boldsymbol{h}_k^{\mathrm{H}} \boldsymbol{Z} \boldsymbol{h}_k + \sigma_k^2}\right] \tag{4.12}$$

is the expected value over the ratio of a quadratic forms in the random channels. For $\boldsymbol{h}_k \sim \mathcal{N}_{\mathbb{C}}(\bar{\boldsymbol{h}}_k, \boldsymbol{C}_k)$, an integral MMSE representation results by applying the complex equivalents of [242, Theorem 6] or [244, Proposition 2] for (4.12):

$$\mathrm{AMSE}_k = 1 - \mathrm{e}^{-\bar{\boldsymbol{h}}_k^{\mathrm{H}} \boldsymbol{C}_k^{-1} \bar{\boldsymbol{h}}_k} \int_0^\infty \mathrm{e}^{-z\sigma_k^2 - \boldsymbol{m}_k^{\mathrm{H}} (\mathbf{I} + z\tilde{\boldsymbol{Q}}_k)^{-1} \boldsymbol{m}_k} \det(\mathbf{I} + z\tilde{\boldsymbol{Q}}_k)^{-1} \times$$
$$\left(\boldsymbol{t}_k^{\mathrm{H}} \boldsymbol{C}_k^{1/2} (\mathbf{I} + z\tilde{\boldsymbol{Q}}_k)^{-1} \boldsymbol{C}_k^{\mathrm{H}/2} \boldsymbol{t}_k + |\boldsymbol{m}_k^{\mathrm{H}} (\mathbf{I} + z\tilde{\boldsymbol{Q}}_k)^{-1} \boldsymbol{C}_k^{\mathrm{H}/2} \boldsymbol{t}_k|^2\right) \mathrm{d}Z, \tag{4.13}$$

where the auxiliary mean and transmit covariance matrices are given by

$$\boldsymbol{m}_k = \boldsymbol{C}_k^{-1/2} \bar{\boldsymbol{h}}_k, \qquad \tilde{\boldsymbol{Q}}_k = \boldsymbol{C}_k^{\mathrm{H}/2} \boldsymbol{Q} \boldsymbol{C}_k^{1/2}.$$

Inserting $\boldsymbol{\tau}_k = \boldsymbol{V}_k^{\mathrm{H}} \boldsymbol{C}_k^{\mathrm{H}/2} \boldsymbol{t}_k$, $\boldsymbol{\nu}_k = \boldsymbol{V}_k^{\mathrm{H}} \boldsymbol{m}_k$, and the eigenvalue decomposition

$$\tilde{\boldsymbol{Q}}_k = \boldsymbol{V}_k \boldsymbol{\Phi} \boldsymbol{V}_k^{\mathrm{H}},$$

with the unitary modal matrix $\boldsymbol{V}_k \in \mathbb{C}^{N \times N}$ and the diagonal matrix $\boldsymbol{\Phi}_k = \mathrm{diag}(\phi_1, \ldots, \phi_N)$ of the eigenvalues $\phi_i \geq 0$, $i = 1, \ldots, N$, (4.13) becomes

$$\mathrm{AMSE}_k = 1 - \mathrm{e}^{-\bar{\boldsymbol{h}}_k^{\mathrm{H}} \boldsymbol{C}_k^{-1} \bar{\boldsymbol{h}}_k} \int_0^\infty \mathrm{e}^{-z\sigma_k^2 - \sum\limits_{i=1}^{N} \frac{|\nu_i|^2}{1+z\phi_i}} \prod_{i=1}^{N} \frac{1}{1+z\phi_i}$$
$$\left(\sum_{i=1}^{N} \frac{|\tau_i|^2}{1+z\phi_i} + \left|\sum_{i=1}^{N} \frac{\nu_i^* \tau_i}{1+z\phi_i}\right|^2\right) \mathrm{d}z.$$

A numerical integration or an infinite series expansion evaluates this integral [244]. For the zero-mean case, i.e., $\boldsymbol{h}_k \sim \mathcal{N}_{\mathbb{C}}(\mathbf{0}, \boldsymbol{C}_k)$, the integral formulation simplifies to

[6] Approximations for moments of ratios of quadratic forms in random variables are shown in [245].

$$\mathrm{AMSE}_k = 1 - \int_0^{\infty} \mathrm{e}^{-z\sigma_k^2} \prod_{i=1}^{N} \frac{|\tau_i|^2}{(1+z\phi_i)^2}\,\mathrm{d}z. \tag{4.14}$$

The above numerical integrations can be inaccurate for increasing system dimensions. An alternative closed-form MMSE expression follows via (4.7) and a known ergodic mutual information expression for zero-mean Gaussian channels. Writing $\mathrm{E}[|s_k|^2] = \sigma_{s_k}^2$ and $\boldsymbol{Q}(\sigma_{s_k}^2) = \sigma_{s_k}^2 \boldsymbol{t}_k \boldsymbol{t}_k^{\mathrm{H}} + \sum_{i \neq k} \boldsymbol{t}_i \boldsymbol{t}_i^{\mathrm{H}}$, the MMSE reads as

$$\mathrm{AMSE}_k = 1 - \frac{\mathrm{d}}{\mathrm{d}\sigma_{s_k}^2} \mathrm{E}\left[\log_2\left(1 + \sigma_k^{-2} \boldsymbol{h}_k^{\mathrm{H}} \boldsymbol{Q}(\sigma_{s_k}^2) \boldsymbol{h}_k\right)\right]\Big|_{\sigma_{s_k}^2 = 1}. \tag{4.15}$$

Kießling [27] and Kang and Alouini [25] have provided closed-form expressions of the ergodic mutual information for point-to-point MIMO channels.[7] In particular, we exploit the representations in [25, Section II.E.2)] and [27, Corollary 9] for a closed form of (4.15). To this end, let $\tilde{\boldsymbol{Q}}(\sigma_{s_k}^2)$ have $\mathrm{rank}\{\tilde{\boldsymbol{Q}}(\sigma_{s_k}^2)\} = K$ and distinct eigenvalues $\phi_1 > \phi_2 > \ldots > \phi_K > 0$ within $\boldsymbol{\Phi}_k = \mathrm{diag}(\phi_1, \ldots, \phi_K)$. Then,

$$\begin{aligned} M_k(\boldsymbol{\Phi}_k) &= \mathrm{E}\left[\log_2(1 + \sigma_k^{-2} \boldsymbol{h}_k^{\mathrm{H}} \boldsymbol{Q} \boldsymbol{h}_k)\right] \\ &= \frac{\det(\boldsymbol{\Phi}_k)^{K-2}}{\ln(2)(-1)^{K-1} \prod_{1 \le i < j \le K} \phi_j - \phi_i} \det(\boldsymbol{\Psi}_k), \end{aligned} \tag{4.16}$$

where the matrix $\boldsymbol{\Psi}_k \in \mathbb{R}^{K \times K}$ has the entries

$$[\boldsymbol{\Psi}_k]_{i,j} = \begin{cases} \int_0^{\infty} \mathrm{e}^{-\frac{z}{\phi_j}} \ln(1 + \sigma_k^{-2} z)\,\mathrm{d}z = \phi_j \mathrm{e}^{\sigma_k^2/\phi_j} \mathrm{E}_1(\sigma_k^2/\phi_j) & i = 1, \\ \phi_j^{-(n-i)} & i > 1. \end{cases}$$

The eigenvalues ϕ_i, $i = 1, \ldots, K$ are linear functions of $\sigma_{s_k}^2$ in (4.16). Therefore, we rewrite (4.15) as a sum of partial derivatives, i.e.,

$$\mathrm{AMSE}_k = 1 - \sigma_{s_k}^2 \sum_{\ell=1}^{K} \frac{\partial M_k(\boldsymbol{\Phi}_k)}{\partial \phi_\ell} \frac{\partial \phi_\ell}{\partial \sigma_{s_k}^2}\Bigg|_{\sigma_{s_k}^2 = 1}. \tag{4.17}$$

The derivations of the distinct eigenvalues ϕ_ℓ, each with multiplicity one, read as[8]

$$\frac{\partial \phi_\ell}{\partial \sigma_{s_k}^2} = \boldsymbol{v}_\ell^{\mathrm{H}} \left(\frac{\partial}{\partial \sigma_{s_k}^2} \tilde{\boldsymbol{Q}}_k(\sigma_{s_k}^2) \right) \boldsymbol{v}_\ell = \left\| \boldsymbol{v}_\ell^{\mathrm{H}} \boldsymbol{C}_k^{1/2} \boldsymbol{t}_k \right\|_2^2,$$

[7] We have exploited [27, Theorem 4] to derive the average MMSE of a point-to-point MIMO channel [157] and for a downlink multi-user MIMO channel [158].

[8] Derivatives of eigenvalues are detailed in [66, Appendix A.14] and the references therein.

where $\boldsymbol{v}_\ell \in \mathbb{C}^N$ is the corresponding eigenvector. This results in the MMSE term

$$\mathrm{AMSE}_k = 1 - M_k(\boldsymbol{\Phi}_k) \sum_{\ell=1}^{K} \left(\frac{1}{\phi_\ell} + \sum_{m \neq \ell} \frac{1}{\phi_\ell - \phi_m} + \frac{\det(\boldsymbol{\Psi}'_{k,\ell})}{\det(\boldsymbol{\Psi}_k)} \right) \left\| \boldsymbol{v}_\ell^{\mathrm{H}} \boldsymbol{C}_k^{1/2} \boldsymbol{t}_k \right\|_2^2, \tag{4.18}$$

where $\boldsymbol{\Psi}'_{k,\ell} = \frac{\partial}{\partial \phi_\ell} \boldsymbol{\Psi}_k \in \mathbb{R}^{K \times K}$ has the entries $[\boldsymbol{\Psi}'_{k,\ell}]_{i,j} = [\boldsymbol{\Psi}_k]_{i,j}$ for $j \neq \ell$ and

$$[\boldsymbol{\Psi}'_{k,\ell}]_{i,\ell} = \begin{cases} 1 + \mathrm{e}^{\sigma_k^2/\phi_\ell}\, \mathrm{E}_1(\sigma_k^2/\phi_\ell)\big(1 + \sigma_k^2/\phi_\ell\big) & i = 1, \\ -(K - i)\phi_\ell^{-(K-i-1)} & i > 1. \end{cases}$$

The advantage of (4.18) is its evaluation based on standard functions and using accurate series expansions for $x\mathrm{e}^x\, \mathrm{E}_1(x)$ [203, Equations 5.1.52, 5.1.56, and 5.1.53].

4.3 Quality-of-Service Optimization

Designing beamformers based on the average MMSE metric remains difficult. Even though we found closed-form expressions for the average MMSE, the MMSE requirements of (4.2) are non-convex in $\boldsymbol{t}$ and convex reformulations are missing. We resolve this issue via alternating between updating the equalizers and beamformers. The beamformer update is for equalizers $f_k(\boldsymbol{h}_k)$, $k = 1, \ldots, K$, that depend on the random channels but are independent of the precoder.[9] The QoS problem reads as

$$\min_{p \geq 0,\, \boldsymbol{t}} \ p \quad \text{s.t.} \quad p^{-1}\boldsymbol{t} \in \mathcal{P}, \quad \mathbf{AMSE}(\boldsymbol{t}) \in \mathcal{M}. \tag{4.19}$$

The beamformers are designed w.r.t. $\mathrm{AMSE}_k(\boldsymbol{t}) = \mathrm{E}[\mathrm{MSE}_k(\boldsymbol{h}_k, \boldsymbol{t}, f_k(\boldsymbol{h}_k))]$ as

$$\mathrm{AMSE}_k(\boldsymbol{t}) = 1 - 2\,\mathrm{Re}\left(\hat{\boldsymbol{h}}_k^{\mathrm{H}} \boldsymbol{t}_k\right) + \sum_{i=1}^{K} \boldsymbol{t}_i^{\mathrm{H}} \boldsymbol{R}_k \boldsymbol{t}_i + \tilde{\sigma}_k^2, \quad k = 1, \ldots, K, \tag{4.20}$$

where we assume knowledge of the first and second order moments

$$\hat{\boldsymbol{h}}_k = \mathrm{E}[\boldsymbol{h}_k f_k(\boldsymbol{h}_k)] \tag{4.21}$$

$$\boldsymbol{R}_k = \mathrm{E}[\boldsymbol{h}_k \boldsymbol{h}_k^{\mathrm{H}} |f_k(\boldsymbol{h}_k)|^2], \tag{4.22}$$

[9] The idea of pre-fixed equalizer functions was also exploited by [235]. Other than [235], $f_k(\boldsymbol{h}_k)$ are herein the MMSE filters for a pre-defined $\boldsymbol{t}'$ and, thereby, approximate the optimal filters.

and the effective noise power

$$\tilde{\sigma}_k^2 = \sigma_k^2 \, \mathrm{E}[|f_k(\boldsymbol{h}_k)|^2]. \tag{4.23}$$

We approximate these terms by the sample average over a large number of channel realizations according to the additive error model (2.1) (cf. Sect. 4.3.4).

Remark 4.1 The above MSE metric upper bounds (4.11), which is for MMSE equalizers, and provides a lower bound for the ergodic rate (4.8). With per-user targets ε_k, the solution of (4.2) is therefore also feasible for the ergodic rate based QoS problem with targets $\rho_k = -\log_2(\varepsilon_k)$, $k = 1, \ldots, K$.[10] The optimum of (4.19) is an upper bound to the minimum of the QoS optimization with constraints $R_k(\boldsymbol{t}) \geq \rho_k$, $k = 1, \ldots, K$.

For the QoS optimization, we exploit that (4.20) is biconvex in $\boldsymbol{t}$ and $f_k(\boldsymbol{h}_k)$, $k = 1, \ldots, K$, that is, it is convex in the precoder for fixed equalizers and vice versa.[11] Hence, an ACS of the precoder and filters converges to a local solution of (4.2). It performs the following two steps repeatedly until convergence (cf. [68]):

1. The equalizers $f_k(\boldsymbol{h}_k)$ from (1.6) are first found in the downlink for fixed $\boldsymbol{t}$, i.e., as functions of $\boldsymbol{h}_k$, $k = 1, \ldots, K$, and the moments $\hat{\boldsymbol{h}}_k$, $\boldsymbol{R}_k$, and $\tilde{\sigma}_k^2$ are evaluated.
2. Second, the downlink beamformers $\boldsymbol{t}_k$, $k = 1, \ldots, K$ are optimized as equalizers in the dual uplink system for given moments $\hat{\boldsymbol{h}}_k$, $\boldsymbol{R}_k$, and $\tilde{\sigma}_k^2$, $k = 1, \ldots, K$.

Repeating these steps iteratively results in a diminishing objective value p in each iteration. This ensures convergence in the objective (cf. [246, Theorem 4.5]). Convergence of the beamformers follows from [246, Theorem 4.9], because the update of the MMSE filter functions is unique for given $\boldsymbol{t}$ and the beamformer update (4.19) is unique for given moments (4.21) and (4.22).

The uplink–downlink duality in Step 2 of the ACS reduces the implementation and computation complexity compared to solvers that handle the generalized power constraints in the downlink, e.g., disciplined convex programming toolboxes [99, 232]. While Joham et al. [162] have introduced such a duality for a single sum power constraint, we derive the uplink formulations of the precoder design step concisely via Lagrangian duality and for the generalized power constraint set $\mathcal{P}$.

To this end, we split the downlink equalizer as $f_k(\boldsymbol{h}_k) = v_j \tilde{f}_k(\boldsymbol{h}_k)$ for $k = 1, \ldots, K$, where $\tilde{f}_k(\boldsymbol{h}_k)$ is the above fixed MMSE filter function and v_j, $j = 1, \ldots, M$ are the additional decision variables for (4.19). Then, the precoder update in Step 2 of the ACS reads as

$$\min_{p \geq 0, \boldsymbol{t}, \boldsymbol{v}} \; p \quad \text{s.t.} \quad p^{-1}\boldsymbol{t} \in \mathcal{P}, \quad \mathbf{AMSE}(\boldsymbol{t}, \boldsymbol{v}) \in \mathcal{M}, \tag{4.24}$$

---

[10] This is in contrast to the QoS optimization with the average SINR from [118], for example.

[11] For example, see [246] for a definition of biconvex functions.

where $\boldsymbol{v} = [v_1, \ldots, v_M]^{\mathrm{T}}$ and the entries $\mathrm{AMSE}_k(\boldsymbol{t}, v_j)$ are given by

$$\mathrm{AMSE}_k(\boldsymbol{t}, v_j) = 1 - 2\,\mathrm{Re}\left(v_j^* \hat{\boldsymbol{h}}_k^{\mathrm{H}} \boldsymbol{t}_k\right) + \sum_{i=1}^{K} |v_j|^2 \boldsymbol{t}_i^{\mathrm{H}} \boldsymbol{R}_k \boldsymbol{t}_i + |v_j|^2 \tilde{\sigma}_k^2, \; k = 1, \ldots, K. \tag{4.25}$$

The virtual filter v_j is a joint variable for the jth constraint in $\mathcal{M}$ [cf. (4.6)], i.e.,

$$\sum_{k \in \mathcal{G}_j} w_k \,\mathrm{AMSE}_k(v_j, \boldsymbol{t}) \le \varepsilon_j. \tag{4.26}$$

Its solution for (4.24) minimizes the left-hand side weighted sum MSE, that is,

$$v_j = \frac{\sum_{k \in \mathcal{G}_j} w_k \hat{\boldsymbol{h}}_k^{\mathrm{H}} \boldsymbol{t}_k}{\sum_{k \in \mathcal{G}_j} w_k (\sum_{i=1}^{K} \boldsymbol{t}_i^{\mathrm{H}} \boldsymbol{R}_k \boldsymbol{t}_i + \tilde{\sigma}_k^2)}. \tag{4.27}$$

Therewith, the sum MSE $\mathrm{WAMSE}_j(\boldsymbol{t}) = \min_{v_j} \sum_{k \in \mathcal{G}_j} w_k \,\mathrm{AMSE}_k(v_j, \boldsymbol{t})$ becomes

$$\mathrm{WAMSE}_j(\boldsymbol{t}) = m_j - \frac{|\hat{\boldsymbol{h}}^{\mathrm{H}} \boldsymbol{G}_j \boldsymbol{t}|^2}{\boldsymbol{t}^{\mathrm{H}} \hat{\boldsymbol{R}}_j \boldsymbol{t} + \hat{\sigma}_j^2}, \tag{4.28}$$

where $m_j = \sum_{k \in \mathcal{G}_j} w_k$, $\hat{\sigma}_j^2 = \sum_{k \in \mathcal{G}_j} w_k \tilde{\sigma}_k^2$, and $\boldsymbol{G}_j = \mathrm{diag}(w_{1,k}, \ldots, w_{M,k}) \otimes \mathbf{I}$ with entries $w_{j,k} = w_k$ for $k \in \mathcal{G}_j$ and $w_{j,k} = 0$ otherwise. Therewith, the precoder update (4.24) is equivalent to the weighted sum MMSE constrained QoS power minimization

$$\begin{aligned} \min_{\boldsymbol{t}} \; p \quad \text{s.t.} \quad & \mathrm{WAMSE}_j(\boldsymbol{t}) \le \varepsilon_j, \quad j = 1, \ldots, M, \\ & \sum_{i=1}^{K} \|\boldsymbol{A}_{i,\ell}^{1/2} \boldsymbol{t}_i\|_2^2 \le p^2 P_\ell, \quad \ell = 1, \ldots, L. \end{aligned} \tag{4.29}$$

This optimization of $\boldsymbol{t}$ features a SOC formulation that serves as a basis for the dual uplink formulation. In the dual problem, the downlink beamformer design becomes an inner filter design and power allocation and an outer worst-case noise covariance matrix search (4.29). We find this worst-case covariance matrix via a PG method and the power allocation via a fixed point search, for example.[12]

Special cases of the duality are obtained if either $\mathcal{M}$ or $\mathcal{P}$ from (4.2) consists of a reduced number of constraints. For example, the dual optimization only contains a worst-case noise search if $\mathcal{M}$ consists of a single average sum MSE constraint [145].

[12] Alternatively, one may employ a convex SDP formulation of the dual problem.

If $\mathcal{P}$ consists of a single sum power constraint, the worst case noise covariance has a closed form and the dual problem degenerates to the formulation from [68, 161].

Remark 4.2 Feasibility of (4.24) is crucial for the initial fixed receiver functions $f_k(\boldsymbol{h}_k)$, $k = 1, \ldots, K$ in Step 2 of the ACS strategy. Only then the power is decreased in each iteration. Otherwise, the ACS strategy fails to find a solution for the QoS precoder design. We describe the achievable region of (4.19) in Sect. 4.4.

4.3.1 MSE Based Uplink–Downlink Dualities

To derive the Lagrangian dual problem of (4.29), we first characterize it as a SOCP:

$$\begin{aligned} &\min_{\boldsymbol{t},p} p^2 \quad \text{s.t.} \quad \|\boldsymbol{A}_\ell^{1/2}\boldsymbol{t}\|_2 \leq p\sqrt{P_\ell}, \quad \ell = 1, \ldots, L, \\ &\|[\hat{\boldsymbol{R}}_j^{\mathrm{H}/2}, \boldsymbol{0}]^{\mathrm{H}}\boldsymbol{t} + \hat{\sigma}_j \mathbf{e}_{N+1}\|_2 \leq \frac{\mathrm{Re}(\hat{\boldsymbol{h}}^{\mathrm{H}}\boldsymbol{G}_j\boldsymbol{t})}{\sqrt{m_j - \varepsilon_j}}, \quad j = 1, \ldots, M. \end{aligned} \tag{4.30}$$

Here, $\hat{\boldsymbol{R}}_j^{1/2}$ and $\boldsymbol{A}_\ell^{1/2}$ are matrix square roots, which we obtain via a Cholesky factorization [119, Section 4.2.10] from the block-diagonal semidefinite matrices

$$\hat{\boldsymbol{R}}_j = \mathbf{I}_K \otimes \sum_{k \in \mathcal{G}_j} w_k \boldsymbol{R}_k \in \mathbb{C}^{NK \times NK}, \quad j = 1, \ldots, M, \tag{4.31}$$

$$\boldsymbol{A}_\ell = \mathrm{bdiag}(\boldsymbol{A}_{1,\ell}, \ldots, \boldsymbol{A}_{K,\ell}) \in \mathbb{C}^{NK \times NK}, \quad \ell = 1, \ldots, L. \tag{4.32}$$

Problems (4.29) and (4.30) are equivalent up to a unit scalar multiple of the beamformers, i.e., $\boldsymbol{t}_k' = \mathrm{e}^{\mathrm{j}\phi_k}\boldsymbol{t}_k$ with $\phi_k \in [0, 2\pi)$, $k = 1, \ldots, K$. Since this variable substitution neither increases the transmit powers nor the interference, we require

$$\mathrm{Re}(\hat{\boldsymbol{h}}_k^{\mathrm{H}}\boldsymbol{t}_k) \leq |\hat{\boldsymbol{h}}_k^{\mathrm{H}}\boldsymbol{t}_k'|, \quad k = 1, \ldots, K \tag{4.33}$$

to hold with equality, i.e., $\mathrm{Im}(\hat{\boldsymbol{h}}_k^{\mathrm{H}}\boldsymbol{t}_k) = 0$, and $\mathrm{Re}(\hat{\boldsymbol{h}}^{\mathrm{H}}\boldsymbol{G}_j\boldsymbol{t}) = |\hat{\boldsymbol{h}}^{\mathrm{H}}\boldsymbol{G}_j\boldsymbol{t}|$ as a consequence. If $\boldsymbol{t}$ denotes the solution of (4.30), any $\boldsymbol{t}'$ with preshifted beamformers $\boldsymbol{t}_k' = \mathrm{e}^{\mathrm{j}\phi_k}\boldsymbol{t}_k$ and $\phi_k \in [0, 2\pi)$, $k = 1, \ldots, K$ is a solution for (4.29) and the minima of (4.30) and (4.29) are equal.

The solution of (4.30) can be obtained with the standard interior-point solvers and the disciplined convex programming toolbox CVX [99, 247].[13] We derive the Lagrangian dual problem of (4.30) and recast it as an uplink QoS optimization.

[13] The problems are similar to the SINR constrained QoS optimization (cf. [71, 73]).

Theorem 4.1 *The strongly dual uplink max–min problem to* (4.30) *reads as*

$$\max_{\boldsymbol{\mu}\geq \mathbf{0}} \min_{\boldsymbol{u},\boldsymbol{\lambda}\geq \mathbf{0}} \sum_{j=1}^{M} \lambda_j \sum_{k\in\mathcal{G}_j} w_k \tilde{\sigma}_k^2 \quad \text{s.t.} \quad \sum_{\ell=1}^{L} \mu_\ell P_\ell \leq 1, \quad \mathbf{AMSE}^{(\mathrm{ul})} \in \mathcal{M}. \tag{4.34}$$

The inner problem designs the entries from $\boldsymbol{u} = \mathrm{vec}([\boldsymbol{u}_1, \ldots, \boldsymbol{u}_M])$ *that minimize*

$$\mathrm{AMSE}_k^{(\mathrm{ul})} = 1 - 2\sqrt{\lambda_j w_k}\,\mathrm{Re}(\hat{\boldsymbol{h}}_k^{\mathrm{H}} \boldsymbol{u}_k) + \boldsymbol{u}_k^{\mathrm{H}}\Big(\sum_{j=1}^{M} \lambda_j \sum_{i\in\mathcal{G}_j} w_i \boldsymbol{R}_i + \sum_{\ell=1}^{L} \mu_\ell \boldsymbol{A}_{k,\ell}\Big)\boldsymbol{u}_k, \tag{4.35}$$

and allocates $\boldsymbol{\lambda} = [\lambda_1, \ldots, \lambda_M]^{\mathrm{T}}$ *to satisfy the sum MSE constraints of* $\mathcal{M}$. *The outer optimization searches the worst-case noise covariance matrix w.r.t. the dual variables* $\boldsymbol{\mu} = [\mu_1, \ldots, \mu_L]^{\mathrm{T}}$ *that correspond to the power constraints of* (4.30).

The derivation of this dual formulation is via similar steps as in [74], which derives duality for SINR constraints. Sect. A.6 shows an alternative derivation for a generic SOC problem and proves the following lemma (cf. [73, Appendix A]).

Lemma 4.1 *When maximizing the concave objective function* $f : \mathbb{R}^n \to \mathbb{R}$, $\boldsymbol{y} \mapsto f(\boldsymbol{y})$ *subject to the SOC conditions* $\boldsymbol{c}_i - \boldsymbol{A}_i^{\mathrm{T}} \boldsymbol{y} \in \mathcal{L}^N$, $i = 1, \ldots, m$, *the structure for the Lagrangian multipliers is* $\boldsymbol{x}_i = \lambda_i\, \mathrm{bdiag}(1, -\mathbf{I})(\boldsymbol{c}_i - \boldsymbol{A}_i^{\mathrm{T}} \boldsymbol{y}) \in \mathcal{L}^N$ *and the Lagrangian function reads as*

$$L(\boldsymbol{y}, \boldsymbol{\lambda}) = f(\boldsymbol{y}) + \sum_{i=1}^{m} \lambda_i \big((c_{i1} - \boldsymbol{a}_{i1}^{\mathrm{T}} \boldsymbol{y})^2 - \|\boldsymbol{c}_{i2} - \boldsymbol{A}_{i2}^{\mathrm{T}} \boldsymbol{y}\|_2^2\big), \tag{4.36}$$

where $\boldsymbol{c}_i = [c_{i1}, \boldsymbol{c}_{i2}^{\mathrm{T}}] \in \mathbb{R}^N$ *and* $\boldsymbol{A}_i = [\boldsymbol{a}_{i1}, \boldsymbol{A}_{i2}] \in \mathbb{R}^{N\times n}$.

This lemma is the basis for the next presented proof of the theorem.

Proof (Proof of Theorem 4.1) Due to Lemma 4.1, we base the duality derivation of (4.30) without loss of generality or strong duality on the Lagrangian function

$$\begin{aligned} L(p, \boldsymbol{t}, \boldsymbol{\lambda}, \boldsymbol{\mu}) = &\sum_{j=1}^{M} \lambda_j \hat{\sigma}_j^2 + p^2\Big(1 - \sum_{\ell=1}^{L} \mu_\ell P_\ell\Big) \\ &+ \boldsymbol{t}^{\mathrm{H}}\Big(\boldsymbol{Y} - \sum_{j=1}^{M} \frac{\lambda_j}{m_j - \varepsilon_j} \boldsymbol{G}_j \hat{\boldsymbol{h}} \hat{\boldsymbol{h}}^{\mathrm{H}} \boldsymbol{G}_j\Big)\boldsymbol{t}, \end{aligned} \tag{4.37}$$

where $\boldsymbol{Y} = \sum_{j=1}^{M} \lambda_j \hat{\boldsymbol{R}}_j + \sum_{\ell=1}^{L} \mu_\ell \boldsymbol{A}_\ell$. Here, $\lambda_j \geq 0$, $j = 1, \ldots, M$ and $\mu_\ell \geq 0$, $\ell = 1, \ldots, L$, are the multipliers to the MSE and power constraints of (4.30), respectively.

While the primal optimization essentially minimizes the supremum of (4.37), the dual problem results from exchanging the min- and maximization, i.e.,[14]

$$\sup_{\boldsymbol{\lambda}\geq 0,\boldsymbol{\mu}} \inf_{p\geq 0,\boldsymbol{t}} L(p,\boldsymbol{t},\boldsymbol{\lambda},\boldsymbol{\mu}), \tag{4.38}$$

and the dual objective function $g : \mathbb{R}^K \times \mathbb{R}^L \to \mathbb{R}$ reads as

$$g(\boldsymbol{\lambda},\boldsymbol{\mu}) = \inf_{p\geq 0,\boldsymbol{t}} L(p,\boldsymbol{t},\boldsymbol{\lambda},\boldsymbol{\mu}).$$

Since $p \in \mathbb{R}_+$ and $\boldsymbol{t} \in \mathbb{C}^{NK}$ are unconstrained, $g(\boldsymbol{\lambda},\boldsymbol{\mu})$ is unbounded below unless both $\sum_{\ell=1}^{L} \mu_\ell P_\ell \leq 1$ and $\boldsymbol{Y} - \sum_{j=1}^{M} \frac{\lambda_j}{m_j-\varepsilon_j} \boldsymbol{G}_j \hat{\boldsymbol{h}}\hat{\boldsymbol{h}}^{\mathrm{H}} \boldsymbol{G}_j \succeq \mathbf{0}$ [cf. (4.37)]. Necessary and sufficient conditions for the latter Linear matrix inequality (LMI) are

$$\boldsymbol{Y} - \frac{\lambda_j}{m_j-\varepsilon_j} \boldsymbol{G}_j \hat{\boldsymbol{h}}\hat{\boldsymbol{h}}^{\mathrm{H}} \boldsymbol{G}_j \succeq \mathbf{0}, \quad j = 1,\ldots,M, \tag{4.39}$$

because the vectors $\boldsymbol{G}_j\hat{\boldsymbol{h}}$, $j = 1,\ldots,M$ are mutually orthogonal. With Schur's complement [102, Appendix A.5.5], we recast these LMIs as[15]

$$m_j - \lambda_j \hat{\boldsymbol{h}}^{\mathrm{H}} \boldsymbol{G}_j \boldsymbol{Y}^{\dagger} \boldsymbol{G}_j \hat{\boldsymbol{h}} \geq \varepsilon_j, \quad j = 1,\ldots,M. \tag{4.40}$$

The constraints (4.39) and (4.40) are equivalent because $\hat{\boldsymbol{h}} \in \mathrm{range}\{\boldsymbol{Y}\}$.[16] Inserting (4.40), a first version of the dual problem of (4.30) reads as

$$\max_{\boldsymbol{\mu}\geq\mathbf{0},\boldsymbol{\lambda}\geq\mathbf{0}} \sum_{j=1}^{M} \lambda_j \hat{\sigma}_j^2 \quad \text{s.t.} \sum_{\ell=1}^{L} \mu_\ell P_\ell \leq 1,\ \ \varepsilon_j \leq m_j - \lambda_j \hat{\boldsymbol{h}}^{\mathrm{H}} \boldsymbol{G}_j \boldsymbol{Y}^{\dagger} \boldsymbol{G}_j \hat{\boldsymbol{h}},$$
$$j = 1,\ldots,M. \tag{4.41}$$

The latter M constraints upper bound the objective in (4.41). Their right-hand sides decrease when scaling $\boldsymbol{\lambda}$ as $(1+\delta)\boldsymbol{\lambda}$, $\delta > 1$, assuming arbitrary but fixed $\boldsymbol{\mu} \geq \mathbf{0}$. These constraints form a M-tuple of interference constraints with respect to $\boldsymbol{\lambda}$, i.e.,

$$\lambda_j \leq (m_j - \varepsilon_j) I_j(\boldsymbol{\lambda};\boldsymbol{\mu}), \quad j = 1,\ldots,M, \tag{4.42}$$

---

[14] The primal objective $\sup_{\boldsymbol{\lambda}\geq 0,\boldsymbol{\mu}} L(p,\boldsymbol{t},\boldsymbol{\lambda},\boldsymbol{\mu})$ upper bounds the dual objective $\inf_{p\geq 0,\boldsymbol{t}} L(p,\boldsymbol{t},\boldsymbol{\lambda},\boldsymbol{\mu})$.

[15] The alternative derivation in Sect. A.6 is up to this point. The remaining steps are the same.

[16] This follows from $\hat{\boldsymbol{h}}_k \in \mathrm{range}\{\boldsymbol{R}_k\}$ and the block entries of the block-diagonal matrix $\boldsymbol{Y}$ are $\boldsymbol{Y}_k = \sum_{j=1}^{M} \lambda_j \sum_{k\in\mathcal{G}_j} w_k \boldsymbol{R}_k + \boldsymbol{C}_k$ with the matrix $\boldsymbol{C}_k = \sum_{\ell=1}^{L} \mu_\ell \boldsymbol{A}_{k,\ell} \succeq \mathbf{0}$.

where $I_j(\cdot;\boldsymbol{\mu}) : \mathbb{R}_+^M \to \mathbb{R}_+$ is the effective interference function

$$I_j(\boldsymbol{\lambda};\boldsymbol{\mu}) = \frac{1}{\hat{\boldsymbol{h}}^{\mathrm{H}} \boldsymbol{G}_j \left(\sum_{i=1}^M \lambda_i \hat{\boldsymbol{R}}_i + \sum_{\ell=1}^L \mu_\ell \boldsymbol{A}_\ell \right)^{\dagger} \boldsymbol{G}_j \hat{\boldsymbol{h}}}. \tag{4.43}$$

These interference functions are positive and sublinear monotonically increasing in $\boldsymbol{\lambda} \in \mathbb{R}_+^M$ for any fixed $\boldsymbol{\mu} \in \mathbb{R}_+^L$. For this reason, the function $\boldsymbol{I} : \mathbb{R}_+^M \to \mathbb{R}_+^M$, with

$$\boldsymbol{I}(\boldsymbol{\lambda};\boldsymbol{\mu}) = \operatorname{diag}(m_1 - \varepsilon_1, \dots, m_M - \varepsilon_M)[I_1(\boldsymbol{\lambda};\boldsymbol{\mu}), \dots, I_M(\boldsymbol{\lambda};\boldsymbol{\mu})]^{\mathrm{T}}, \tag{4.44}$$

defines a standard interference function in $\boldsymbol{\lambda} \in \mathbb{R}_+^M$. Thus, the maximum of (4.41) is the unique fixed point $\boldsymbol{\lambda} = \boldsymbol{I}(\boldsymbol{\lambda};\boldsymbol{\mu})$. This solution remains valid when jointly reversing the maximization over $\boldsymbol{\lambda}$ into a minimization and the inequality of the interference constraints. This reversed optimization reads as

$$\max_{\boldsymbol{\mu}\geq \mathbf{0}} \min_{\boldsymbol{\lambda}\geq \mathbf{0}} \sum_{j=1}^M \lambda_j \hat{\sigma}_j^2 \text{ s.t. } \sum_{\ell=1}^L \mu_\ell P_\ell \leq 1,$$
$$\varepsilon_j \geq m_j - \lambda_j \hat{\boldsymbol{h}}^{\mathrm{H}} \boldsymbol{G}_j \boldsymbol{Y}^{\dagger} \boldsymbol{G}_j \hat{\boldsymbol{h}},$$
$$j = 1, \dots, M. \tag{4.45}$$

The uplink constraint formulation from (4.45) now follows via exploiting the block-diagonal structure of $\boldsymbol{G}_j$, $\hat{\boldsymbol{R}}_j$, and $\boldsymbol{A}_\ell$ and the equalizer optimization

$$m_j - \lambda_j \hat{\boldsymbol{h}}^{\mathrm{H}} \boldsymbol{G}_j \boldsymbol{Y}^{\dagger} \boldsymbol{G}_j \hat{\boldsymbol{h}} = \sum_{k \in \mathcal{G}_j} w_k \min_{\boldsymbol{u}_k} \mathrm{AMSE}_k^{(\mathrm{ul})} \tag{4.46}$$

with the uplink (M)MSE equalizers $\boldsymbol{u}_k \in \mathbb{C}^N$ from (4.55) as its solution. □

Strong duality holds between (4.30) and (4.34), because the proof starts from a convex SOC formulation of the primal problem and involves transformations to the Lagrangian function that preserve the KKT conditions. The solution of (4.34) allows a reconstruction of the primal optimization variables of (4.30). Furthermore, the dual problems for a set $\mathcal{M}$ with either only per-user constraints (4.4) or a single sum-MSE requirement (4.5) are special cases of the dual formulation from Theorem 4.1.

Corollary 4.1 *If the primal QoS problem* (4.24) *has the per-user MSE target set* (4.4), *i.e.,* $w_k = 1$ *and* $\mathcal{G}_k = \{k\}$, $k = 1, \dots, K$, *the strongly dual problem to* (4.30) *becomes* (cf. [164])

$$\max_{\boldsymbol{\mu}\geq \mathbf{0}} \min_{\boldsymbol{u},\boldsymbol{\lambda}\geq \mathbf{0}} \sum_{k=1}^K \lambda_k \tilde{\sigma}_k^2 \quad \text{s.t.} \quad \sum_{\ell=1}^L P_\ell \mu_\ell \leq 1, \quad \mathrm{AMSE}_k^{(\mathrm{ul})} \leq \varepsilon_k, \quad k = 1, \dots, K, \tag{4.47}$$

and the uplink (average) MSEs read as

$$\mathrm{AMSE}_k^{(\mathrm{ul})} = 1 - 2\sqrt{\lambda_k}\,\mathrm{Re}(\hat{\boldsymbol{h}}_k^{\mathrm{H}}\boldsymbol{u}_k) + \boldsymbol{u}_k^{\mathrm{H}}\Big(\sum_{i=1}^{K}\lambda_i\boldsymbol{R}_i + \sum_{\ell=1}^{L}\mu_\ell\boldsymbol{A}_{k,\ell}\Big)\boldsymbol{u}_k. \tag{4.48}$$

Computing the solution of (4.47) can be intractable for practical implementations if the number of users K becomes large (cf. Sect. 4.3.2). A single (weighted) sum MSE constraint reduces this computation complexity (e.g., see [68, 116]).

Corollary 4.2 *With a sum MSE constraint* (4.5) *and a strictly feasible target* $\varepsilon \in (\sum_{k=1}^{K} w_k\,\mathrm{AMSE}_k^{(\mathrm{ul})}|_{\boldsymbol{\mu}'=\mathbf{0}}, \sum_{k=1}^{K} w_k)$*, the dual problem of* (4.24) *is (cf.* [145]*)*

$$\min_{\boldsymbol{u},\boldsymbol{\mu}'\geq\mathbf{0}} \frac{1}{\sum_{\ell=1}^{L}\mu'_\ell P_\ell} \quad \text{s.t.} \quad \sum_{k=1}^{K} w_k\,\mathrm{AMSE}_k^{(\mathrm{ul})} \leq \varepsilon \tag{4.49}$$

and the uplink MSEs only depend on the noise parameters $\boldsymbol{\mu}$*, i.e.,*

$$\mathrm{AMSE}_k^{(\mathrm{ul})} = 1 - 2\sqrt{w_k}\,\mathrm{Re}(\hat{\boldsymbol{h}}_k^{\mathrm{H}}\boldsymbol{u}_k) + \boldsymbol{u}_k^{\mathrm{H}}\Big(\sum_{i=1}^{K}w_i\boldsymbol{R}_i + \sum_{i=1}^{K}w_i\tilde{\sigma}_i^2\sum_{\ell=1}^{L}\mu'_\ell\boldsymbol{A}_{k,\ell}\Big)\boldsymbol{u}_k. \tag{4.50}$$

Proof For a single weighted sum MSE requirement, $M = 1$ and $\mathcal{G}_1 = \{1, \ldots, K\}$, the inner minimization in (4.34) is only over a single scalar $\lambda_1 \in \mathbb{R}_+$:

$$\max_{\boldsymbol{\mu}\geq\mathbf{0}}\min_{\lambda_1\geq 0} \lambda_1\sum_{k=1}^{K} w_k\tilde{\sigma}_k^2 \quad \text{s.t.} \quad \sum_{\ell=1}^{L} P_\ell\mu_\ell \leq 1, \quad \sum_{k=1}^{K} w_k\,\mathrm{AMSE}_k^{(\mathrm{ul})} \leq \varepsilon. \tag{4.51}$$

Restricting to a positive λ_1[17] and inserting the equalizers (4.55), the MSEs are

$$\mathrm{AMSE}_k^{(\mathrm{ul})} = 1 - \hat{\boldsymbol{h}}_k^{\mathrm{H}}\Big(\sum_{i=1}^{K}w_i\boldsymbol{R}_i + \sum_{\ell=1}^{L}\frac{\mu_\ell}{\lambda_1}\boldsymbol{A}_{k,\ell}\Big)^{\dagger}\hat{\boldsymbol{h}}_k.$$

The substitution $\boldsymbol{\mu}' = (\lambda_1\sum_{i=1}^{K} w_i\tilde{\sigma}_i^2)^{-1}\boldsymbol{\mu}$ leads to the MSE representation in (4.50) and the restriction $\sum_{\ell=1}^{L} P_\ell\mu'_\ell \leq (\lambda_1\sum_{i=1}^{K} w_i\tilde{\sigma}_i^2)^{-1}$ for the outer optimization. Since any solution $\boldsymbol{\mu}'$ satisfies this condition with equality, because the uplink MSEs are monotonically increasing in the entries of $\boldsymbol{\mu}'$, we can replace $\lambda_1\sum_{i=1}^{K} w_i\tilde{\sigma}_i^2$ by $(\sum_{\ell=1}^{L} P_\ell\mu'_\ell)^{-1}$ if λ_1 remains finite. Finite λ_1 is guaranteed for a target $\varepsilon > \sum_{k=1}^{K} w_k\,\mathrm{AMSE}_k^{(\mathrm{ul})}|_{\boldsymbol{\mu}'=\mathbf{0}}$. □

[17] This restriction is without loss of generality for $\varepsilon < \sum_{k=1}^{K} w_k$ and $(\lambda_1, \boldsymbol{\mu}) = (0, \mathbf{0})$ for $\varepsilon \geq \sum_{k=1}^{K} w_k$.

The weighted sum MSE metric may result in situations, where transmission to a user is switched off because the channel conditions are too bad. A QoS formulation with the original MSE constraints is preferred for this situation. Computational tractability for a large system may then be achieved by simplifying the power constraint set $\mathcal{P}$ to a single sum power restriction, for example, [68, 161, 162].

Corollary 4.3 *If the power constraint set* $\mathcal{P}$ *of the primal problem* (4.24) *contains only the sum power limit* $\|\boldsymbol{t}\|_2 \leq P_1$*, the strongly dual problem of* (4.34) *reads as*

$$\min_{\boldsymbol{u},\boldsymbol{\lambda}\geq\boldsymbol{0}} \sum_{j=1}^{M} \lambda_j \sum_{k\in\mathcal{G}_j} w_k \tilde{\sigma}_k^2 \quad \text{s.t.} \quad \sum_{k\in\mathcal{G}_j} w_k \,\mathrm{AMSE}_k^{(\mathrm{ul})} \leq \varepsilon_j, \quad j=1,\ldots,M, \tag{4.52}$$

where the uplink MSE terms become

$$\mathrm{AMSE}_k^{(\mathrm{ul})} = 1-2\sqrt{\lambda_j w_k}\,\mathrm{Re}(\hat{\boldsymbol{h}}_k^{\mathrm{H}}\boldsymbol{u}_k)+\boldsymbol{u}_k^{\mathrm{H}}\Big(\sum_{j=1}^{M}\lambda_j\sum_{i\in\mathcal{G}_j} w_i\boldsymbol{R}_i+P_1^{-1}\mathbf{I}\Big)\boldsymbol{u}_k. \tag{4.53}$$

Proof Let μ_1 be the Lagrangian multiplier to the sum power constraint $\|\boldsymbol{t}\|_2 \leq \sqrt{P_1}$. Then $\mu_1 P_1 \leq 1$ for the dual problem. Moreover, since the objective $g(\boldsymbol{\lambda}, \mu_1)$ increases with μ_1, because $I_j(\boldsymbol{\lambda}, \mu_1)$ increases and $\lambda_j \geq (m_j - \varepsilon_j)I_j(\boldsymbol{\lambda}, \mu_1)$ is required (cf. proof of Theorem 4.1), $\mu_1 = P_1^{-1}$ solves the outer maximization in (4.34). □

In contrast to the dualities for generalized power limitations, which are only tight at the solution $\boldsymbol{\mu}$ of the dual problem, the dual formulation (4.52) is always tight because the maximizer μ_1 is known a priori. A feasible precoder can be reconstructed for any feasible $\boldsymbol{\lambda}$ via the transformation in Sect. 4.3.3, which ensures equality between the downlink and uplink weighted sum MSEs and power constraints $\|\boldsymbol{t}\|_2^2 = P_1 \sum_{j=1}^{M} \hat{\sigma}_j^2 \lambda_j$ [68, 161].[18]

Another set $\mathcal{P}$ with per-beamformer constraints, i.e., $\|\boldsymbol{A}_{\ell,\ell}^{1/2}\boldsymbol{t}_\ell\|_2 \leq \sqrt{P_\ell}$ with $\boldsymbol{A}_{\ell,\ell} = \mathbf{I}$ for $\ell = 1,\ldots,K$, was focused by [248]. For this case, (4.34) trades off the users' experienced noise powers because the uplink MSE

$$\mathrm{AMSE}_k^{(\mathrm{ul})} = 1-2\sqrt{\lambda_j w_k}\,\mathrm{Re}(\hat{\boldsymbol{h}}_k^{\mathrm{H}}\boldsymbol{u}_k)+\boldsymbol{u}_k^{\mathrm{H}}\Big(\sum_{j=1}^{M}\lambda_j\sum_{i\in\mathcal{G}_j} w_i\boldsymbol{R}_i+\mu_k\mathbf{I}\Big)\boldsymbol{u}_k$$

only contains the Lagrangian multiplier μ_k to the kth primal power constraint.

[18] For the duality results in [68, 161], the power allocation is substituted by $\boldsymbol{\lambda}' = P_1\boldsymbol{\lambda}$.

4.3.2 Power Allocation and Worst-Case Noise Search

The dual problem formulations reduce the search of the $NM \times 1$ complex-valued precoder $\boldsymbol{t}$ to the $M + L$ non-negative reals of $\boldsymbol{\lambda}$ and $\boldsymbol{\mu}$. This decreases the optimization complexity, especially if L or M is small. The interpretation as an uplink power minimization enables the fast iterative solutions from [68, 89, 90]. This power allocation depends on the uplink noise covariance matrices $\sum_{\ell=1}^{L} \mu_\ell \boldsymbol{A}_{k,\ell}$, $k = 1, \ldots, K$. We may find $\boldsymbol{\mu}$ and $\boldsymbol{\lambda}$ jointly by exploiting the SDP [19]

$$\max_{\boldsymbol{\mu}\geq\boldsymbol{0},\boldsymbol{\lambda}\geq\boldsymbol{0}} \sum_{j=1}^{M} \lambda_j \hat{\sigma}_j^2 \text{ s.t. } \sum_{\ell=1}^{L} \mu_\ell P_\ell \leq 1,$$

$$\sum_{j=1}^{M} \lambda_j \hat{\boldsymbol{R}}_j + \sum_{\ell=1}^{L} \mu_\ell \boldsymbol{A}_\ell - \sum_{j=1}^{M} \frac{\lambda_j}{m_j - \varepsilon_j} \boldsymbol{G}_j \hat{\boldsymbol{h}} \hat{\boldsymbol{h}}^{\mathrm{H}} \boldsymbol{G}_j \succeq \boldsymbol{0}. \tag{4.54}$$

Alternatively, we search the $M + L$ variables via an inner fixed point iteration and an outer PG algorithm, respectively. The uplink equalizers $\boldsymbol{u}_k \in \mathbb{C}^N$ are

$$\boldsymbol{u}_k = \sqrt{\lambda_j w_k} \Big(\sum_{i=1}^{M} \lambda_i \sum_{n \in \mathcal{G}_j} w_n \boldsymbol{R}_n + \sum_{\ell=1}^{L} \mu_\ell \boldsymbol{A}_{k,\ell} \Big)^{\dagger} \hat{\boldsymbol{h}}_k, \quad k = 1, \ldots, K. \tag{4.55}$$

Inserting them into (4.35), the interference constraints $\lambda_j \geq (m_j - \varepsilon_j) I_j(\boldsymbol{\lambda}; \boldsymbol{\mu})$, with ... from (4.43), $j = 1, \ldots, M$, bound the objective of the dual problem (4.34) and form the entries of the standard interference function $\boldsymbol{I}(\boldsymbol{\lambda}; \boldsymbol{\mu})$ from (4.44). Given $\boldsymbol{\mu}$, the unique solution $\boldsymbol{\lambda}^\star$ is, therefore, the result of the globally converging sequence

$$\lambda_j^{(n+1)} = (m_j - \varepsilon_j) I_j(\boldsymbol{\lambda}^{(n)}, \boldsymbol{\mu}), \quad j = 1, \ldots, M. \tag{4.56}$$

It remains to find the solution of the outer maximization from (4.34), which is a convex program (see Lemma A.5 in Sect. A.7). We suggest an iterative PG algorithm for this task, where the gradient $\boldsymbol{\delta} = [\delta_1, \ldots, \delta_L]^{\mathrm{T}}$ has the entries

$$\delta_i = \sum_{j=1}^{M} \frac{\partial \lambda_j}{\partial \mu_m} \sum_{k \in \mathcal{G}_j} w_k \tilde{\sigma}_k^2, \quad i = 1, \ldots, L. \tag{4.57}$$

Its computation requires the derivative of $\boldsymbol{\lambda}$, which we obtain by the satisfied fixed point equations (4.56). These derivatives are affine in $\frac{\partial \lambda_i}{\partial \mu_m}$, $i = 1, \ldots, M$, i.e.,

$$\frac{\partial \lambda_j}{\partial \mu_m} = (m_j - \varepsilon_j) \frac{\partial I_j(\boldsymbol{\lambda}(\boldsymbol{\mu}); \boldsymbol{\mu})}{\partial \mu_m}, \tag{4.58}$$

[19]The LMI constraint stems from (4.39) within the proof of Theorem 4.1.

where the derivative of the interference function is

$$\frac{\partial I_j(\boldsymbol{\lambda}(\boldsymbol{\mu}); \boldsymbol{\mu})}{\partial \mu_m} = \frac{\hat{\boldsymbol{h}}^{\mathrm{H}} \boldsymbol{G}_j \boldsymbol{Y}^{\dagger} \boldsymbol{A}_m \boldsymbol{Y}^{\dagger} \boldsymbol{G}_j \hat{\boldsymbol{h}} + \sum_{i=1}^{M} \hat{\boldsymbol{h}}^{\mathrm{H}} \boldsymbol{G}_j \boldsymbol{Y}^{\dagger} \hat{\boldsymbol{R}}_i \boldsymbol{Y}^{\dagger} \boldsymbol{G}_j \hat{\boldsymbol{h}} \frac{\partial \lambda_i}{\partial \mu_m}}{(\hat{\boldsymbol{h}}^{\mathrm{H}} \boldsymbol{G}_j \boldsymbol{Y}^{\dagger} \boldsymbol{G}_j \hat{\boldsymbol{h}})^2} \tag{4.59}$$

and $\boldsymbol{Y} = \sum_{i=1}^{M} \lambda_i^{(n)} \hat{\boldsymbol{R}}_i + \sum_{\ell=1}^{L} \mu_\ell \boldsymbol{A}_\ell$. Thus, $\frac{\partial \lambda_i}{\partial \mu_m}$ are solutions of the equation

$$(\mathbf{I} - \boldsymbol{\Lambda} \boldsymbol{\Psi}) \frac{\partial \boldsymbol{\lambda}}{\partial \mu_m} = \boldsymbol{\Lambda} \boldsymbol{n}_m,$$

where $\boldsymbol{n}_m = [\tilde{n}_{m,1}, \ldots, \tilde{n}_{m,M}]^{\mathrm{T}} \in \mathbb{R}_+^M$ and $\boldsymbol{\Lambda} = \mathrm{diag}(\tilde{\lambda}_1, \ldots, \tilde{\lambda}_M)$ have the entries

$$\begin{aligned} \tilde{n}_{m,j} &= \hat{\boldsymbol{h}}^{\mathrm{H}} \boldsymbol{G}_j \boldsymbol{Y}^{\dagger} \boldsymbol{A}_m \boldsymbol{Y}^{\dagger} \boldsymbol{G}_j \hat{\boldsymbol{h}} \\ \tilde{\lambda}_j &= (m_j - \varepsilon_j)(\hat{\boldsymbol{h}}^{\mathrm{H}} \boldsymbol{G}_j \boldsymbol{Y}^{\dagger} \boldsymbol{G}_j \hat{\boldsymbol{h}})^{-2} = \lambda_j (\hat{\boldsymbol{h}}^{\mathrm{H}} \boldsymbol{G}_j \boldsymbol{Y}^{\dagger} \boldsymbol{G}_j \hat{\boldsymbol{h}})^{-1}, \end{aligned} \tag{4.60}$$

respectively, and the matrix $\boldsymbol{\Psi} \in \mathbb{R}_+^{M \times M}$ consists of the non-negative entries

$$[\boldsymbol{\Psi}]_{j,i} = \hat{\boldsymbol{h}}^{\mathrm{H}} \boldsymbol{G}_j \boldsymbol{Y}^{\dagger} \hat{\boldsymbol{R}}_i \boldsymbol{Y}^{\dagger} \boldsymbol{G}_j \hat{\boldsymbol{h}}, \quad i, j = 1, \ldots, M. \tag{4.61}$$

We remark that $\mathbf{I} - \boldsymbol{\Lambda}\boldsymbol{\Psi}$ has non-positive off-diagonal and positive diagonal elements. In particular, $\mathbf{I} - \boldsymbol{\Lambda}\boldsymbol{\Psi}$ is a non-singular M-matrix according to [229, Theorem 6.2.3, (M36)][20] because $\mathrm{diag}(\lambda_1, \ldots, \lambda_M)^{-1}(\mathbf{I} - \boldsymbol{\Lambda}\boldsymbol{\Psi})\,\mathrm{diag}(\lambda_1, \ldots, \lambda_M)$ is diagonally dominant, i.e., the row sums of this matrix are

$$1 - \frac{\hat{\boldsymbol{h}}^{\mathrm{H}} \boldsymbol{G}_j \boldsymbol{Y}^{\dagger} \sum_{i=1}^{M} \lambda_i \hat{\boldsymbol{R}}_i \boldsymbol{Y}^{\dagger} \boldsymbol{G}_j \hat{\boldsymbol{h}}}{\hat{\boldsymbol{h}}^{\mathrm{H}} \boldsymbol{G}_j \boldsymbol{Y}^{\dagger} \boldsymbol{G}_j \hat{\boldsymbol{h}}} 1 \geq 0, \quad j = 1, \ldots, M.$$

Due to this property, the inverse matrix $(\mathbf{I} - \boldsymbol{\Lambda}\boldsymbol{\Psi})^{-1}$ exists and all its entries are positive. Hence, also the derivatives (4.58) are non-negative and $\boldsymbol{\delta} \in \mathbb{R}_+^L$.

Being aware of the gradient (4.57), an update for $\boldsymbol{\mu}$ reads as

$$\boldsymbol{\mu}^{(n'+1)} = \boldsymbol{\Pi}_{\mathcal{C}}\left(\boldsymbol{\mu}^{(n')} + a_{n'} \boldsymbol{\delta}\right),$$

where $a_{n'} \in \mathbb{R}_+$ is the step size in iteration n'. For example, we chose $a_{n'}$ via Armijo's rule (e.g., see [227, Section 2.3] or [249, Section 7.2]) for the simulations.

The orthogonal projection $\boldsymbol{\Pi}_{\mathcal{C}} : \mathbb{R}_+^L \to \mathcal{C}$ is again onto the simplex

$$\mathcal{C} = \left\{\boldsymbol{\mu} \in \mathbb{R}_+^L : \sum_{\ell=1}^{L} P_\ell \mu_\ell \leq 1\right\}.$$

[20] Berman and Plemmons [229, Chapter 6] provide an extensive discussion about non-singular M-matrices including 50 equivalent statements for their definition [229, Theorem 6.2.3].

It maps any vector $\boldsymbol{x} = \boldsymbol{\mu}^{(n')} + a_{n'}\boldsymbol{\delta} \in \mathbb{R}_+^L, \boldsymbol{x} \notin \mathcal{C}$ onto the upper right boundary $\partial\mathcal{C}$, where equality holds for the limiting half-space constraint. Therewith, all sequence elements satisfy $\boldsymbol{\mu}^{(n')} \in \partial\mathcal{C}$ when starting from a boundary point $\boldsymbol{\mu}^{(0)} \in \partial\mathcal{C}$. Since also the solution resides at $\partial\mathcal{C}$, this behavior is without loss of optimality. An explicit representation of this projection onto $\partial\mathcal{C}$ reads as (cf. [228])

$$\boldsymbol{\Pi}_{\mathcal{C}}(\boldsymbol{x}) = [\boldsymbol{x} - \gamma \boldsymbol{p}]^+, \tag{4.62}$$

where $\boldsymbol{p} = [P_1, \ldots, P_L]^{\mathrm{T}}$, $[\cdot]^+$ is an elementwise mapping into the non-negative orthant,[21] and γ ensures equality for $\sum_{\ell=1}^L P_\ell[x_\ell - \gamma P_\ell] \leq 1$.

4.3.3 *Primal Reconstruction of Beamformers*

Given the solutions $\boldsymbol{\mu}$ and $\boldsymbol{\lambda}$ and the equalizers $\boldsymbol{u}_k$, $k = 1, \ldots, K$ of (4.34), the downlink beamformers read as (cf. [161])

$$\boldsymbol{t}_k = \beta'_j \boldsymbol{u}_k, \quad k \in \mathcal{G}_j, \quad j = 1, \ldots, M, \tag{4.63}$$

where the scaling factors $\beta'_j \in \mathbb{R}_+$ are the solution of a linear equation system. Hunger et al. [161] proved this for the uplink–downlink duality with a sum-MSE, per-user MSEs, and per-stream MSEs but only for a sum power constraint. The next paragraphs show that this relation even holds for multiple power constraints.

Given the dual solution, reconstructing the primal variables spatial structure is by the zero-derivative condition of the Lagrangian function (4.37). Then, we find the norm of the beamformers $\boldsymbol{t}_k$, $k = 1, \ldots, K$ with the weighted sum MSE constraints, which hold with equality at the optimum, and the zero duality gap between the optima of the dual and primal problem. The zero-derivative condition for $\boldsymbol{t}$ reads as

$$\frac{\partial}{\partial \boldsymbol{t}^*} L(p, \boldsymbol{t}, \boldsymbol{\mu}, \boldsymbol{\lambda}) = \Big(\boldsymbol{Y} - \sum_{j=1}^M \frac{\lambda_j}{m_j - \varepsilon_j} \boldsymbol{G}_j \hat{\boldsymbol{h}} \hat{\boldsymbol{h}}^{\mathrm{H}} \boldsymbol{G}_j\Big) \boldsymbol{t} = \boldsymbol{0}. \tag{4.64}$$

Therefore, the solution of the precoder has the characteristics

$$\boldsymbol{t} = \boldsymbol{Y}^\dagger \sum_{j=1}^M \beta_j \boldsymbol{G}_j \hat{\boldsymbol{h}}, \tag{4.65}$$

where $\beta_j = \frac{\lambda_j}{m_j - \varepsilon_j} \hat{\boldsymbol{h}}^{\mathrm{H}} \boldsymbol{G}_j \boldsymbol{t}$. The block-diagonal structure of $\boldsymbol{Y}$ and $\boldsymbol{G}_j$ allows a separation of (4.65) into the individual beamformers with the suggested structure (4.55):

[21]The scalar projection is onto $\mathbb{R}_+$ and reads as $[z]^+ = \max(0, z)$ for $z \in \mathbb{R}$.

$$t_k = \beta_j \sqrt{w_k \lambda_j^{-1}} u_k, \quad k \in \mathcal{G}_j, \quad j = 1, \ldots, M. \tag{4.66}$$

Due to positivity of the solution $\boldsymbol{\lambda}$, equality also holds for the primal constraints:

$$\sum_{k \in \mathcal{G}_j} w_k \, \text{AMSE}_k = \sum_{k \in \mathcal{G}_j} w_k \, \text{AMSE}_k^{(\text{ul})} = \varepsilon_j, \quad j = 1, \ldots, M. \tag{4.67}$$

Inserting the beamformers (4.65) and exploiting $\boldsymbol{Y}^\dagger \boldsymbol{G}_j = \boldsymbol{G}_j \boldsymbol{Y}^\dagger$ and $\boldsymbol{G}_i \boldsymbol{G}_j = \mathbf{0}$ for $i \neq j$,[22] these equality requirements after some algebraic reformulations in

$$\beta_j^2 = \frac{\lambda_j}{\hat{\boldsymbol{h}}^{\text{H}} \boldsymbol{G}_j \boldsymbol{Y}^\dagger \boldsymbol{G}_j \hat{\boldsymbol{h}}} \Big(\sum_{i=1}^{M} \hat{\boldsymbol{h}}^{\text{H}} \boldsymbol{G}_i \boldsymbol{Y}^\dagger \hat{\boldsymbol{R}}_j \boldsymbol{Y}^\dagger \boldsymbol{G}_i \hat{\boldsymbol{h}} \beta_i^2 + \hat{\sigma}_j^2 \Big), \quad j = 1, \ldots, M.$$

Hence, $\boldsymbol{\beta} = [\beta_1^2, \ldots, \beta_M^2]^{\text{T}}$ is the solution of the linear equation system

$$\big(\mathbf{I} - \boldsymbol{\Lambda} \boldsymbol{\Psi}^{\text{T}}\big) \boldsymbol{\beta} = \boldsymbol{\Lambda} \boldsymbol{\sigma}$$

with $\boldsymbol{\Psi}$ from (4.61) and $\boldsymbol{\sigma} = [\hat{\sigma}_1^2, \ldots, \hat{\sigma}_M^2]^{\text{T}}$. Also $\mathbf{I} - \boldsymbol{\Lambda} \boldsymbol{\Psi}^{\text{T}}$ is a non-singular M-matrix [229, Theorem 6.2.3], which assures that $\boldsymbol{\beta} = (\mathbf{I} - \boldsymbol{\Lambda} \boldsymbol{\Psi}^{\text{T}})^{-1} \boldsymbol{\Lambda} \boldsymbol{\sigma}$ is valid.

Remark 4.3 For the QoS optimization with a single sum-MSE constraint, i.e., $M = 1$ and $w_k = 1$ for $k = 1, \ldots, K$, the scalar transformation factor reads as

$$\beta^2 = \Big(1 - \frac{\lambda \hat{\boldsymbol{h}}^{\text{H}} \boldsymbol{Y}^\dagger \hat{\boldsymbol{R}}_1 \boldsymbol{Y}^\dagger \hat{\boldsymbol{h}}}{\hat{\boldsymbol{h}}^{\text{H}} \boldsymbol{Y}^\dagger \hat{\boldsymbol{h}}} \Big)^{-1} \frac{\lambda^2 \hat{\sigma}_1^2}{K - \varepsilon_1} \tag{4.68}$$

and λ is found via a root search for the implicit function

$$F(\lambda) = \lambda - \frac{K - \varepsilon_1}{\hat{\boldsymbol{h}}^{\text{H}} (\lambda \hat{\boldsymbol{R}}_1 + \sum_{\ell=1}^{L} \mu_\ell \boldsymbol{A}_\ell)^\dagger \hat{\boldsymbol{h}}} = 0.$$

4.3.4 *Expectation Evaluation for Alternating Convex Optimization*

With the solution $\boldsymbol{t}$ of (4.29), the ACS recomputes the moments (4.21)–(4.23) in Step 1 and repeats the QoS optimization of Step 2. The computational extensive part of the ACS besides the beamformer update is the evaluation of the effective channels' moments and the mean noise variance with the updated downlink MMSE

[22] This follows from the block-diagonal structure of $\boldsymbol{Y}$.

filter functions. Numerical integrations for this task can be based on first order derivatives of $\mathrm{RQ}_k(\boldsymbol{t}_k, \boldsymbol{Q})$ (cf. [158]). The derivatives for $\hat{\boldsymbol{h}}_k$ and $\tilde{\sigma}_k^2$ are

$$\hat{\boldsymbol{h}}_k = \frac{\partial}{\partial \boldsymbol{x}^*} \mathrm{RQ}_k(\boldsymbol{x}, \boldsymbol{Q})\big|_{\boldsymbol{x}=\boldsymbol{t}_k} = \mathrm{E}\left[\frac{(\boldsymbol{h}_k^{\mathrm{H}} \boldsymbol{t}_k)\boldsymbol{h}_k}{\boldsymbol{h}_k^{\mathrm{H}} \boldsymbol{Q} \boldsymbol{h}_k + \sigma_k^2}\right], \tag{4.69}$$

$$\tilde{\sigma}_k^2 = -\sigma_k^2 \frac{\partial}{\partial \sigma_k^2} \mathrm{RQ}_k(\boldsymbol{t}_k, \boldsymbol{Q}) = \sigma_k^2 \,\mathrm{E}\left[\frac{|\boldsymbol{h}_k^{\mathrm{H}} \boldsymbol{t}_k|^2}{(\boldsymbol{h}_k^{\mathrm{H}} \boldsymbol{Q} \boldsymbol{h}_k + \sigma_k^2)^2}\right], \tag{4.70}$$

respectively, and the effective channel's second order moment has the form

$$\boldsymbol{R}_k = -\frac{\partial}{\partial \boldsymbol{Z}^*} \mathrm{RQ}_k(\boldsymbol{t}_k, \boldsymbol{Z})\big|_{\boldsymbol{Z}=\boldsymbol{Q}} = \mathrm{E}\left[\frac{|\boldsymbol{h}_k^{\mathrm{H}} \boldsymbol{t}_k|^2 \boldsymbol{h}_k \boldsymbol{h}_k^{\mathrm{H}}}{(\boldsymbol{h}_k^{\mathrm{H}} \boldsymbol{Q} \boldsymbol{h}_k + \sigma_k^2)^2}\right]. \tag{4.71}$$

Expressions for the Gaussian channel model can be based on (4.13) and (4.14).[23]

However, a numerical evaluation of the integrals has only limited accuracy and becomes intractable when increasing the system dimensions. The numerical results are therefore computed with a sample average approach. With a tractable number of randomly generated samples for the additive fading channels, the expectations for $\hat{\boldsymbol{h}}_k$, $\boldsymbol{R}_k$, and $\tilde{\sigma}_k$ are approximated by the sample means [250].

4.4 Quality-of-Service Feasibility Region

Feasibility of the MMSE target region $\mathcal{M}$ is crucial for the QoS beamformer design. Unfortunately, this feasible region is unknown for imperfect transmitter and perfect receiver CSI. The half-space constraint $\sum_{k=1}^{K} \varepsilon_k \geq K - N$ of the perfect CSI feasibility region $\mathcal{F}_{\mathrm{MMSE}}$ from (1.30) is only a necessary condition for feasible MSE targets ε_k if the transmitter CSI is imperfect. The achievable MMSE region has also the interval bounds $\varepsilon_k \in [0, 1]$, $k = 1, \ldots, K$. However, the zero lower bound of

$$\mathrm{AMSE}_k(\boldsymbol{t}) = 1 - \mathrm{E}\left[\frac{|\boldsymbol{h}_k^{\mathrm{H}} \boldsymbol{t}_k|^2}{\sum_{i=1}^{K} |\boldsymbol{h}_k^{\mathrm{H}} \boldsymbol{t}_i|^2 + \sigma_k^2}\right] \geq 0$$

is only reached for $\boldsymbol{t} = \alpha \boldsymbol{t}'$, $\alpha \to \infty$ if $|\boldsymbol{h}_k^{\mathrm{H}} \boldsymbol{t}_i'|^2 = \boldsymbol{0}$ for $i \neq k$ holds with probability one. This requires $\boldsymbol{t}_i' = 0$ for the additive channel error model (2.1) with a full-rank covariance matrix $\boldsymbol{C}_k$. Hence, the sum MSE remains strictly positive also for $N \geq K$ because the interference of all $K - 1$ receivers is positive with non-zero probability. In the worst case, the sum-bound even degrades to [114, Remark 1]

[23] In [158], we have suggested alternative derivations based on the expectation in (4.18).

$$\sum_{k=1}^{K} \varepsilon_k \geq K - 1. \tag{4.72}$$

Another achievable region $\widehat{\mathcal{F}}_{\text{MMSE}}$ for imperfect transmitter but perfect receiver CSI follows with the fixed filter functions $f_k(\boldsymbol{h})$, $k = 1, \ldots, K$ from Sect. 4.3 and the dual formulation of the MSEs for a total sum power constraint (cf. [65, 251]):[24]

$$\text{AMSE}_k^{(\text{ul})}(\boldsymbol{\lambda}) = 1 - \lambda_k \hat{\boldsymbol{h}}_k^{\text{H}} \Big(\sum_{i=1}^{K} \lambda_i \boldsymbol{R}_i + P_1^{-1} \mathbf{I} \Big)^{-1} \hat{\boldsymbol{h}}_k, \quad k = 1, \ldots, K. \tag{4.73}$$

These MSE expressions are obviously lower bounded by the minima

$$\varepsilon_{k,\min} = 1 - \hat{\boldsymbol{h}}_k^{\text{H}} \boldsymbol{R}_k^{\dagger} \hat{\boldsymbol{h}}_k = \big(1 + \hat{\boldsymbol{h}}_k^{\text{H}} \hat{\boldsymbol{C}}_k^{\dagger} \hat{\boldsymbol{h}}_k\big)^{-1}, \quad k = 1, \ldots, K, \tag{4.74}$$

and the sum-MSE bound is the optimum of a convex problem:

$$\varepsilon_{\text{sum},\min} = K - \max_{\boldsymbol{\lambda}' \in \partial\mathcal{C}} \operatorname{tr}\Big(\hat{\boldsymbol{H}} \boldsymbol{\Lambda}' \hat{\boldsymbol{H}}^{\text{H}} \Big(\sum_{i=1}^{K} \lambda_i' \boldsymbol{R}_i \Big)^{-1} \Big), \tag{4.75}$$

where $\partial\mathcal{C} = \{\boldsymbol{\lambda} \in \mathbb{R}_+^K : \sum_{k=1}^K \lambda_k = 1\}$. A complete characterization of the lower Pareto boundary $\partial\widehat{\mathcal{F}}_{\text{MMSE}}$ follows by the equations $\varepsilon_k = \text{AMSE}_k^{(\text{ul})}(\boldsymbol{\lambda})$, $k = 1, \ldots, K$ and the results from [11, 252]. The equalities result in [75, Corollary 1]

$$\boldsymbol{\lambda} = \boldsymbol{I}(\boldsymbol{\lambda}; \boldsymbol{\varepsilon}), \tag{4.76}$$

where $\boldsymbol{I}(\cdot; \boldsymbol{\varepsilon})$ comprises the concave functions $(1 - \varepsilon_k) I_k(\boldsymbol{\lambda})$ with [cf. (4.43)][25]

$$I_k(\boldsymbol{\lambda}) = \left(\hat{\boldsymbol{h}}_k^{\text{H}} \Big(\sum_{i=1}^{K} \lambda_i \boldsymbol{R}_i + P_1^{-1} \mathbf{I} \Big)^{-1} \hat{\boldsymbol{h}}_k \right)^{-1}. \tag{4.77}$$

The fixed point (4.76) exists and is unique if the targets reside in the interior of the feasible region, i.e., $\boldsymbol{\varepsilon} \in \operatorname{int}(\widehat{\mathcal{F}}_{\text{MMSE}})$. Targets at the lower Pareto boundary $\partial\widehat{\mathcal{F}}_{\text{MMSE}}$ are only achieved in the limit, i.e., for $\boldsymbol{\lambda} = \alpha \boldsymbol{\lambda}'$ with $\alpha \to \infty$. Then, (4.77) degenerates to the *linear* interference function (cf. [252, Definition 1])

[24] Using a sum power constraint for this characterization is without loss of generality. Since the transmit power is unbounded, $\mathcal{P}' = \{\boldsymbol{t} \in \mathbb{C}^{NK} : \|\boldsymbol{t}\|_2^2 \leq P_\ell'\} \subseteq \mathcal{P}$ can replace $\mathcal{P}$ from (1.32).

[25] Here, $\boldsymbol{I}(\boldsymbol{\lambda}; \boldsymbol{\varepsilon})$ is the entrywise minimum of the function $\boldsymbol{I}_{\boldsymbol{\varepsilon}}(\boldsymbol{\lambda}, \boldsymbol{u}) = \boldsymbol{D}_{\boldsymbol{\varepsilon}} \boldsymbol{M}(\boldsymbol{u})\boldsymbol{\lambda} + \boldsymbol{n}(\boldsymbol{u})$, where $\boldsymbol{D}_{\boldsymbol{\varepsilon}} = \operatorname{diag}(\varepsilon_1^{-1} - 1, \ldots, \varepsilon_K^{-1} - 1)$, $[\boldsymbol{M}(\boldsymbol{u})]_{k,i} = |\hat{\boldsymbol{h}}_k^{\text{H}} \boldsymbol{u}_k|^{-2} \boldsymbol{u}_k^{\text{H}} \boldsymbol{R}_i \boldsymbol{u}_k$, and $[\boldsymbol{n}]_k = P^{-1} |\hat{\boldsymbol{h}}_k^{\text{H}} \boldsymbol{u}_k|^{-2} \|\boldsymbol{u}_k\|^2$. This reformulation supports the results and algorithms from [89, 90] and many results from [11].

$$I_k^{\infty}(\boldsymbol{\lambda}') = \frac{1}{\hat{\boldsymbol{h}}_k^{\mathrm{H}}\left(\sum_{i=1}^{K}\lambda_i'\boldsymbol{R}_i\right)^{-1}\hat{\boldsymbol{h}}_k}. \tag{4.78}$$

The fixed point $\boldsymbol{\lambda}' = \boldsymbol{I}^{\infty}(\boldsymbol{\lambda}'; \boldsymbol{\varepsilon})$ also exists if I_k^{∞} is strict monotonic in $\boldsymbol{\lambda}'$ [252, Corollary 14], e.g., if the second order moments $\boldsymbol{R}_k$ are full-rank matrices.[26]

These properties enable a characterization of the feasible region $\widehat{\mathcal{F}}_{\mathrm{MMSE}}$ via a max–min SIR optimization [11, 252]. A target tuple $\boldsymbol{\varepsilon}$ is feasible, i.e., $\boldsymbol{\varepsilon} \in \widehat{\mathcal{F}}_{\mathrm{MMSE}}$, if

$$C = \min_{\boldsymbol{\lambda}' \in \mathcal{C}} \max_{k} \frac{(1-\varepsilon_k) I_k^{\infty}(\boldsymbol{\lambda}')}{\lambda_k'} \tag{4.79}$$

is smaller or equal to one (e.g., see [252, Section 2]). If $C < 1$, the solution satisfies $\boldsymbol{\lambda}' > \boldsymbol{I}^{\infty}(\boldsymbol{\lambda}'; \boldsymbol{\varepsilon})$ and $\boldsymbol{\varepsilon} \in \mathrm{int}(\widehat{\mathcal{F}}_{\mathrm{MMSE}})$. In other words, a finite $\boldsymbol{\lambda} \in \mathbb{R}_+^K$ exists that meets the MSE requirements $\mathrm{AMSE}_k^{(\mathrm{ul})}(\boldsymbol{\lambda}) = \varepsilon_k$, $k = 1, \ldots, K$. For $C = 1$, the target resides at the Pareto boundary, i.e., $\boldsymbol{\varepsilon} \in \partial\widehat{\mathcal{F}}_{\mathrm{MMSE}}$, and the downlink transmitter requires infinite power to achieve this target. For $C > 1$, the requirements $\boldsymbol{\varepsilon}$ are infeasible, i.e., no fixed point exists for (4.76).

Figure 4.1a depicts the Pareto boundary for the feasible MSE regions of a two user BC with $N = K = 2$ and perfect transmitter CSI, no transmitter CSI, and three examples with imperfect transmitter CSI and fixed filter functions. The boundaries are found with the targets $\varepsilon_k' = 1 - C^{-1}(1-\varepsilon_k)$, $k = 1, \ldots, K$, where $\boldsymbol{\varepsilon} \in \mathcal{C}$ and $C \in \mathbb{R}_+$ is the optimum of (4.79). While example 1 is for symmetric channel conditions, the error variance of channel two is reduced for example 2. This results in an asymmetric feasibility region. The region of example 3 is for channels that are close to each other, i.e., the vectors are collinear up to a small error.

Figure 4.1b shows the Pareto boundaries for users 1 and 2 of a three user vector BC with 2 antennas and fixed MSE target for user 3. Decreasing this fixed MSE to 0.6, the feasible region of the other users is smaller. In particular, $\partial\widehat{\mathcal{F}}_{\mathrm{MMSE}}$ approaches the no transmitter CSI bound because the interference of user 3 is dominant in this case.

4.5 Average Mean Square Error Balancing

MSE minimizations have been studied for multi-antenna transmitters, receivers, multiple streams per-user, and even for imperfect CSI, e.g., [68, 116, 136, 161, 248]. Average MSE balancing is focused because it provides a lower bound for ergodic

[26] Since $\boldsymbol{I}^{\infty}(\boldsymbol{\lambda}'; \boldsymbol{\varepsilon}) = \min_{\boldsymbol{u}} \boldsymbol{D}_{\boldsymbol{\varepsilon}} \boldsymbol{M}(\boldsymbol{u}) \boldsymbol{\lambda}'$, the fixed point $\boldsymbol{\lambda}'^{,\star} = \boldsymbol{I}_{\boldsymbol{\varepsilon}}^{\infty}(\boldsymbol{\lambda}'^{,\star})$ is the eigenvector of $\boldsymbol{D}_{\boldsymbol{\varepsilon}} \boldsymbol{M}(\boldsymbol{u})$ with the SIR maximizing equalizers $\boldsymbol{u}_k = (\sum_{i=1}^{K} \lambda_i \boldsymbol{R}_i)^{-1} \hat{\boldsymbol{h}}_k$, $k = 1, \ldots, K$ [252, Section 6]. Alternatively, we find $\boldsymbol{\lambda}'^{,\star}$ as the limit point of the converging sequence $\boldsymbol{\lambda}'^{,(n+1)} = \|\boldsymbol{I}_{\boldsymbol{\varepsilon}}^{\infty}(\boldsymbol{\lambda}'^{,(n)})\|_1^{-1} \boldsymbol{I}_{\boldsymbol{\varepsilon}}^{\infty}(\boldsymbol{\lambda}'^{,(n)})$ [92, 226].

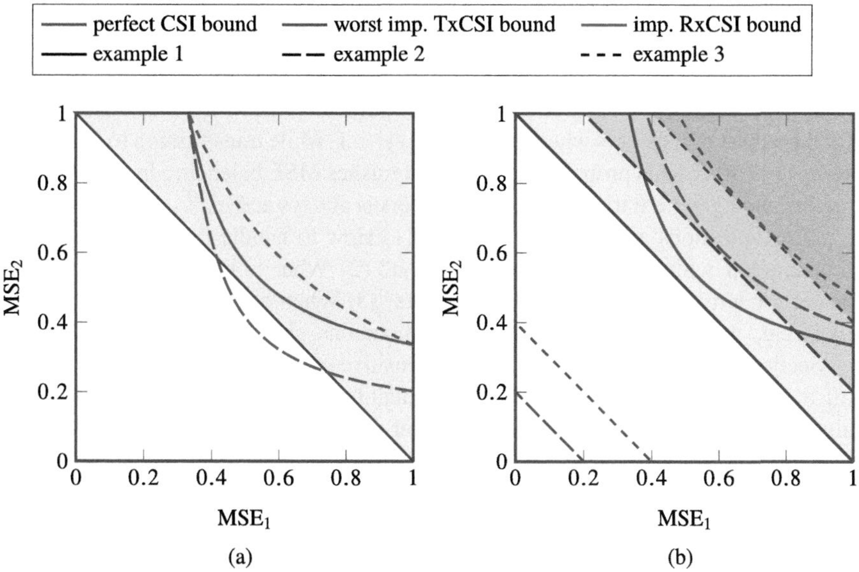

Fig. 4.1 Inner bounds for the MSE feasibility region in the K user BC with $N = 2$ antennas, imperfect CSI, channels $\hat{\boldsymbol{h}}_1 = [1, 0]^{\mathrm{T}}$, $\hat{\boldsymbol{h}}_2 = \frac{1}{2}[-\sqrt{3}, 1]^{\mathrm{T}}$, and $\hat{\boldsymbol{h}}_3 = \frac{1}{2}[-\sqrt{3}, -1]^{\mathrm{T}}$, and $\hat{\boldsymbol{C}}_i = \frac{1}{2}\mathbf{I}$ [example 1 of subfigure (**a**)]. For example 2, $\hat{\boldsymbol{C}}_2 = \frac{1}{4}\mathbf{I}$, and $\hat{\boldsymbol{h}}_2 = \frac{1}{\sqrt{10}}[-3, 1]^{\mathrm{T}}$ and $\hat{\boldsymbol{h}}_3 = \frac{1}{\sqrt{10}}[-3, -1]^{\mathrm{T}}$ for example 3. In subfigure (**b**), the third target is $\varepsilon_3 = 1$ (solid line), $\varepsilon_3 = 0.8$ (dashed line), and $\varepsilon_3 = 0.6$ (dotted line). (**a**) $N = K = 2$ with distinct channels. (**b**) $N = 2$, $K = 3$ with distinct $\varepsilon_3 = 1, 0.8, 0.6$

rate balancing (cf. Sect. 4.1) and, thus, ensures reliable data transmission. In contrast to this work, the imperfect CSI is often the same at the transmitter and receivers. Then, MSE balancing is similar to SINR balancing with perfect CSI [67, 73, 78, 79].

With perfect CSI at the receivers and imperfect CSI at the transmitter, a similar ACS as for QoS optimization finds a local solution of the MSE balancing problem (4.3). The filter update remains, but the precoder update now reads as

$$\min_{\varepsilon, \boldsymbol{t}} \varepsilon \quad \text{s.t.} \quad \boldsymbol{t} \in \mathcal{P}, \quad \varepsilon^{-1} \mathbf{AMSE}(\boldsymbol{t}) \in \mathcal{M}. \tag{4.80}$$

Section 4.5.2 provides a dual approach to transform the downlink precoder updates into an uplink filter design and power allocation. Such a MSE duality was missing. While former dualities are mainly for a sum power limitation (cf. [68, 136, 162, 241]), the duality framework herein is for the generalized power constraints (1.32). Similarly, Shi et al.[248] focused on per-beamformer constraints and perfect CSI and Bogale and Vandendorpe [116] only consider the sum MSE as an objective for the optimization with multiple power constraints.[27] These are special cases of the duality framework for the weighted sum MSE balancing in this section.

[27] Both works base their duality derivation on an algorithmic perspective.

A further novelty is the discussion about unequal MSE targets. Distinct targets allow a flexible prioritization of receivers by lowering their MSE targets in comparison to other users. Transmission to receivers with low prioritization, i.e., large MSE targets, can even be switched off. Generally, the MSEs are balanced only for the subset of receivers where ε satisfies $\varepsilon\varepsilon_k < 1$, while transmission to users with $\varepsilon\varepsilon_k \geq 1$ is off.[28] This property strictly distinguishes MSE balancing from SINR and rate balancing, where transmission to all users is always active.[29]

Three questions arise from this effect: (1) How to handle the (de)activation of transmission within an algorithmic solution? (2) What is the minimum transmit power for active transmission to all receivers? (3) When do other problem formulations, e.g., variations of SINR and RB optimizations, share this property?

Section 4.5.1 proposes an ACS that is sensitive with respect to switching on and off transmission to receivers and exploits duality for the beamformer update. The algorithmic solution for the dual beamformer update takes the unequal MSE targets into account and activates transmission to receivers dependent on the balancing level. From the corresponding discussion follows that the minimum power for active transmission to all receivers is the solution of a QoS optimization with targets

$$\varepsilon_k{}' = \varepsilon_k \max\left\{\varepsilon \in \mathbb{R} : \varepsilon \leq \tfrac{1}{\varepsilon_k}\right\}$$

if $\varepsilon_k, k = 1, \ldots, K$ are the per-user MSE targets of the balancing optimization.

To answer the third question, Fig. 4.2 depicts the curves for balanced rates, SINRs, and MSEs for specific targets in the MSE and rate region for a two user setup. It even sketches the switching effect for MSE balancing. Transmission can also be switched off for balancing the rate bounds $R_k^{(\mathrm{B})}(\boldsymbol{t}) = r_k^{(\mathrm{B})}(\boldsymbol{t}) + \mu_k^{(\mathrm{B})}$ with offset $\mu_k^{(\mathrm{B})} \in \mathbb{R}_+$. When jointly maximizing the ratios

$$\frac{r_k^{(\mathrm{B})}(\boldsymbol{t}) + \mu_k^{(\mathrm{B})}}{\rho_k}, \quad k = 1, \ldots, K$$

at a common level $\rho \in \mathbb{R}_+$, transmission is active only to those receivers where $\rho\rho_k \geq \mu_k^{(\mathrm{B})}$. Section 4.6 discusses the relation between MSE and ergodic RB with unequal targets in detail. Therein, the ACS is extended by an adaptive weighting strategy for approximating the RB solution similar to [155, 156, 237].

Remark 4.4 The effect of soft switching transmission to users on (and off) may be exploited for scheduling in higher layers. The prioritization of users allows to distinguish between primary users with low MSE targets and secondary users with

[28] Only for equal MSE targets, the MSEs for all the users are always equal at the optimum.

[29] The minimum for SINR and rate balancing is zero and corresponds to zero transmit power. As soon as the transmit power is positive, transmission to all the users is active, even for unequal targets [11, 166, 167]. This stands in contrast to MSE balancing with unequal MSE targets [164].

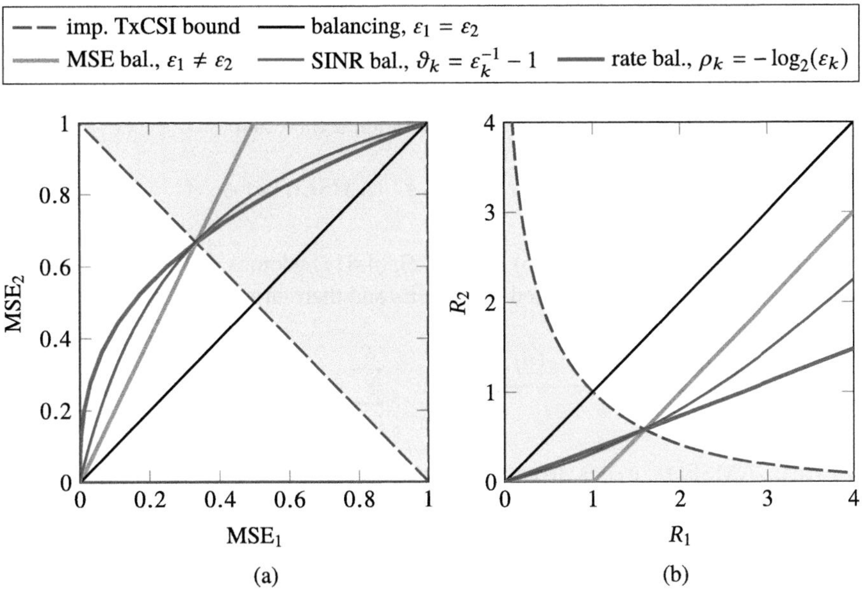

Fig. 4.2 Difference between the MSE, SINR, and rate balancing curves of a $K = 2$ user vector BC for equal and unequal targets, ε_k, $\vartheta_k = \varepsilon_k^{-1} - 1$, and $\rho_k = -\log_2(\varepsilon_k)$. (**a**) Mean square error region. (**b**) Ergodic rate region

larger targets. Transmission to secondary users is then only active if the primary users achieve a threshold level $\varepsilon^\star \leq \max\{\varepsilon \in \mathbb{R}_+ : \varepsilon\varepsilon_k \leq 1, k = 1, \ldots, K\}$.

4.5.1 Alternating Convex Search for Balancing

A local solution for the min–max MSE problem (4.3) with imperfect transmitter and perfect receiver CSI is also found via an ACS that alternates between the update of the moments $\hat{\boldsymbol{h}}_k$, $\boldsymbol{R}_k$, and $\tilde{\sigma}_k^2$ and the precoder update. In contrast to the iterations from Sect. 4.3, Step 2 shall now solve a max–min MSE optimization that (de)activates transmission to receivers dependent on the balancing level ε.[30]

1. Again, the downlink filters $f_k(\boldsymbol{h}_k)$ are updated for given $\boldsymbol{t}$. For the set of active beamformers $\mathcal{A} = \{k \in \{1, \ldots, K\} : \varepsilon^{(n)}\varepsilon_k < 1\}$, with $\boldsymbol{t}_k \neq \boldsymbol{0}$, the filter functions are from (1.6). Generic filters replace these functions for beamformers from the complement set $\mathcal{A}^c$, where $\boldsymbol{t}_i = \boldsymbol{0}$. Therewith, we compute the channels moments $\hat{\boldsymbol{h}}_k$ and $\boldsymbol{R}_k$ and the effective noise $\tilde{\sigma}_k^2$ from (4.69)–(4.23).

[30]For the general case, we again consider set of weighted sum MSE constraints (4.6).

2. Then, the downlink beamformers $\boldsymbol{t}_k$, $k = 1, \ldots, K$ are jointly optimized as MMSE equalizers for the dual uplink MSE balancing formulation.

The beamformer design in Step 2 again exploits the helping filter variables v_j, $j = 1, \ldots, M$ for the average MSE optimization [cf. (4.24)]:

$$\min_{\varepsilon, \boldsymbol{t}, \boldsymbol{v}} \varepsilon \quad \text{s.t.} \quad \boldsymbol{t} \in \mathcal{P}, \quad \varepsilon^{-1} \mathbf{AMSE}(\boldsymbol{t}, \boldsymbol{v}) \in \mathcal{M}. \tag{4.81}$$

Inserting the constraint sets (4.6) and (1.32), (4.81) becomes a min–max optimization for the ratios of the weighted sum MSEs and their targets, i.e.,[31]

$$\min_{\boldsymbol{v}, \boldsymbol{t}} \max_j \frac{\sum_{k \in \mathcal{G}_j} w_k \, \mathrm{AMSE}_k(v_j, \boldsymbol{t})}{\varepsilon_j} \quad \text{s.t.} \quad \sum_{i=1}^{K} \| \boldsymbol{A}_{i,\ell}^{1/2} \boldsymbol{t}_i \|_2^2 \leq P_\ell, \quad \ell = 1, \ldots, L. \tag{4.82}$$

Here, $\mathrm{AMSE}_k(\boldsymbol{t}, v_j)$ is from (4.25) and includes the first and second order moments of $\boldsymbol{h}_k f_k(\boldsymbol{h}_k)$, i.e., $\hat{\boldsymbol{h}}_k$ and $\boldsymbol{R}_k$, and $\tilde{\sigma}_k^2$ from (4.21)–(4.23). Therewith, (4.82) is unable to activate any previously inactive beamformer $\boldsymbol{t}_i$ if $\hat{\boldsymbol{h}}_i = \mathbf{0}$ because $f_i(\boldsymbol{h}_i) = 0$. For this case, the moments are alternatively calculated with

$$\tilde{f}_i(\boldsymbol{h}_i) = \frac{\tilde{\boldsymbol{t}}_i^{\mathrm{H}} \boldsymbol{h}_i}{\sum_{k=1}^{K} |\boldsymbol{h}_i^{\mathrm{H}} \boldsymbol{t}_k|^2 + \sigma_i^2}, \quad i \in \mathcal{A}^{\mathrm{c}} \tag{4.83}$$

and a unit-norm $\tilde{\boldsymbol{t}}_i \in \mathbb{C}^N$. Ideally, $\tilde{\boldsymbol{t}}_i$ would be the beamformer at the boundary point where transmission to receiver i becomes active. Since this term is unknown, we approximate it with the normalized MRT filters $\tilde{\boldsymbol{t}}_i = \alpha_i \bar{\boldsymbol{h}}_i$, $\alpha_i \in \mathbb{R}_+$.[32] The joint optimization of v_j and $\boldsymbol{t}_k$, $k = 1, \ldots, K$ compensates the scaling mismatch of $\hat{\boldsymbol{h}}_k$ due to $\tilde{\boldsymbol{t}}_k$, such that the ACS still decreases the objective in each iteration.

With the weighted sum-MSEs (4.28), which follow from inserting the MMSE equalizers v_j from (4.27), the precoder update optimization (4.82) reads as

$$\min_{\boldsymbol{t}} \max_j \frac{\mathrm{WAMSE}_j(\boldsymbol{t})}{\varepsilon_j} \quad \text{s.t.} \quad \sum_{i=1}^{K} \| \boldsymbol{A}_{i,\ell}^{1/2} \boldsymbol{t}_i \|_2^2 \leq P_\ell, \quad \ell = 1, \ldots, L. \tag{4.84}$$

With the upper bound $m_j = \sum_{k \in \mathcal{G}_j} w_j$ for $\mathrm{WAMSE}_j(\boldsymbol{t})$, only that ratios are balanced where $\varepsilon \varepsilon_j < m_j$ at the optimum ε. This distinguishes the above optimization from formulations with equal targets, where the MSEs are equal at the optimum.

[31] Equivalence holds for $\varepsilon > 0$. While (4.82) is even valid for $\varepsilon = 0$, this case is excluded in (4.81).

[32] This beamformer is also the initialization for equal targets $\varepsilon_k = c$ and optimal for very low SNR.

Problem (4.84) is a quasiconvex program. The power constraints are convex and the objective function is quasiconvex.[33] Particularly, the superlevel sets $\mathcal{M}_j(\varepsilon)$, $j = 1, \dots, M$ of the ratios $\text{WAMSE}_j(\boldsymbol{t})/\varepsilon_j$ have the convex SOC form[34]

$$\mathcal{M}_j(\varepsilon) = \left\{ \boldsymbol{t} \in \mathbb{C}^{NK} : \left\| [\hat{\boldsymbol{R}}_j^{\mathrm{H}/2}, \boldsymbol{0}]^{\mathrm{H}} \boldsymbol{t} + \hat{\sigma}_j \mathbf{e}_{N+1} \right\|_2 \le \frac{\mathrm{Re}(\hat{\boldsymbol{h}}^{\mathrm{H}} \boldsymbol{G}_j \boldsymbol{t})}{\sqrt{m_j - \varepsilon \varepsilon_j}} \right\}.$$

The objective is the maximum of these ratios and, thus, quasiconvex [102, Section 3.4].

Being a quasiconvex optimization problem, (4.84) can be solved in the downlink via a sequence of QoS optimizations. Scaling the power constraint of (4.84) by $p^{-1} \in \mathbb{R}_+$, the minimum of (4.84) becomes the monotonic increasing function $\varepsilon^{(\mathrm{dl})} : \mathbb{R}_+ \to \mathbb{R}_+$:

$$\varepsilon^{(\mathrm{dl})}(p) = \min \left\{ \varepsilon \in \mathbb{R}_+ : \; p^{-1}\boldsymbol{t} \in \mathcal{P}, \; \boldsymbol{t} \in \mathcal{M}_j(\varepsilon), \; j = 1, \dots, M \right\}. \tag{4.85}$$

Its inverse function $p^{(\mathrm{dl})} : \mathbb{R}_+ \to \mathbb{R}_+$ is the optimum of (4.29), but with the inserted slack variable $\varepsilon^{-1} \in \mathbb{R}_+$:

$$p^{(\mathrm{dl})}(\varepsilon) = \min \left\{ p \in \mathbb{R}_+ : \; p^{-1}\boldsymbol{t} \in \mathcal{P}, \; \boldsymbol{t} \in \mathcal{M}_j(\varepsilon), \; j = 1, \dots, M \right\}. \tag{4.86}$$

With this notation, the relation of the functions $p^{(\mathrm{dl})}$ and $\varepsilon^{(\mathrm{dl})}$ reads as

$$\begin{aligned} p^{(\mathrm{dl})}(\varepsilon^{(\mathrm{dl})}(p)) &= p, \\ \varepsilon^{(\mathrm{dl})}(p^{(\mathrm{dl})}(\varepsilon)) &= \varepsilon. \end{aligned} \tag{4.87}$$

This relation follows by monotonicity and equivalence of the constraint sets for (4.85) and (4.30). The value $\varepsilon^{(\mathrm{dl})}(p)$ decreases monotonically with increasing p. The other way round, $p^{(\mathrm{dl})}(\varepsilon)$ decreases when increasing ε. Moreover, a solution $\boldsymbol{t}^\star$ for (4.85) also satisfies the equivalent MSE constraints of (4.86).

For this reason, we can find a solution for (4.85) via a line search, e.g., a bisection over $\varepsilon \in (\varepsilon_{\min}, \varepsilon_{\max}]$ with the lower and upper bounds

$$\begin{aligned} \varepsilon_{\min} &= \min\{\varepsilon \in \mathbb{R}_+ : \varepsilon\varepsilon_j \ge \varepsilon_{j,\min}, \, j = 1, \dots, M\} \\ \varepsilon_{\max} &= \max\{\varepsilon \in \mathbb{R}_+ : \varepsilon\varepsilon_j \le m_j, \, j = 1, \dots, M\}, \end{aligned}$$

and the weighted sum MSE bounds $\varepsilon_{j,\min} = m_j - \hat{\boldsymbol{h}}^{\mathrm{H}} \boldsymbol{G}_j \hat{\boldsymbol{R}}_j^{\dagger} \boldsymbol{G}_j \hat{\boldsymbol{h}} \le \text{WAMSE}_j(\boldsymbol{t})$. A QoS optimization is solved for each sample ε', until $p^{(\mathrm{dl})}(\varepsilon') = 1$ is met with the

[33] Assuming fixed $\boldsymbol{v}$, the optimization problem is convex, because $\text{AMSE}_k(v_j, \boldsymbol{t})$ is convex quadratic in $\boldsymbol{t}$ for fixed v_j and the maximum of convex functions is convex [102, Section 3.2.3].

[34] Requiring $\hat{\boldsymbol{h}}_k^{\mathrm{H}} \boldsymbol{t}_k$ to be real valued does not change the objectives minimum value (cf. Sect. 4.3.1).

desired accuracy. If $p^{(\mathrm{dl})}(\varepsilon') < 1$, ε' lies above the optimum of (4.82), and likewise ε' lies below the optimum if $p^{(\mathrm{dl})}(\varepsilon') > 1$ or (4.30) is infeasible.[35] Constraints with $\varepsilon'\varepsilon_j \geq m_j$ can be discarded from the QoS optimization as the beamformers are $\boldsymbol{t}_k = \boldsymbol{0}$, $k \in \mathcal{G}_j$ for this case.

This iterative balancing solution for (4.81) is still computationally complex for a large number of power and MSE constraints and maladaptive to changes in the channel moments $\hat{\boldsymbol{h}}$ and $\hat{\boldsymbol{R}}_j$, the noise variance $\hat{\sigma}_j^2$, or the weights $w_k \in \mathbb{R}_+$. This motivates the dual uplink formulation and iterative solution in the next sections.

4.5.2 Uplink–Downlink Dualities for Balancing

The main result is the following dual uplink formulation for MSE balancing (4.85).

Theorem 4.2 *A strongly dual uplink formulation for problem* (4.82) *is*

$$\max_{\boldsymbol{\mu}\geq\boldsymbol{0}} \min_{\boldsymbol{u},\boldsymbol{\lambda}\geq\boldsymbol{0},\varepsilon} \varepsilon \quad \text{s.t.} \quad \sum_{j=1}^{M}\lambda_j\sum_{k\in\mathcal{G}_j} w_k\tilde{\sigma}_k^2 \leq 1, \quad \sum_{\ell=1}^{L}\mu_\ell P_\ell \leq 1, \qquad (4.88)$$
$$\varepsilon^{-1}\,\mathbf{AMSE}^{(\mathrm{ul})} \in \mathcal{M},$$

where the MSEs are the same as in (4.35). *While the inner filter design and allocation of* $\boldsymbol{\lambda} = [\lambda_1, \ldots, \lambda_M]^{\mathrm{T}}$ *minimize the maximum ratio* $\varepsilon_j^{-1}\sum_{k\in\mathcal{G}_j} w_k\,\mathrm{AMSE}_k^{(\mathrm{ul})}$, *the outer search w.r.t.* $\boldsymbol{\mu} = [\mu_1, \ldots, \mu_L]^{\mathrm{T}}$ *maximizes this value.*

For the proof, we first establish the connection of this max–min MSE formulation to the QoS optimization of Theorem 4.1. Inserting (4.6) and the equalizers (4.55), the dual uplink MSE minimization (4.88) is equivalent to

$$\max_{\boldsymbol{\mu}\geq\boldsymbol{0}} \min_{\boldsymbol{\lambda}\geq\boldsymbol{0},\varepsilon} \varepsilon \quad \text{s.t.} \quad \mathrm{WAMSE}_j^{(\mathrm{ul})}(\boldsymbol{\lambda},\boldsymbol{\mu}) \leq \varepsilon\varepsilon_j, \quad j = 1, \ldots, M,$$
$$\sum_{\ell=1}^{L}\mu_\ell P_\ell \leq 1, \quad \sum_{j=1}^{M}\lambda_j\hat{\sigma}_j^2 \leq p, \qquad (4.89)$$

where $p = 1$ and the (minimum) uplink weighted sum MSE reads as [cf. (4.46)]

$$\mathrm{WAMSE}_j^{(\mathrm{ul})}(\boldsymbol{\lambda},\boldsymbol{\mu}) = m_j - \lambda_j\hat{\boldsymbol{h}}^{\mathrm{H}}\boldsymbol{G}_j\left(\sum_{i=1}^{M}\lambda_i\hat{\boldsymbol{R}}_i + \sum_{\ell=1}^{L}\mu_\ell\boldsymbol{A}_\ell\right)^{\dagger}\boldsymbol{G}_j\hat{\boldsymbol{h}} \in [\varepsilon_{j,\min}, m_j]. \qquad (4.90)$$

[35] The same functionality has the test in [79, Algorithm 2], which is for SINR balancing.

The bounds for the uplink weighted sum MSE are the same as in the downlink. Here, the upper bound is tight for $\lambda_j = 0$ and the lower bound is tight for $\boldsymbol{\mu} = \mathbf{0}$ and $\lambda_i = 0$ for all $i \neq j$. Again, only a subset of the MSE constraints is satisfied by $\lambda_j = 0$ if the objective ε of (4.89) is too large. We refer to the remaining MSE constraints by the index set

$$\mathcal{J} = \{j \in \{1, \dots, M\} : m_j \geq \varepsilon \varepsilon_j\}.$$

Furthermore, the optimum of (4.89), i.e., $\varepsilon^{(\mathrm{ul})} : \mathbb{R}_+ \to (\varepsilon_{\min}, \varepsilon_{\max}]$ with $p \mapsto \varepsilon^{(\mathrm{ul})}(p)$, monotonically decreases for increasing p. The weighted sum MSEs jointly decrease when scaling $\boldsymbol{\lambda} \in \mathbb{R}_+^M$, $\boldsymbol{\lambda} \neq \mathbf{0}$ with a factor larger than one and the inner minimization of (4.89) satisfies the uplink power constraint with equality at the optimum. This property holds for all $\boldsymbol{\mu} \in \mathbb{R}_+^L$, hence, also for the solution of (4.89).

Lemma 4.2 *Let the inverse mapping of $\varepsilon^{(\mathrm{ul})}$ be $p^{(\mathrm{ul})} : (\varepsilon_{\min}, \varepsilon_{\max}] \to \mathbb{R}_+$ with $\varepsilon \mapsto p^{(\mathrm{ul})}(\varepsilon)$. This function is the optimum of the uplink QoS optimization* (4.34) *with targets $\varepsilon'_j = \varepsilon \varepsilon_j$.*

Proof The uplink QoS optimization (4.34) equivalently reads as

$$\max_{\boldsymbol{\mu} \geq \mathbf{0}} \min_{\boldsymbol{\lambda} \geq \mathbf{0}, p} p \quad \text{s.t.} \quad \mathrm{WAMSE}_j^{(\mathrm{ul})}(\boldsymbol{\lambda}, \boldsymbol{\mu}) \leq \varepsilon \varepsilon_j, \quad j = 1, \dots, M,$$
$$\sum_{\ell=1}^{L} \mu_\ell P_\ell \leq 1, \quad \sum_{j=1}^{M} \lambda_j \hat{\sigma}_j^2 \leq p \tag{4.91}$$

when inserting (4.55) into the MSEs and scaling them as $\varepsilon^{-1}\,\mathbf{AMSE} \in \mathcal{M}$. Let the QoS optimal variables of this problem be $\boldsymbol{\mu}^\star$ and $\boldsymbol{\lambda}^\star$ for a given MSE factor $\varepsilon \in (\varepsilon_{\min}, \varepsilon_{\max}]$. We prove that these variables are optimal for (4.89) to conclude that $\varepsilon^{(\mathrm{ul})}(p) = \varepsilon$ if $p = p^{(\mathrm{ul})}(\varepsilon)$. The solutions of both problems, the QoS optimization (4.91) and balancing formulation (4.89), satisfy

$$\mathrm{WAMSE}_j^{(\mathrm{ul})}(\boldsymbol{\lambda}^\star, \boldsymbol{\mu}^\star) = \varepsilon \varepsilon_j, \qquad j \in \mathcal{J}, \tag{4.92a}$$

$$\mathrm{WAMSE}_j^{(\mathrm{ul})}(\boldsymbol{\lambda}^\star, \boldsymbol{\mu}^\star) = m_j \leq \varepsilon \varepsilon_j, \quad j \notin \mathcal{J}, \tag{4.92b}$$

$$\sum_{\ell=1}^{L} \mu_\ell^\star P_\ell = 1, \qquad \sum_{j \in \mathcal{J}} \lambda_j^\star \hat{\sigma}_j^2 = p. \tag{4.92c}$$

The latter equalities hold because $\mathrm{WAMSE}_j^{(\mathrm{ul})}(\delta \boldsymbol{\lambda}, \gamma \boldsymbol{\mu})$ decreases with increasing $\delta \in \mathbb{R}_+$ and increases with $\gamma \in \mathbb{R}_+$ if $\lambda_j > 0$. The equalities (4.92a) and (4.92b) are valid because $\mathrm{WAMSE}_j^{(\mathrm{ul})}(\boldsymbol{\lambda}, \boldsymbol{\mu})$ decreases with increasing interference powers λ_j and increases with $\lambda_i, i \neq j$ (see Lemma A.6). For $j \notin \mathcal{J}$, the solution is $\lambda_j^\star = 0$, because this minimizes the uplink power.

As a consequence of (4.92a) and (4.92b), the solution $\boldsymbol{\lambda}^\star$ to both (4.91) and (4.89) is a fixed point of the standard interference function $\boldsymbol{I}_\varepsilon(\boldsymbol{\lambda}^\star, \boldsymbol{\mu})$, that is,

$$\lambda_j^\star = [m_j - \varepsilon\varepsilon_j]^+ I_j(\boldsymbol{\lambda}^\star, \boldsymbol{\mu}), \quad j = 1, \ldots, M, \tag{4.93}$$

where $[z]^+ = \max\{z, 0\}$ and $I_j(\boldsymbol{\lambda}, \boldsymbol{\mu})$ is given by (4.43). Since this fixed point is unique, it remains to prove that the optimizer $\boldsymbol{\mu}^\star$ of (4.91) is a maximizer of (4.89).

Achievability, i.e., that $\boldsymbol{\mu}^\star$ is feasible for (4.89), follows from (4.92c). Therefore, the optimum of (4.89) is $\varepsilon^{(\mathrm{ul})}(p) \geq \varepsilon^\star$. The converse is via contradiction. Assume there is an optimizer $\boldsymbol{\mu}'$ of (4.89) with $\varepsilon' = \varepsilon^{(\mathrm{ul})}(p) > \varepsilon$. Since such a pair $(\boldsymbol{\mu}', \varepsilon')$ is feasible but strictly suboptimal for (4.91), the corresponding $\boldsymbol{\lambda}'$ requires only the reduced power $p' = \sum_{j \in \mathcal{J}} \hat{\sigma}_j^2 \lambda_j' < p^{(\mathrm{ul})}(\varepsilon) = p$. This contradicts optimality of $\boldsymbol{\lambda}'$, $\boldsymbol{\mu}'$, and ε' for (4.89). Thus, the converse inequality relation $\varepsilon^{(\mathrm{ul})}(p) \leq \varepsilon^\star$ must hold.

Together with the achievability result, this shows that the solutions $\boldsymbol{\mu}^\star$ and $\boldsymbol{\lambda}^\star$ of (4.91) are indeed optimizers for (4.89), such that the optimum is ε. Proving the reverse, i.e., $p = p^{(\mathrm{ul})}(\varepsilon)$ if ε is the optimum of (4.89), is via similar steps. □

With Lemma 4.2, we next prove strong duality between (4.82) and (4.88) via a detour through the strongly dual QoS problems (4.29) and (4.34).

Proof (of Theorem 4.2) Let $(\varepsilon^\star, \boldsymbol{t}^\star)$ denote the solution to the primal max–min MSE optimization (4.82), where $\varepsilon^\star\varepsilon_j < m_j$ for $j \in \mathcal{J} \subseteq \{1, \ldots, M\}$. MSE constraints with indices $i \notin \mathcal{J}$ are neglected because $\boldsymbol{t}_k^\star = \boldsymbol{0}$ for $k \in \mathcal{G}_i$.

By the inversion property of Sect. 4.5.1, the optimum of the corresponding QoS problem (4.29), with the MSE constraint $\varepsilon^{\star-1}\,\mathbf{AMSE}(\boldsymbol{t}) \in \mathcal{M}$, is $p^{(\mathrm{dl})}(\varepsilon^\star) = 1$. It is achieved for any $\boldsymbol{t}$ with $\boldsymbol{t}_k = \mathrm{e}^{\mathrm{j}\phi_k}\boldsymbol{t}_k^\star$ and $\phi_k \in [0, 2\pi)$, $k = 1, \ldots, K$. Vice versa, $p^{(\mathrm{dl})}(\varepsilon^\star) = 1$ is an identifier to declare $\varepsilon^\star$ and $\boldsymbol{t}^\star$ as a solution of (4.82).

The minima of (4.29) and (4.34) are the same, i.e., $p^{(\mathrm{dl})}(\varepsilon^\star) = p^{(\mathrm{ul})}(\varepsilon^\star) = 1$, due to zero duality gap (cf. Theorem 4.1). This duality result remains valid also for the reduced number of uplink constraints $\mathrm{WAMSE}_j^{(\mathrm{ul})}(\boldsymbol{\lambda}, \boldsymbol{\mu}) \leq \varepsilon^\star\varepsilon_j$, $j \in \mathcal{J}$. Therefore, $p^{(\mathrm{ul})}(\varepsilon^\star) = 1$ also identifies $\varepsilon^\star$ as the optimum of the downlink weighted sum MSE balancing problem (4.82).

According to Lemma 4.2, $p^{(\mathrm{ul})}(\varepsilon^\star) = 1$ also identifies $\varepsilon^\star$ as the minimum of the uplink max–min MSE optimization (4.88). Hence, the optima of the two optimizations (4.88) and (4.82) are the same, that is, strong duality holds. □

All the special cases for the uplink–downlink duality from the QoS problem (see Sect. 4.3.1) are also available for the MSE balancing optimization.

Corollary 4.4 *If the MSE balancing problem* (4.82) *has per-user MSE targets* (4.4), *i.e.,* $w_k = 1$ *and* $\mathcal{G}_k = \{k\}$, $k = 1, \ldots, K$, *the strongly dual problem becomes*

$$\max_{\boldsymbol{\mu} \geq \boldsymbol{0}} \min_{\boldsymbol{u}, \boldsymbol{\lambda} \geq \boldsymbol{0}} \max_k \frac{\mathrm{AMSE}_k^{(\mathrm{ul})}}{\varepsilon_k} \quad \text{s.t.} \quad \sum_{k=1}^{K} \lambda_k \tilde{\sigma}_k^2 \leq 1, \quad \sum_{\ell=1}^{L} \mu_\ell P_\ell \leq 1 \tag{4.94}$$

and the uplink MSEs are given by (4.48). *The inner filter design and allocation of* $\boldsymbol{\lambda}$ *jointly minimize the ratios between the users' performance and their demands, and the outer worst-case uplink noise search w.r.t.* $\boldsymbol{\mu}$ *maximizes this minimum value.*

The other special cases are for a single sum MSE restriction within $\mathcal{M}$ (4.5) [116] or a simple sum power restriction for $\mathcal{P}$ [68, 162]. For the previous example, the dual optimization only consists of the outer noise covariance matrix search [145], and for the latter example, the worst case noise covariance matrix is $P^{-1}\mathbf{I}$ [68, 161].

Corollary 4.5 *When minimizing a single weighted sum MSE, that is,* $\mathcal{M}$ *has the form of* (4.5)*, the strongly dual problem of* (4.82) *simplifies to (cf.* [145]*)*

$$\max_{\boldsymbol{\mu}\geq\mathbf{0}} \sum_{k=1}^{K} w_k \min_{\boldsymbol{u}_k} \mathrm{AMSE}_k^{(\mathrm{ul})} \quad \text{s.t.} \quad \sum_{\ell=1}^{L} P_\ell \mu_\ell \leq 1, \tag{4.95}$$

and the uplink MSEs are those from (4.50).

Proof Due to the sole MSE requirement of (4.88), i.e., $M = 1$, ε is directly replaced with the sum MSE. Moreover, the tight uplink power constraint results in $\lambda_1 = \tilde{\sigma}_1^{-2}$. Inserting this λ_1 and the equalizer (4.55), only the outer search of $\boldsymbol{\mu}$ remains. □

Corollary 4.6 *If* $\mathcal{P}$ *only contains the requirement* $\|\boldsymbol{t}\|_2 \leq P_1$, (4.88) *reduces to the inner power allocation and equalizer design*

$$\min_{\boldsymbol{u},\boldsymbol{\lambda}\geq\mathbf{0},\varepsilon} \varepsilon \quad \text{s.t.} \sum_{j=1}^{M} \lambda_j \sum_{k\in\mathcal{G}_j} w_k \tilde{\sigma}_k^2 \leq 1, \ \sum_{k\in\mathcal{G}_j} w_k\, \mathrm{AMSE}_k^{(\mathrm{ul})} \leq \varepsilon_j, \ j = 1, \ldots, M. \tag{4.96}$$

and the uplink MSEs are given by (4.53). *The solution of the outer problem is simply* $\mu_1 = P^{-1}$.

4.5.3 Iterative Minimum Mean Square Error Search

In advantage to optimizing the $NM \times 1$ complex entries of $\boldsymbol{t}$ with (4.82), the uplink optimization (4.88) is only w.r.t. the $M+L$ non-negative reals for $\boldsymbol{\lambda}$ and $\boldsymbol{\mu}$ besides ε. Solutions for this balancing problem are closely related to the QoS solutions in Sect. 4.3.2, using either convex programming solvers or an iterative search.

While the uplink QoS optimization features a SDP formulation (4.54), finding the max–min MSE optimizers $\boldsymbol{\mu}$, $\boldsymbol{\lambda}$, and ε of (4.88) requires a sequence of SDPs. When jointly inverting the inner minimization into a maximization and the inequalities of the MSE requirements, the problem has the same solution $(\varepsilon, \boldsymbol{\lambda}, \boldsymbol{\mu})$, but reads as

$$\max_{\boldsymbol{\mu}\geq\mathbf{0},\boldsymbol{\lambda}\geq\mathbf{0},\varepsilon} \varepsilon \quad \text{s.t.} \quad \sum_{j=1}^{M}\lambda_j\hat{\sigma}_j^2 \leq 1, \quad \sum_{\ell=1}^{L}\mu_\ell P_\ell \leq 1,$$
$$\sum_{j=1}^{M}\lambda_j\hat{\boldsymbol{R}}_j + \sum_{\ell=1}^{L}\mu_\ell\boldsymbol{A}_\ell - \sum_{j=1}^{M}\frac{\lambda_j}{m_j-\varepsilon_j\varepsilon}\boldsymbol{G}_j\hat{\boldsymbol{h}}\hat{\boldsymbol{h}}^{\mathrm{H}}\boldsymbol{G}_j \succeq \mathbf{0}. \tag{4.97}$$

The latter constraint is non-convex in ε, but the problem can be solved via a line search over the objective value, e.g., using a QoS feasibility test in each iteration.

The alternative solution consists of two nested loops. Similar to Sect. 4.3.2, the inner loop solves the power allocation with the globally convergent update

$$\lambda_j^{(n+1)} = [m_j - \varepsilon_j\varepsilon^{(n+1)}]^+ I_j(\boldsymbol{\lambda}^{(n)}, \boldsymbol{\mu}), \quad j = 1, \ldots, M, \tag{4.98}$$

which is motivated by (4.92a)–(4.92c). The normalization with $[m_j - \varepsilon_j\varepsilon^{(n+1)}]^+$ is to exploit full transmit power, i.e., $\sum_{j\in\mathcal{J}}\hat{\sigma}_j^2\lambda_j^{(n+1)} = 1$. Hence, $\varepsilon^{(n+1)}$ is the root of

$$\varepsilon^{(n+1)} = \min\left\{\varepsilon \in \mathbb{R}_+ : f(\varepsilon) = \sum_{j=1}^{M}\hat{\sigma}_j^2[m_j - \varepsilon_j\varepsilon^{(n+1)}]^+ I_j(\boldsymbol{\lambda}^{(n)}, \boldsymbol{\mu}) - 1 = 0\right\}, \tag{4.99}$$

which is unique because $f(\varepsilon)$ increases monotonically with ε.[36]

With knowledge of $\mathcal{J}$, an algebraic reformulation of (4.99) reads as

$$\varepsilon^{(n+1)} = \frac{\sum_{j\in\mathcal{J}}\hat{\sigma}_j^2 m_j I_j(\boldsymbol{\lambda}^{(n)}, \boldsymbol{\mu}) - 1}{\sum_{j\in\mathcal{J}}\hat{\sigma}_j^2\varepsilon_j I_j(\boldsymbol{\lambda}^{(n)}, \boldsymbol{\mu})}. \tag{4.100}$$

The choice for $\mathcal{J}$ is found via testing at most M candidate sets. Let the ratios $\varepsilon_j^{-1}m_j$ be in descending order, i.e., $\varepsilon_1^{-1}m_1 \geq \varepsilon_2^{-1}m_2 \geq \ldots \geq \varepsilon_M^{-1}m_M$. Then, the solution is $\mathcal{J} = \{1, \ldots, M'\}$, where M' is the largest integer satisfying $\varepsilon^{(n+1)}\varepsilon_{M'} < m_{M'}$.[37] Here, users with small ratios ε_k/m_k are strictly preferred.

Remark 4.5 Shi et al. [253] provide another method for the power allocation with individual MSEs. By splitting the uplink equalizers as $\boldsymbol{u}_k = \tilde{\boldsymbol{u}}_k\lambda_k^{-1/2}$, the power allocation can be formulated as an eigenvalue problem. Then, the results for ε and $\boldsymbol{\lambda}$ from a repeated filter update and eigenvector calculation [89].

[36]From (4.98), $\lambda_j^{(n+1)}$ clearly decreases for increasing $\varepsilon^{(n+1)}$ until $\lambda_j^{(n+1)} = 0$ for $\varepsilon_j\varepsilon^{(n+1)} \geq m_j$.

[37]The search for M' requires less than M' iterations for a bisection over the indices $i \in \{1, \ldots, M''\}$ that halves the number of candidates in each iteration.

The outer optimization searches the matrix $\sum_{\ell=1}^{L} \mu_\ell \boldsymbol{A}_\ell$, which maximizes

$$\varepsilon(\boldsymbol{\mu}) = \frac{\sum_{j\in\mathcal{J}} \hat{\sigma}_j^2 m_j I_j(\boldsymbol{\lambda}(\boldsymbol{\mu}), \boldsymbol{\mu}) - 1}{\sum_{j\in\mathcal{J}} \hat{\sigma}_j^2 \varepsilon_j I_j(\boldsymbol{\lambda}(\boldsymbol{\mu}), \boldsymbol{\mu})}. \tag{4.101}$$

This objective is quasiconcave in $\boldsymbol{\mu}$ because $I_j(\boldsymbol{\lambda}(\boldsymbol{\mu}), \boldsymbol{\mu})$ is concave in $\boldsymbol{\mu}$ for $\varepsilon \in \mathbb{R}_+$ (cf. Proof of Lemma A.5 in Sect. A.7) and the constraint set is a convex polytope. Hence, the outer maximization of (4.88) is a quasiconvex problem.

We find a solution with a PG algorithm similar to that of Sect. 4.3.2. To compute the gradient $\boldsymbol{\delta} = [\delta_1, \ldots, \delta_L]^{\mathrm{T}}$ with $\delta_m = \frac{\partial \varepsilon}{\partial \mu_m}$, we exploit that

$$\lambda_j = (m_j - \varepsilon\varepsilon_j) I_j(\boldsymbol{\lambda}, \boldsymbol{\mu}), \quad j \in \mathcal{J}, \qquad \sum_{j=1}^{M} \hat{\sigma}_j^2 \lambda_j = 1$$

as a result of the inner minimization. The derivatives of these equations are

$$0 = \sum_{j=1}^{M} \hat{\sigma}_j^2 \frac{\partial \lambda_j}{\partial \mu_m}, \qquad \frac{\partial \lambda_j}{\partial \mu_m} = \varepsilon_j \frac{\partial \varepsilon}{\partial \mu_m} I_j(\boldsymbol{\lambda}, \boldsymbol{\mu}) + (m_j - \varepsilon_j \varepsilon) \frac{\partial I_j(\boldsymbol{\lambda}(\boldsymbol{\mu}), \boldsymbol{\mu})}{\partial \mu_m}, \tag{4.102}$$

respectively, where the derivative of $I_j(\boldsymbol{\lambda}(\boldsymbol{\mu}), \boldsymbol{\mu})$ with respect to μ_m is given by (4.59). These derivatives provide us the equation system

$$\begin{bmatrix} 0 & \hat{\boldsymbol{\sigma}}^{\mathrm{T}} \\ \boldsymbol{D}\boldsymbol{I}(\boldsymbol{\lambda}, \boldsymbol{\mu}) & \mathbf{I} - \boldsymbol{\Lambda}\boldsymbol{\Psi} \end{bmatrix} \begin{bmatrix} \frac{\partial \varepsilon}{\mu_m} \\ \frac{\partial \boldsymbol{\lambda}}{\mu_m} \end{bmatrix} = \begin{bmatrix} 0 \\ \boldsymbol{\Lambda}\boldsymbol{n}_m \end{bmatrix}, \tag{4.103}$$

where $\hat{\boldsymbol{\sigma}} = [\hat{\sigma}_1^2, \ldots, \hat{\sigma}_{|\mathcal{J}|}]^{\mathrm{T}}$, the diagonal matrix $\boldsymbol{D} = \mathrm{diag}(-\varepsilon_1, \ldots, -\varepsilon_{|\mathcal{J}|})$, and $\boldsymbol{\Lambda}$, $\boldsymbol{\Psi}$, and $\boldsymbol{n}_m$ are from (4.60) and (4.61). The solution from (4.103) reads as

$$\delta_m = \frac{\partial \varepsilon}{\partial \mu_m} = -\frac{\hat{\boldsymbol{\sigma}}^{\mathrm{T}}(\mathbf{I} - \boldsymbol{\Lambda}\boldsymbol{\Psi})^{-1}\boldsymbol{\Lambda}\boldsymbol{n}_m}{\hat{\boldsymbol{\sigma}}^{\mathrm{T}}(\mathbf{I} - \boldsymbol{\Lambda}\boldsymbol{\Psi})^{-1}\boldsymbol{D}\boldsymbol{I}(\boldsymbol{\lambda}, \boldsymbol{\mu})}, \qquad m = 1, \ldots, L. \tag{4.104}$$

Using this closed-form gradient formulation, the PG method updates $\boldsymbol{\mu}$ with the same projection and step size selection as for the QoS optimization in Sect. 4.3.2. After convergence of the suggested PG algorithm, the primal beamformers are reconstructed as detailed in Sect. 4.3.3.

4.5.3.1 Uplink Weighted Sum-MSE Minimization

Only the inner iteration remains for solving the MSE balancing optimization with a single sum power constraint (cf. Corollary 4.3) and only the outer Projected gradient (PG) iteration needs to be performed for the solution of a single weighted sum MSE minimization (cf. Corollary 4.2). The latter problem is moreover convex in $\boldsymbol{\mu}$ in contrast to the general uplink max–min MSE formulation (4.97). Inserting the equalizer (4.55), the maximization of (4.95) transforms to

$$\max_{\boldsymbol{\mu}\geq\mathbf{0}} m_1 - \hat{\boldsymbol{h}}^{\mathrm{H}}\boldsymbol{G}_j\Big(\hat{\boldsymbol{R}}_1 + \hat{\sigma}_1^2\sum_{\ell=1}^{L}\mu_\ell\boldsymbol{A}_\ell\Big)^{-1}\boldsymbol{G}_j\hat{\boldsymbol{h}} \quad \text{s.t.} \quad \sum_{\ell=1}^{L}\mu_\ell P_\ell \leq 1. \tag{4.105}$$

The objective is concave, elementwise increasing with $\boldsymbol{\mu} \in \mathbb{R}_+^L$, and limited by an affine half-space constraint. Hence, (4.105) can be solved via a disciplined convex programming solver [99] after an appropriate problem reformulation. In particular, using Schur's complement, the standard SDP formulation of (4.105) reads as[38]

$$\max_{\varepsilon,\boldsymbol{\mu}\geq\mathbf{0}} \ \varepsilon \quad \text{s.t.} \quad \begin{bmatrix} m_1 - \varepsilon_1\varepsilon & \hat{\boldsymbol{h}}^{\mathrm{H}}\boldsymbol{G}_1 \\ \boldsymbol{G}_1\hat{\boldsymbol{h}} & \hat{\boldsymbol{R}}_1 + \hat{\sigma}_1^2\sum_{\ell=1}^{L}\mu_\ell\boldsymbol{A}_\ell \end{bmatrix} \succeq \mathbf{0}, \quad \sum_{\ell=1}^{L}\mu_\ell P_\ell \leq 1. \tag{4.106}$$

The alternative PG algorithm for (4.105) is similar to Sect. 3.2.3. However, this algorithm now uses the gradient $\boldsymbol{\delta} = [\delta_1, \ldots, \delta_L]^{\mathrm{T}}$, with

$$\delta_\ell = \hat{\sigma}_1^2\hat{\boldsymbol{h}}^{\mathrm{H}}\boldsymbol{G}_1\boldsymbol{X}^{\dagger}\boldsymbol{A}_\ell\boldsymbol{X}^{\dagger}\boldsymbol{G}_1\hat{\boldsymbol{h}} \tag{4.107}$$

and $\boldsymbol{X} = \hat{\boldsymbol{R}} + \hat{\sigma}_1^2\sum_{\ell=1}^{L}\mu_\ell\boldsymbol{A}_\ell$, for a successive update of the decision variable $\boldsymbol{\mu}$.

4.6 Ergodic Rate Balancing Approximations

As mentioned in Sect. 4.1, the max–min weighted (sum-)MSE optimization also approximates the robust ergodic rate maximization with additive channel fading:

$$\max_{\rho,\,\boldsymbol{t}} \ \rho \quad \text{s.t.} \quad \boldsymbol{t} \in \mathcal{P}, \quad \rho^{-1}\boldsymbol{R}(\boldsymbol{t}) \in \mathcal{R}. \tag{4.108}$$

For ergodic rate balancing, i.e., $\mathcal{R} = \{\boldsymbol{r} \in \mathbb{R}_+^K : r_k \geq \rho_k, k = 1, \ldots, K\}$, with equal targets $\rho_k = c$, the approximation (4.8) suggests solving the MSE balancing optimization (4.80) with per-user targets $\varepsilon_k = 2^{-c}$ instead. The predicted data rate is then given by $c\rho = -c\log_2(\varepsilon)$ if ε is the minimum of the MSE optimization.

For unequal targets ρ_k, the lower bound (4.8) results in the balancing problem

$$\max_{\rho,\,\boldsymbol{t}} \ \rho \quad \text{s.t.} \quad \boldsymbol{t} \in \mathcal{P}, \quad \mathrm{AMSE}_k(\boldsymbol{t}) \geq \varepsilon_k(\rho), \quad k = 1, \ldots, K, \tag{4.109}$$

[38]The LMI constraint is affine in the decision variables $\boldsymbol{\mu}$ and ε.

where the target functions are $\varepsilon_k : \mathbb{R}_+ \to (0, 1]$ with $\varepsilon_k(\rho) = 2^{-\rho_k \rho}$, $k = 1, \dots, K$. Unlike for the min–max MSE optimization with distinct but fixed (MSE) targets ε_k, all MSE constraints of this problem hold with equality in the optimum.

Again, a local solution of (4.109) results from the ACS strategy in Sect. 4.5. The dual uplink problem for the precoder update of the ACS now reads as [cf. (4.88)]

$$\min_{\boldsymbol{\mu} \geq \mathbf{0}} \max_{\rho,\, \boldsymbol{\lambda} \geq \mathbf{0}} \rho \quad \text{s.t.} \quad \boldsymbol{t} \in \mathcal{P}, \quad \text{AMSE}_k^{(\text{ul})}(\boldsymbol{\lambda}, \boldsymbol{\mu}) \geq \varepsilon_k(\rho), \quad k = 1, \dots, K, \tag{4.110}$$

where $\text{AMSE}_k^{(\text{ul})}(\boldsymbol{\lambda}, \boldsymbol{\mu})$ is from (4.48). We find a solution of (4.110) by the iterative procedure in Sect. 4.5.3. For the inner maximization, $\varepsilon_k(\rho^{(n+1)})$ replaces $\varepsilon^{(n+1)}\varepsilon_j$ in (4.98) and (4.99). Now, the update $\rho^{(n+1)}$ requires an iterative search and the gradient $\boldsymbol{\delta} = [\delta_1, \dots, \delta_L]^{\mathrm{T}}$ of the outer PG search has the entries [cf. (4.104)]

$$\delta_m = \frac{\partial \rho}{\partial \mu_m} = -\frac{\tilde{\boldsymbol{\sigma}}^{\mathrm{T}}(\mathbf{I} - \boldsymbol{\Lambda}\boldsymbol{\Psi})^{-1}\boldsymbol{\Lambda}\boldsymbol{n}_m}{\tilde{\boldsymbol{\sigma}}^{\mathrm{T}}(\mathbf{I} - \boldsymbol{\Lambda}\boldsymbol{\Psi})^{-1}\boldsymbol{D}\boldsymbol{I}(\boldsymbol{\lambda}, \boldsymbol{\mu})}, \quad m = 1, \dots, L,$$

but $\boldsymbol{D} = \text{diag}(d_1, \dots, d_K)$ has the entries $d_k = \rho_k \log_2(\mathrm{e})2^{-\rho_k \rho}$, while the other terms, e.g., $\tilde{\boldsymbol{\sigma}} = [\tilde{\sigma}_1^2, \dots, \tilde{\sigma}_K^2]^{\mathrm{T}}$, are the per-user equivalent of (4.60) and (4.61).

An alternative approach for ergodic rate balancing provides the weighted MSE approximation of [155, 156, 237]. This approach also uses an ACS, but with changing weights $w_k(\boldsymbol{h}_k)$, $k = 1, \dots, K$ that even depend on the channel realization. Thereby, the ACS sequentially approximates the ergodic rates by affine functions of the MMSEs (4.10). For the BC at hand, this approximation reads as [cf. (4.10)]

$$R_k(\boldsymbol{t}) \geq \bar{R}_k(\boldsymbol{t}) = \mathrm{E}[\log_2(w_k(\boldsymbol{h}_k))] + \log_2(\mathrm{e})\big(1 - \text{WAMSE}_k(\boldsymbol{t})\big), \tag{4.111}$$

where the average weighted MSE function is $\text{WAMSE}_k(\boldsymbol{t}) = \mathrm{E}[w_k(\boldsymbol{h}_k)\text{MMSE}_k(\boldsymbol{t}, \boldsymbol{h}_k)]$ for perfect CSI MMSE receivers (1.6). To satisfy (4.111) with equality, the weights $w_k(\boldsymbol{h}_k)$ obviously must be the inverse MMSEs, that is, (cf. [155])

$$w_k(\boldsymbol{h}_k) = \text{MMSE}_k^{-1}(\boldsymbol{t}, \boldsymbol{h}_k) = 1 + \frac{|\boldsymbol{h}_k^{\mathrm{H}}\boldsymbol{t}_k|^2}{\sum_{i \neq k} |\boldsymbol{h}_k^{\mathrm{H}}\boldsymbol{t}_i|^2 + \sigma_k^2}, \quad k = 1, \dots, K. \tag{4.112}$$

Since these weights are random, we update them for the given precoder $\boldsymbol{t}^{(n-1)}$ in Step 1 of the ACS (cf. [155, 156, 237]). Thereby, we approximate the rates $\bar{R}_k(\boldsymbol{t})$ for the beamformer optimization in Step 2 of the ACS by the MSE bound [cf. (4.111)]

$$\text{WAMSE}_k(\boldsymbol{t}) \leq \bar{w}_k\, \text{AMSE}_k(\boldsymbol{t}), \tag{4.113}$$

with $\text{AMSE}_k(\boldsymbol{t})$ from (4.25), but with the following moments for the effective channels $w_k(\boldsymbol{h}) f_k(\boldsymbol{h}_k)\boldsymbol{h}_k$, $k = 1, \dots, K$ and the noise powers:

$$\hat{\boldsymbol{h}}_k = \bar{w}_k^{-1}\, \mathrm{E}\big[\boldsymbol{h}_k \tilde{f}_k(\boldsymbol{h}_k) w_k(\boldsymbol{h}_k)\big], \tag{4.114}$$

$$\boldsymbol{R}_k = \bar{w}_k^{-1} \,\mathrm{E}\big[\boldsymbol{h}_k \boldsymbol{h}_k^{\mathrm{H}} w_k(\boldsymbol{h}_k) |\tilde{f}_k(\boldsymbol{h}_k)|^2\big], \tag{4.115}$$

$$\tilde{\sigma}_k^2 = \bar{w}_k^{-1} \sigma_k^2 \,\mathrm{E}\big[w_k(\boldsymbol{h}_k) |\tilde{f}_k(\boldsymbol{h}_k)|^2\big]. \tag{4.116}$$

Given sufficient accurate computations for (4.114)–(4.116), the weighted MSE balancing problem for the precoder update in the ACS reads as

$$\max_{\rho,\, \boldsymbol{t}} \ \rho \quad \text{s.t.} \quad \boldsymbol{t} \in \mathcal{P}, \quad \bar{w}_k \,\mathrm{AMSE}_k(\boldsymbol{t}) \le \varepsilon_k(\rho), \quad k = 1, \ldots, K. \tag{4.117}$$

The target functions are now affine in ρ, namely

$$\varepsilon_k(\rho) = \mathrm{E}[\ln(w_k(\boldsymbol{h}_k))] + 1 - \ln(2)\rho_k \rho, \quad k = 1, \ldots, K. \tag{4.118}$$

This allows us to solve (4.117) again in the dual uplink. Transmission to some users may be switched off also in this case if the optimum is $\varepsilon_k(\rho) \ge 1$. This effect is due to the limited approximation accuracy and vanishes when (4.111) becomes tight.

The channel dependent weights (4.112) promise performance advantages in terms of data rate compared to the approximation for (4.109). However, sufficiently accurate computing $\mathrm{E}[\ln(w_k(\boldsymbol{h}_k))]$, $\bar{w}_k = \mathrm{E}[w_k(\boldsymbol{h}_k)]$, and (4.114)–(4.116) within Step 1 of the ACS increases the computational complexity. Computing the channels' moments $\hat{\boldsymbol{h}}_k$, $\boldsymbol{R}_k$, and the noise power $\tilde{\sigma}_k^2$ is even more complicated than with deterministic weights. Even though $w_k(\boldsymbol{h}_k)$ and $f_k(\boldsymbol{h}_k)$ incorporate the same precoder $\boldsymbol{t}' \in \mathcal{P}$, closed-form expressions and tractable integral representations for these expectations are missing. Therefore, their evaluation requires Monte Carlo sampling already for the additive Gaussian channel error model, which results in small inaccuracies of the solution.

Razaviyayn et al. [34] present an alternative sample average approximation strategy to overcome this problem (e.g., see [254, Chapter 5]). He adds one sample channel in each iteration of the ACS and computes the moments by the values of the previous iteration and the terms for the added sample. Convergence is assured only in the long term and after a large number of ACS iterations, because the previous realizations of the channel moments and the noise are based on suboptimal precoders. This further reduces the convergence speed of the ACS.

4.7 Numerical Results for Mean Square Error Optimizations

The simulation setup for the MSE based optimizations is for $K = 2$ or $K = 4$ users and $N = 4$ or $N = 8$ antennas. The noise variances are fixed at $\sigma_k^2 = 1$ and the channel means are created for a Gaussian and an exponential power profile model.[39]

[39]The exponential power profile is amongst others used for multi-spotbeam SatCom [255].

For the Gaussian model, the channel means $\bar{\boldsymbol{h}}_k$ are drawn from a standard Gaussian distribution and the channels' covariance matrices are $\boldsymbol{C}_k = \frac{\kappa}{N}\mathbf{I}_N$, where κ equals either 0 dB, −10 dB, or −20 dB to model Rician-like fading. For each of the channels' means, another 1.000 channels $\boldsymbol{h}_k = \bar{\boldsymbol{h}}_k + \boldsymbol{e}_k$ are drawn for the sample average approximations of $\hat{\boldsymbol{h}}_k$, $\boldsymbol{R}_k$, and $\tilde{\sigma}_k^2$ and to compute the average MSE for imperfect transmitter CSI. The norm bound $\|\boldsymbol{t}^{(n+1)} - \boldsymbol{t}^{(n)}\|_2^2 \leq \epsilon$, with $\epsilon = 10^{-4}$, serves as an abort criteria for the alternating convex beamformer search.

The latter channel model is based on an exponential power profile with i.i.d. uniformly distributed phase coefficients and log-normal shadow fading. The corresponding channel model reads as $\bar{\boldsymbol{h}}_k = \frac{1}{\sqrt{\tau_k}}\boldsymbol{b}_k$, where

$$b_{ki} = \mathrm{e}^{\mathrm{j}\phi_{i,k}}\sqrt{\rho^{|i-k|}}$$

with $\phi_{i,k} \in [0, 2\pi)$ and $\ln(\tau_k^{\mathrm{dB}}) \sim \mathcal{N}(0, 1)$ and $\boldsymbol{C}_k = \frac{\kappa}{N\tau_k}\operatorname{diag}(\rho^{|1-k|}, \ldots, \rho^{|N-k|})$ for the vector channels. Due to the exponential profile, the transmitter serves the kth user mainly by allocating power to the kth antenna element. The factor τ_k is a measure for the users' link quality. This model suits for coordinated multi-point systems and for multi-spotbeam SatCom, where the distance and antenna directivity, respectively, result in the above power profile [255]. An extreme case is $\rho = 0$. Then, each transmit antenna serves one user and the channels are orthogonal. Moreover, per-antenna constraints result in strictly larger MSEs than a sum power constraint if the attenuation values are distinct. The simulations have been performed for $\rho = 0.1$.

The max–min MSE balancing results in Sect. 4.7.1 are for the Gaussian channel setup and per-antenna power constraints with $P_\ell = P/N$. An analysis for distinct power constraints and the performance comparison for the exponential channel setup are then provided in Sect. 4.7.2. Finally, Sect. 4.7.3 discusses the ergodic RB results via the weighted MSE approximations.

4.7.1 Max–Min Mean Square Error Optimization Results

The numerical results for the max-min MSE optimizations is divided into three parts. The simulations with per-receiver MSE constraints and per-antenna power constraints in the first part are for the targets ε_k, $k = 1, \ldots, K$ from Table 4.1. For $K = 4$, only receiver 3 is active when $\varepsilon \geq 4/3$. Receivers 2 and 4 are served if the optimum level is $\varepsilon < 4/3$. If the optimum is $\varepsilon < 1$, also receiver 1 is served. For $K = 2$, only receiver 1 is served if $\varepsilon \geq 1$ and both users are served for $\varepsilon < 1$. Simulation results for weighted sum-MSE optimizations in the second part use functions of these values for their targets. The third part then provides a short analysis of the complexity for optimizing either individual or weighted sum-MSE constraints.

Table 4.1 Targets for MSE balancing with per-receiver constraints

# MSE constraints	MSE targets
$K = 4$	$\varepsilon_1 = 1, \varepsilon_2 = 3/4, \varepsilon_3 = 1/2, \varepsilon_4 = 3/4$
$K = 2$	$\varepsilon_1 = 1, \varepsilon_2 = 1/2$

4.7.1.1 Per-User Mean Square Error Balancing Results

This part analyzes the balancing results for $J = K$ Mean square error (MSE) constraints. Figure 4.3 compares the MSE balancing optimum of the ACS and the dual PG precoder optimization with that of the following schemes: balancing for perfect transmitter and receiver CSI, balancing with imperfect transmitter and receiver CSI, and perfect receiver and imperfect transmitter CSI with pre-fixed ZF beamforming, MRT beamforming, and RZF beamforming. The pre-fixed beamforming has been improved by post-processing power allocation. The figure shows the average performance for 100 randomly drawn channel mean realizations and 1000 channel realizations for each of the channel means to compute the perfect transmitter CSI results and the sample estimates for imperfect CSI.

Clearly, the balancing level ε for imperfect transmitter but perfect receiver CSI resides between the perfect transmitter CSI and imperfect receiver CSI curves. The perfect CSI curves fall unbounded for sufficiently accurate computations. In contrast, the imperfect CSI curves saturate due to the non-vanishing interference. This saturation level varies with the accuracy of the CSI, that is, it decreases with decreasing κ (see Fig. 4.3d). The performance of the fixed beamforming schemes with optimal power allocation is below the performance of the proposed ACS. While the performance for RZF beamforming is still close to the ACS result, the MRT design becomes only close for low SNR, and ZF beamforming is only tight in the asymptotic limit $P \to \infty$.

The saturation level for imperfect CSI also decreases when increasing the ratio N/K. This is especially visible for pre-fixed MRT beamforming. The reason is that the channel means are drawn from an Identically and independent distributed (i.i.d.) complex normal distribution. These vectors are asymptotically orthogonal when increasing N for fixed K. In other words, the MRT beamforming results asymptotically approach the ZF results for $N/K \to \infty$. The decreasing interference due to this effect is already visible for a moderate increase of N.

Only a subset of the per-antenna constraints is tight when the SNR is sufficiently high. This is seen in Fig. 4.4, which depicts the relative number of inactive per-antenna constraints for 100 channel mean realizations. A power constraint is markted as inactive if $\sum_{i=1}^{K} \boldsymbol{t}_i^{\mathrm{H}} \boldsymbol{A}_{i\ell} \boldsymbol{t}_i < 0.98 P_\ell$. All power constraints are active for $P \leq 10\,\mathrm{dB}$, while only one constraint is active for $P = 30\,\mathrm{dB}$. Increasing the transmit power for antennas with inactive constraints would cause unwanted interference in the high SNR regime.

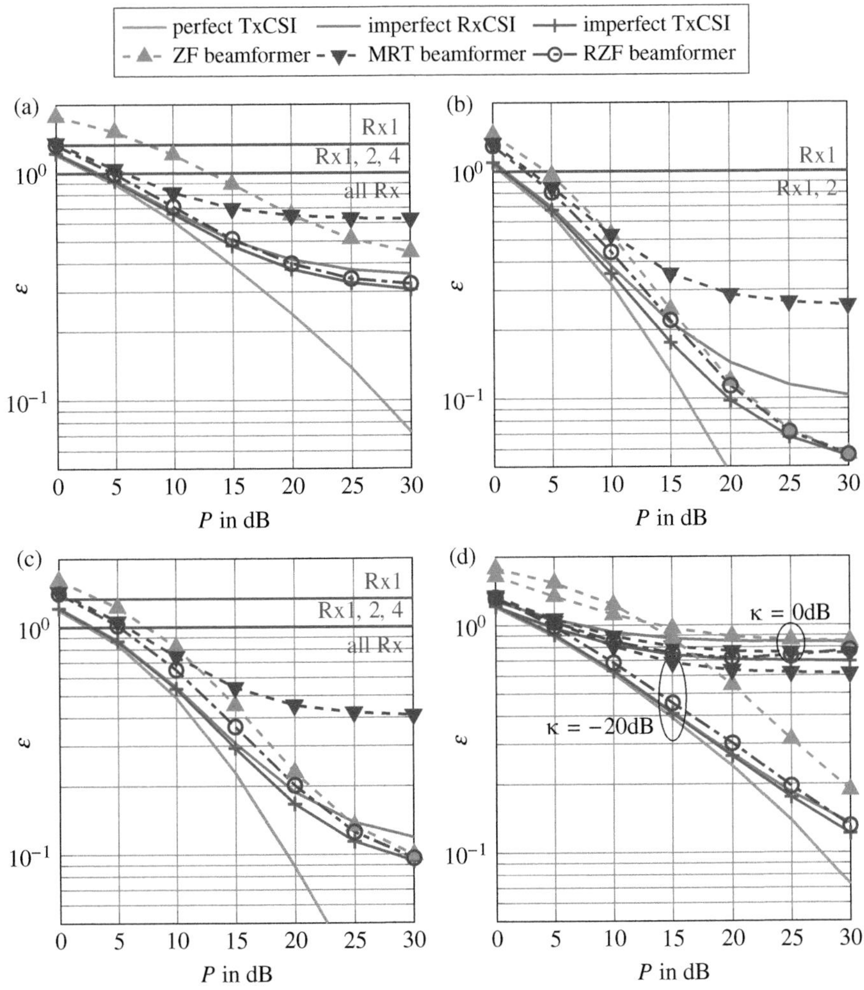

Fig. 4.3 Average MSE balancing result with per-antenna power constraints, K MSE constraints, and Rician-like fading channels, i.e., $\boldsymbol{h}_k \sim \mathcal{N}_{\mathbb{C}}(\bar{\boldsymbol{h}}_k, \boldsymbol{C}_k)$ with $\boldsymbol{C}_k = \frac{\kappa}{N}\mathbf{I}$. (**a**) $\kappa = -10$ dB, $K = 4$, and $N = 4$ antennas. (**b**) $\kappa = -10$ dB, $K = 2$, and $N = 4$ antennas. (**c**) $\kappa = -10$ dB, $K = 4$, and $N = 8$ antennas. (**d**) $\kappa = 0$ dB, -20 dB, $K = 4$, and $N = 4$ antennas

4.7.1.2 Per-user MSE Balancing Versus Weighted Sum-MSE Balancing

Next follows a comparison of the above per-user MSE balancing with an optimization of the weighted sums of the MSEs. The weighted sum MSE balancing with $J < K$ focuses on two scenarios for the targets $\varepsilon_j^{(J)}$, $j = 1, \ldots, J$ and weights $w_k^{(J)}$, $k = 1, \ldots, K$. Either the ratios of the weights and their associated targets are

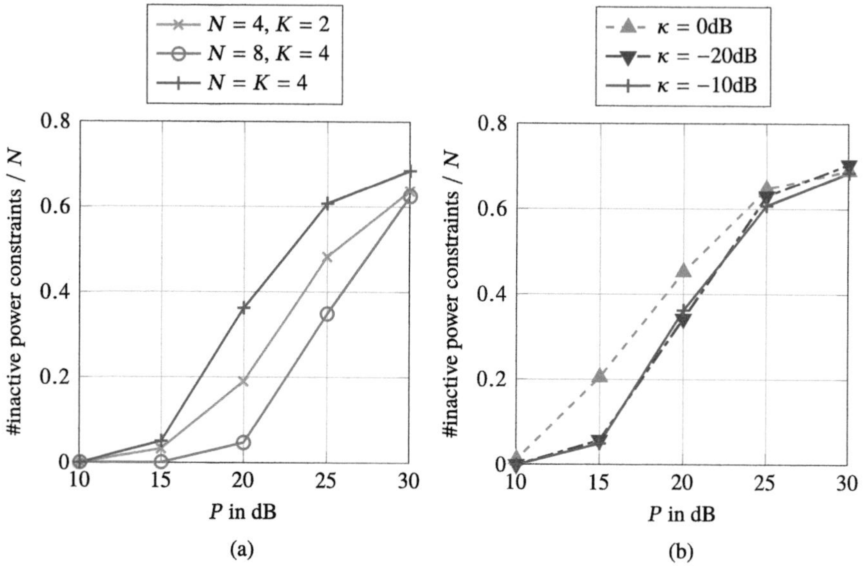

Fig. 4.4 Average percentage of inactive per-antenna constraints for the ACS results. (**a**) $\kappa = -10\,\text{dB}$. (**b**) $N = K = 4$

$$w_k^{(J)} \varepsilon_j^{(J),-1} = \varepsilon_k^{-1}, \quad k \in \mathcal{G}_j, \quad j = 1, \ldots J,$$

where the superscripts (J) indicate the number of MSE constraints and ε_k are the values of Table 4.1, or the weights are $w_k^{(J)} = 1$ for $k = 1, \ldots, K$ and

$$\varepsilon_j^{(J)} = K/J \sum_{k \in \mathcal{G}_j} \varepsilon_k, \quad j = 1, \ldots J.$$

The simulations are with $K = 4$ and $J = 2$, where the receivers are grouped as $\mathcal{G}_1^{(2)} = \{1, 2\}$ and $\mathcal{G}_2^{(2)} = \{3, 4\}$, and for $J = 1$ with $\mathcal{G}_1^{(1)} = \{1, 2, 3, 4\}$. Furthermore, the $K = 2$ two user scenario has been simulated with $J = 1$ and $\mathcal{G}_1^{(1)} = \{1, 2\}$. For both configurations, the MSE balancing optima for $J < K$ are approximately K/J times the result for $J = K$. To account for this fact, Fig. 4.5 presents the optimized balancing level $\varepsilon \cdot J/K$. The left plot depicts the $K = 4$ user scenario with $\kappa = -10\,\text{dB}$, while the right plot shows the $K = 2$ user scenario with $\kappa = 0\,\text{dB}$ and $\kappa = -10\,\text{dB}$.

The per-receiver balancing constraints with $J = K$ are stricter than the weighted sum MSE constraints for $J < K$. In particular, the individual MSE constraints for $J = K = 4$ are stricter than the constraints for $J = 2$, which are in turn stricter than the single weighted sum MSE constraint $J = 1$. Therefore, the weighted sum MSE optimization curves provide lower bounds for the performance of the per-user MSE balancing results. The optimization gain with (weighted) sum MSE

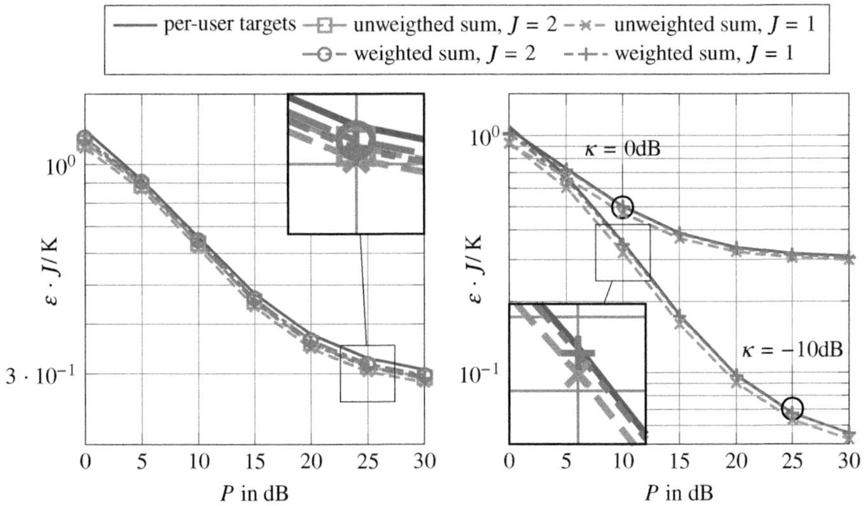

Fig. 4.5 Average weighted sum MSE balancing result with per-antenna power constraints. The number J of sum MSE constraints is varied. (**a**) $\kappa = -10\,\text{dB}$, $K = 4$, and $N = 4$ antennas. (**b**) $\kappa = 0\,\text{dB}, -10\,\text{dB}$, $K = 2$, and $N = 4$ antennas

constraints over per-receiver constraints is small and changes only slightly with the user configurations. This indicates that the sum-MSE optimization provides a good approximation for the balancing performance. The reason is that the channel conditions do not prefer any receiver on average and the targets ε_k are sufficiently close. The approximation for the weighted sum MSEs is slightly tighter than that with the unweighted sum MSEs. The two approximations are even identical if the targets ε_k are equal. However, only the weighted MSE approximation curve remains close if the differences between the individual targets ε_k are large, because it accounts for the influence of these targets.

4.7.1.3 Complexity Comparison for MSE Balancing

Using weighted sum MSE constraints instead of per-user constraints can reduce the computational complexity of the balancing optimization. The number of ACS iterations increases for these sum MSE optimizations and the required PG steps of the precoder updates remain almost the same compared to MSE balancing (see Fig. 4.6). However, finding the dual variable $\boldsymbol{\lambda} \in \mathbb{R}_+^J$ simplifies for decreasing J, e.g., $\lambda_1 = \tilde{\sigma}_1^{-2}$ for $J = 1$. Since this dual power allocation is performed for each evaluation of the objective function, this decreases the complexity of the PG precoder update. Hence, the sum MSE optimization is attractive for a low complex approximation of the MSE balancing performance when the number of users and antennas is large and the direct optimization is untractable.

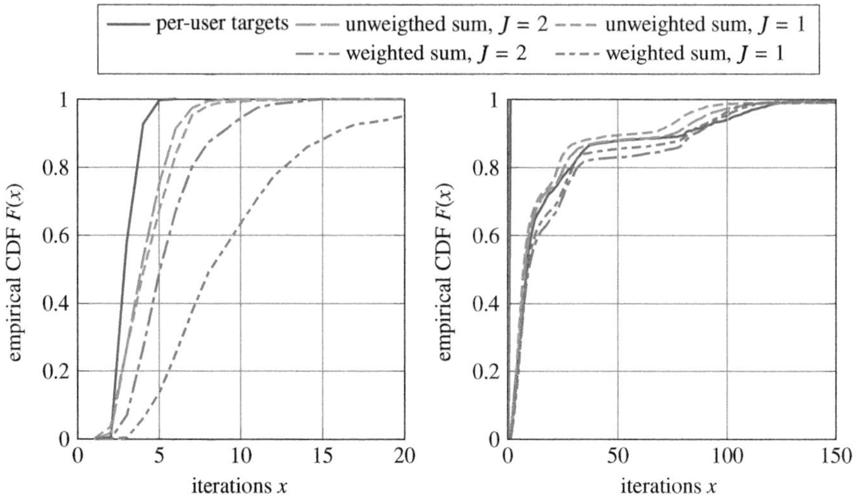

Fig. 4.6 Empirical CDFs of the number of iterations until convergence for the weighted sum MSE minimization with Rician fading $\kappa = -10\,\text{dB}$, $K = 4$, and $N = 4$. (**a**) Alternating convex search. (**b**) Precoder update

We compare the performance of the dual PG based beamformer update also with other precoder updates, particularly, the bisection over ε with a primal SOC power minimization in each iteration (cf. Sect. 4.3.1) and a dual fixed point method from [118] for finding $\boldsymbol{\mu}$. The fixed point approach updates the entries of $\boldsymbol{\mu}$ as

$$\mu_\ell^{(n+1)} = \frac{P_\ell^{-1} \mu_\ell^{(n)} \sum_{i=1}^K \boldsymbol{t}_i^{\mathrm{H}} \boldsymbol{A}_{i,\ell} \boldsymbol{t}_i}{\sum_{m=1}^L P_m^{-1} \mu_m^{(n)} \sum_{i=1}^K \boldsymbol{t}_i^{\mathrm{H}} \boldsymbol{A}_{i,m} \boldsymbol{t}_i}. \tag{4.119}$$

This update ensures strong duality for the dual beamformer update optimization in the convergence point, where $\mu_\ell \big(P_\ell - \sum_{i=1}^K \boldsymbol{t}_i^{\mathrm{H}} \boldsymbol{A}_{i,\ell} \boldsymbol{t}_i\big) = 0$.

All beamformer updates result approximately in the same average performance for the ACS (see Fig. 4.7), but the number of iterations for the dual update of $\boldsymbol{\mu}$ changes. For a comparison, all the methods had to achieve the optimum $\varepsilon^\star$ with an accuracy $\epsilon \leq 10^{-4}$. Figure 4.7b shows the empirical CDFs for the number of iterations until convergence of the beamformer updates. While the number of outer ACS steps is almost identical for the three beamformer updates, the iteration number for the beamformer methods strictly differs. The fixed point search with (4.119) requires a much larger number of iterations until convergence. In fact, convergence of the fixed point method fails for about 20% of the cases if the maximum number of iterations is 100. In contrast, convergence of the PG method is ensured by the step size selection. This makes the dual PG method an attractive alternative for the direct primal SOCP in the downlink.

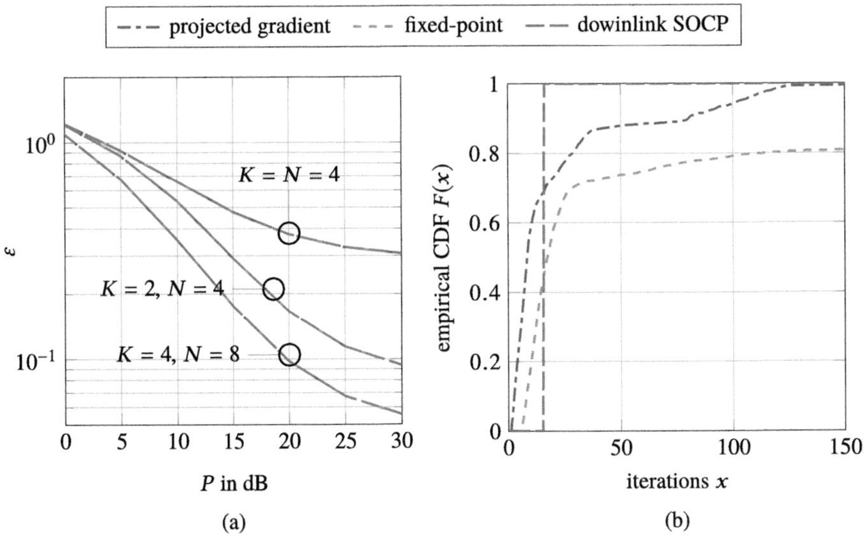

Fig. 4.7 Performance and empirical CDF of the number of outer iterations for the precoder update of the MSE balancing algorithm with $\kappa = -10\,\text{dB}$ and $K = N = 4$. (**a**) Alternating convex search. (**b**) Precoder update

4.7.2 Sum Mean Square Error Minimization Analysis

An analysis of the performance for different power constraints is based on the sum MSE minimization. This optimization was performed for per-antenna, per-array, and a sum power restriction, where $P = \sum_\ell P_\ell$ is varied between $-10\,\text{dB}$ and $20\,\text{dB}$.

Figure 4.8 depicts the minimum sum MSE for the Gaussian channel realizations and the exponential channel model with $N = 4$ and $N = 8$ antennas and $K = 2$ and $K = 4$ users. Four lines are seen for each configuration: for limited sum power P, per-array constraints with two antennas per array and $P_\ell = 2P/N, \ell = 1, \ldots, N/2$, per-antenna constraints with $P_\ell = P/N$, and for imperfect receiver CSI and the same per-antenna constraints. We computed the results with the ACS and the dual PG method.[40]

With the above power limits, the per-antenna constraints are stricter than the per-array constraints and the sum power constraint, which results in an increased sum MSE. This affects the exponential channel model, but results in only small performance differences for the Gaussian model (see Fig. 4.8a, b). For the Gaussian model, the sum power limitation provides a tight lower bound for the minimum MSE with per-array or per-antenna constraints. The performance curve for imperfect receiver CSI upper bounds all the curves with only imperfect transmitter CSI.

[40]The alternative updates via a dual SDP formulation provided the same performance.

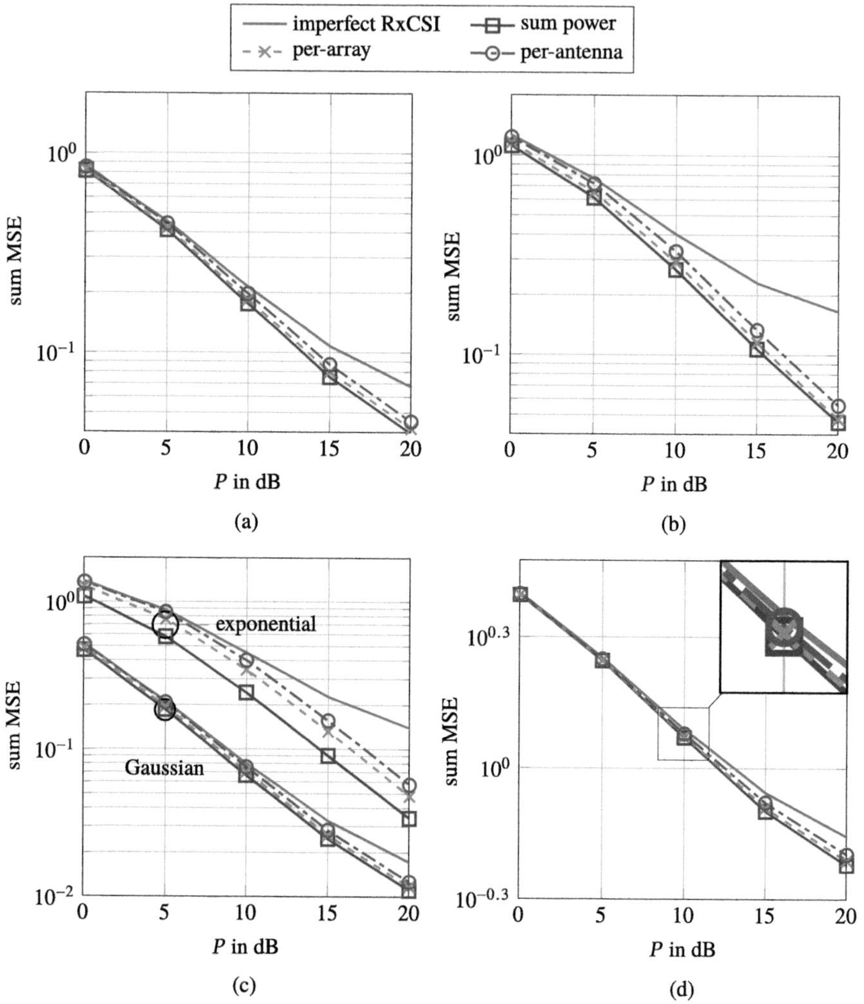

Fig. 4.8 Minimum sum MSE versus $P = \sum_{\ell=1}^{L} P_\ell$ with $P_\ell = L^{-1}P$ for per-antenna, per-array, and a single sum power constraint and Rician fading with $\kappa = -10$ dB. (**a**) Rician fading $N = 4$ antennas, $K = 2$ users. (**b**) Exponential profile $N = 4$ antennas, $K = 2$ users. (**c**) Both models $N = 8$ antennas, $K = 2$ users. (**d**) Rician fading $N = K = 4$ antennas and users

The figures show that the larger the K (for constant N), the tighter the sum power curve approximates the sum MSE for per-array or per-antenna constraints. On average, the Gaussian channel mean characteristics do not prefer any antenna. For the exponential channel model, the difference between the MSE curves for the sum power constraint and the per-antenna and per-array constraints increases with N (for constant K), while this difference decreases for the Gaussian channel model.

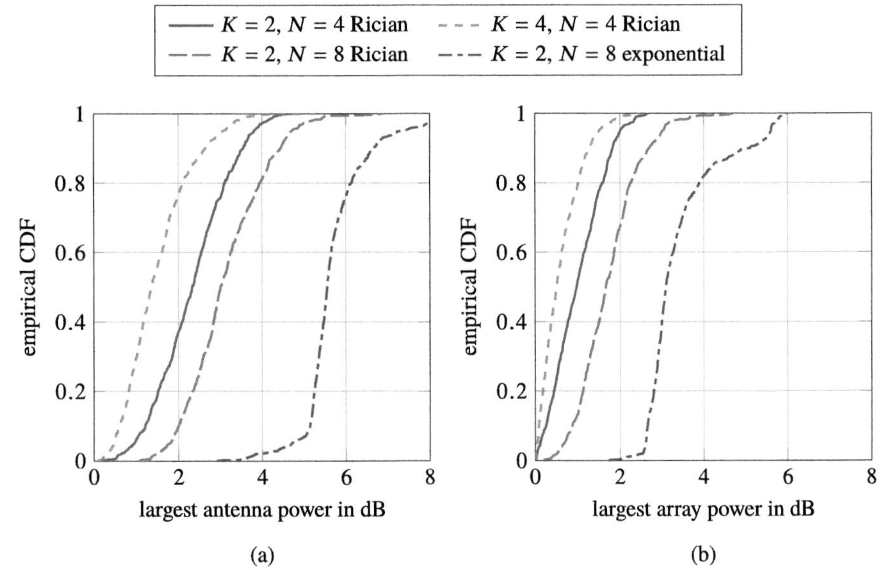

Fig. 4.9 Empirical CDFs of the largest per-antenna and per-array power in dB (normalized to P_ℓ) of the sum MSE minimization with a sum power constraint and either Rician fading with $\kappa = -10\,\text{dB}$ or an exponential channel model. (**a**) Sum vs. per-antenna power. (**b**) Sum vs. per-array power

Even though the sum MSE optimization with the less complex sum power constraint provides a good approximation for the results with the more realistic constraints, it highly increases the dynamic range of the per-antenna and per-array gains. Figure 4.9 depicts the empirical Cumulative distribution function (CDF) of the largest per-antenna and per-array power when minimizing the sum MSE based on a sum power constraint. The power in dB is normalized such that P_ℓ represents 0 dB.

Figure 4.9a focuses on the largest per-antenna power for the optimization with a sum power constraint. For $N = 4$ antennas and $K = 2$ users, a double of the per-antenna bound P_ℓ (i.e., 3 dB) is surpassed in more than 20% of the channel realizations for the Rician fading model. This 20% bound decreases to 2 dB for increasing the number of receivers to $K = 4$. The observation is in accordance with the conclusion that the sum power constrained MSE well approximates that with per-antenna constraints for sufficiently large K (cf. Fig. 4.9d). In contrast, the normalized 20% largest per-antenna powers of the sum power constrained optimization increase when increasing the number of transmit antennas to $N = 8$. While the increase is only to about 4 dB for the Gaussian model, 6 dB is surpassed with the exponential model. Then, the transmitter focuses most of the power to the two main antennas for serving the two users. Since the channels are additionally subject to shadow fading, the channel gains differ. This results in an unbalanced

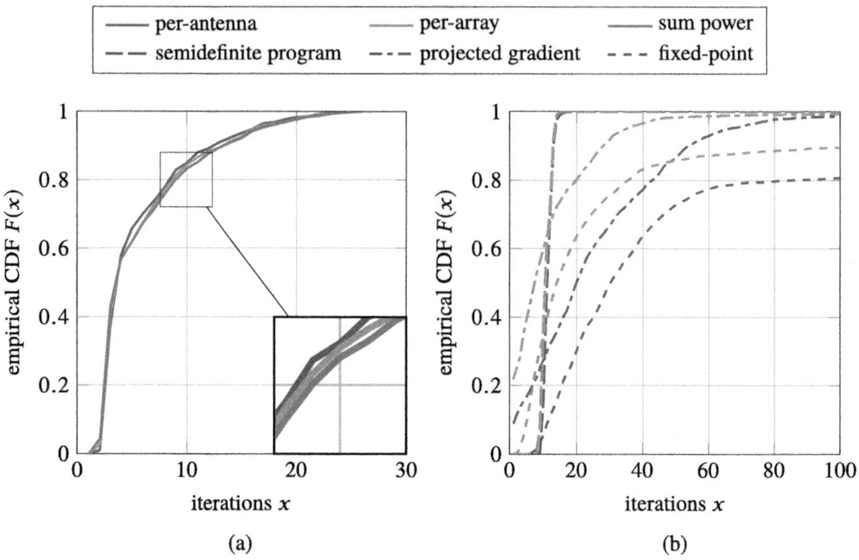

Fig. 4.10 Empirical CDFs of the number of ACS and beamformer update iterations for the sum MSE minimization with Rician fading $\kappa = -10\,\text{dB}$, $N = 4$, and $K = 2$. (**a**) Alternating convex search. (**b**) Precoder update

power distribution for the users' main serving antennas. The unbalanced power is the reason for the increased sum MSE gap between the results for the sum power and the per-antenna constraints in Fig. 4.8b.

The same ordering of the curves shows Fig. 4.9b for the largest per-array power. The x-axis values are compressed to about half of that from the largest per-antenna curves because the power can freely be distributed between the antenna pairs with a common constraint. Using per-array constraints instead of per-antenna constraints provides itself a good performance approximation and still limits the dynamic range of the transmit power per RF chain.

Nevertheless, the per-array constraints reduce the computational complexity for the beamformer optimizations compared to per-antenna constraints. This is important when scaling up the number of RF chains and transmit antennas, because the complexity of the precoder update increases with the number of constraints. In this context, Fig. 4.10 shows the empirical CDF for the number of iterations until convergence of the outer ACS and the inner precoder update for the various power limitations. The CDFs are for a channel setup with $N = 4$, $K = 2$, and the Rician-like channel model with $\kappa = -10\,\text{dB}$.

The number of ACS iterations is almost independent with respect to the following power constraints (see Fig. 4.10a): $L = 4$ per-antenna constraints, $L = 2$ per-array constraint, or a single sum power constraint. However, the complexity of the dual beamformer update, i.e., the search for $\boldsymbol{\mu}$, increases with the number of constraints L. Even though the dual SDP solution requires about the same number of interior-point iterations for per-antenna and per-array constraints (see Fig. 4.10b), the complexity per iteration increases in the number of variables [256], which is only $L + 1$ for (4.95).

For the dual PG method, the number of iterations increases from the per-array to the per-antenna constraints (see Fig. 4.10). For example, the PG algorithm converges in less than 20 iterations for 80% of the performed simulations with the per-array constraints, but requires more than 40 iterations for the per-antenna constraints. Thus, the computational complexity of this dual PG search for $\boldsymbol{\mu}$ increases at least linearly with the number of primal power requirements. The number of outer iterations for finding $\boldsymbol{\mu}$ is even larger for the reference fixed point update from [116]. This update for the outer search of the dual sum MSE minimization reads as[41]

$$\mu_\ell^{(n+1)} = \frac{\boldsymbol{p}^{\mathrm{T}}\boldsymbol{\mu}^{(n)}}{\boldsymbol{\delta}^{(n),\mathrm{T}}\boldsymbol{\mu}^{(n)}} \frac{\delta_\ell^{(n)}}{P_\ell} \mu_\ell^{(n)}, \tag{4.120}$$

where $\boldsymbol{p} = [P_1, \ldots, P_L]^{\mathrm{T}}$ and $\boldsymbol{\delta}$ is given by (4.107). Its convergence speed is much slower than that of the PG method, especially if there are inactive power constraints at the solution. More than a hundred iterations are required in 10% of the simulations for per-array constraints and this number increases to about 19% for per-antenna constraints. For this reason, the dual PG method is faster than the fixed point method [116], even though a fixed point update is computationally less complex than a PG update.

4.7.3 Approximate Rate Balancing Using MSE Optimizations

Section 4.6 showed that the dual uplink max–min MSE precoder update can also be exploited for RB with imperfect transmitter CSI. The associated performance results are shown in this section. The simulation setup consists of Gaussian channel, per-antenna power constraints, and rate constraints with targets $\rho_k = -\log_2(\varepsilon_k / \sum_{i=1}^{K} \varepsilon_i)$, $k = 1, \ldots, K$ and ε_k from Table 4.1.

Figure 4.11 depicts the RB optima as an average over 100 realizations for the channel characteristics, each corresponding to 1000 channel realizations for perfect CSI and the sample average approximations of the expectations. Besides the proposed ACS, the following schemes were simulated: the approximate balancing optimization, MSE balancing, pre-fixed schemes based on ZF, MRT, or RZF beamformers with optimized power allocation, and RB for perfect CSI.

Clearly, the performance for ergodic RB via weighted MSE balancing lies above the approximate MSE balancing result and below RB for perfect transmitter CSI. The approximate MSE balancing curves are tight for low to medium transmit power,

[41] This normalized power iteration is motivated by the complementary slackness conditions for the power constraints, i.e., $\mu_\ell P_\ell = \mu_\ell \boldsymbol{t}^{\mathrm{H}} \boldsymbol{A}_\ell \boldsymbol{t}$, $\ell = 1, \ldots, L$ need to be satisfied for any local solution.

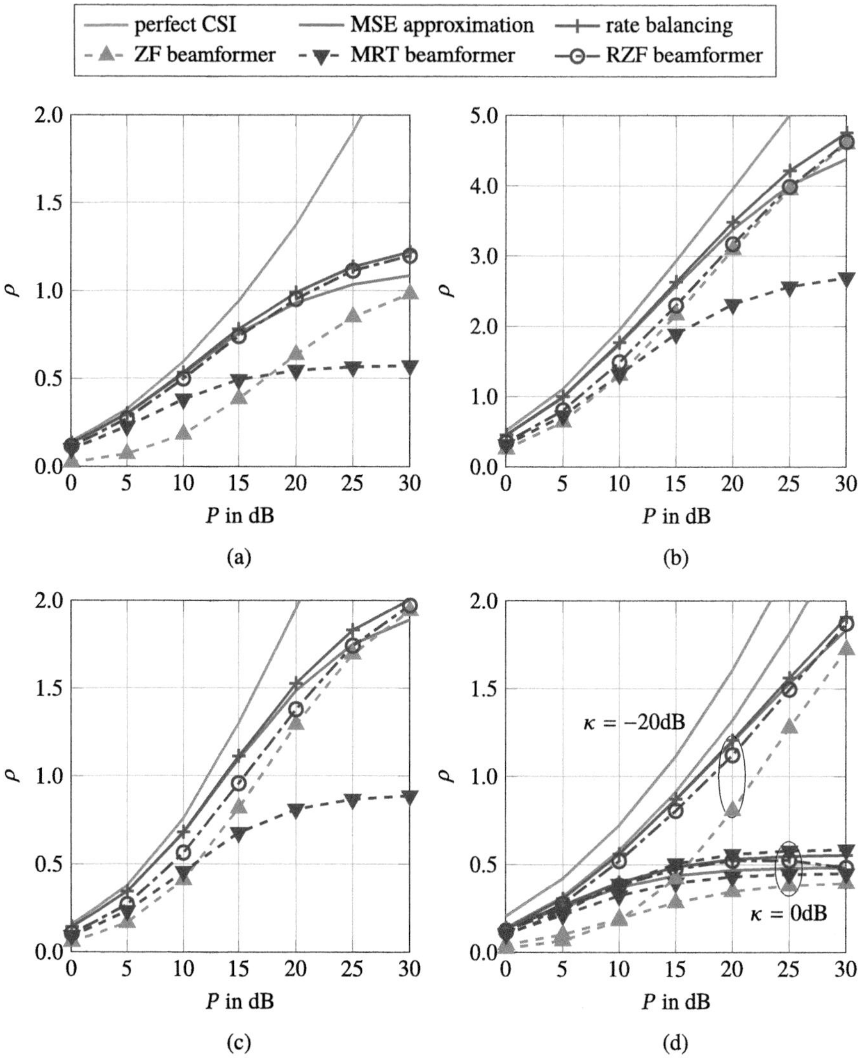

Fig. 4.11 Ergodic rate balancing and approximate rate balancing results with per-antenna power constraints for Rician-like fading, i.e., $\boldsymbol{h}_k \sim \mathcal{N}_{\mathbb{C}}(\bar{\boldsymbol{h}}_k, \boldsymbol{C}_k)$ with $\boldsymbol{C}_k = \frac{\kappa}{N}\mathbf{I}$. (**a**) $\kappa = -10\,\text{dB}$, $K = 4$, and $N = 4$ antennas. (**b**) $\kappa = -10\,\text{dB}$, $K = 2$, and $N = 4$ antennas. (**c**) $\kappa = -10\,\text{dB}$, $K = 4$, and $N = 8$ antennas. (**d**) $\kappa = 0\,\text{dB}$, $-20\,\text{dB}$ $K = 4$, and $N = 4$ antennas

e.g., for $P \leq 15\,\text{dB}$ in Fig. 4.11a, c. In the high transmit power regime, the gap between the MSE approximation and ergodic RB increases. The fixed beamforming schemes with optimized power allocation reside below the alternating beamformer searches. Similar to the results for MSE balancing (cf. Sect. 4.7.1), the performance for RZF beamforming is still close to the ACS design, while the MRT and ZF scheme are only close for low and very high transmit power, respectively.

All imperfect CSI curves saturate due to the non-vanishing interference with the additive channel error model. The saturation level increases with decreasing fading factor κ (see Fig. 4.11d) and when increasing the ratio N/K, due to the asymptotic orthogonality of the randomly drawn channel means. In contrast to the direct MSE balancing results for fixed weights, transmission to all users is active at the optimum of the RB optimization, even for very low transmit power.

Chapter 5
Outage Constrained Beamformer Design

While Chaps. 3 and 4 discuss ergodic robust beamforming, which implies a fast-fading channel model, this chapter focuses on slower fading. The coherence time shall be in the range of the transmit phase, such that the channel is either quasistatic or each transmit code word experiences only a few channel realizations. Transmitter CSI can still be imperfect, e.g., due to limited feedback and delayed CSI usage. In this case, non-robust transmit strategies are unable to ensure reliable data transmission—an unknown number of outages occur when the transmitter sends information at the imposed rates. An *outage* defines the event that the channel is too bad to decode error-free at the transmit data rate. The chance of such an event is the outage probability [1, Section 5.4.1]:

$$\Pr(r_k(\boldsymbol{t}, \boldsymbol{h}_k) < \rho_k). \tag{5.1}$$

Here, $r_k(\boldsymbol{t}, \boldsymbol{h}_k)$ is a random variable due to the random channel $\boldsymbol{h}_k$ and ρ_k is the imposed transmit data rate. Restricting (5.1) to lie below a threshold $\epsilon_k \in (0, 1)$ provides a statistically robust formulation for the maximum supported data rate [257]—the ϵ-outage rate:

$$\varrho_k(\boldsymbol{t}) = \max\{\varrho \in \mathbb{R}_+ : \Pr(r_k(\boldsymbol{t}, \boldsymbol{h}_k) \geq \varrho) \geq 1 - \epsilon_k\}, \qquad k = 1, \ldots, K. \tag{5.2}$$

Relying on (5.2) for QoS and RB optimization is appropriate when reliable communication is a central issue (cf. [1, Section 5.4]) and ergodic rate constraints are insufficient to restrict the frequency for outages. Similar to [36, 44, 47, 51, 138], the QoS based beamformer design under this ϵ-outage rate metric reads as

$$\min_{p \geq 0,\, \boldsymbol{t}} p \quad \text{s.t.} \quad p^{-1}\boldsymbol{t} \in \mathcal{P}, \quad \boldsymbol{\varrho}(\boldsymbol{t}) \in \mathcal{R}, \tag{5.3}$$

A. Gründinger, *Statistical Robust Beamforming for Broadcast Channels and Applications in Satellite Communication*, Foundations in Signal Processing, Communications and Networking 22, https://doi.org/10.1007/978-3-030-29578-3_5

with $\boldsymbol{\varrho}(\boldsymbol{t}) = [\varrho_1(\boldsymbol{t}), \ldots, \varrho_K(\boldsymbol{t})]^{\mathrm{T}}$, when the rate targets are mandatory. If the transmit power is limited instead, the focus is on the joint maximization of the ϵ-outage data rates:

$$\max_{\rho,\, \boldsymbol{t}} \rho \quad \text{s.t.} \quad \boldsymbol{t} \in \mathcal{P}, \quad \rho^{-1}\boldsymbol{\varrho}(\boldsymbol{t}) \in \mathcal{R}. \tag{5.4}$$

After repeating basic results on outage constrained programming in Sect. 5.1, Sect. 5.2 discusses the multiplicative fading assumption for the random vector channels. Then follows a study of the outage probability for additive fading vector channels in Sect. 5.3. A variable power allocation strategy is proposed to improve the performance for pre-fixed beamforming. The power allocation is also a tool to test QoS feasibility of various beamforming strategies. Tractable convex approximations for the beamformer design with chance constraints are discussed in Sects. 5.5 and 5.6. The numerical results in Sect. 5.7 compare these strategies.

5.1 Chance-Constrained Optimization

Mathematical literature (e.g., [137, 254, 258]) denotes the probability constraints (5.2) as chance constraints and optimizations with such restrictions as chance constrained programs. Some basic results for chance constrained programming and applications for downlink communication are reviewed next.

5.1.1 *Basic Mathematical Background*

Consider the event $\mathcal{A}_j(\boldsymbol{x}) = \{g_j(\boldsymbol{x}, \boldsymbol{\xi}) \geq 0\}$, where $g_j : \mathbb{R}^p \times \mathbb{R}^q \to \mathbb{R}$ is a continuous function of the decision vector $\boldsymbol{x} \in \mathbb{R}^p$ and the random $\boldsymbol{\xi} \in \mathbb{R}^q$. The chance constraint for satisfying $\mathcal{A}_j(\boldsymbol{x})$ at least with probability $1 - \epsilon_j$ reads as[1]

$$\mathrm{p}_j(\boldsymbol{x}) = \Pr\left(g_j(\boldsymbol{x}, \boldsymbol{\xi}) \geq 0\right) \geq 1 - \epsilon_j \quad \epsilon_j \in (0, 1). \tag{5.5}$$

A chance constrained program with M such probability constraints has the form [254, Chapter 4]:[2]

$$\min_{\boldsymbol{x} \in C} f(\boldsymbol{x}) \quad \text{s.t.} \quad \mathrm{p}_j(\boldsymbol{x}) \geq 1 - \epsilon_j, \quad j = 1, \ldots, M, \tag{5.6}$$

where $f : \mathbb{R}^p \to \mathbb{R}, \boldsymbol{x} \mapsto f(\boldsymbol{x})$ and C shall be convex if not stated otherwise.

---

[1]The formulation is short hand for the probability $\Pr(\mathcal{A}_j(\boldsymbol{x}))$ with respect to the distribution of $\boldsymbol{\xi}$.

[2]The emphasis is here on probability constraints for scalar stochastic inequalities (5.5), because the downlink beamformer design considers unicast data transmission to independent receivers. An analysis of chance constraints based on vector inequalities is provided by [254, Chapter 4].

Solutions for stochastic programs strongly depend on the probability measure and the properties of the stochastic inequality. Therefore, [137, 254, 258] categorize chance constraints with respect to the type of the stochastic function $g_j(\boldsymbol{x}, \boldsymbol{\xi})$ and the distribution of $\boldsymbol{\xi}$. Deterministic reformulations are available for *separable functions*

$$g_j(\boldsymbol{x}, \boldsymbol{\xi}) = g_{j1}(\boldsymbol{x}) - g_{j2}(\boldsymbol{\xi}), \tag{5.7}$$

which separate the decision variable $\boldsymbol{x}$ from $\boldsymbol{\xi}$ via the functions $g_{j1} : \mathbb{R}^p \to \mathbb{R}$ and $g_{j2} : \mathbb{R}^q \to \mathbb{R}$ [137, Section 3.1]. Then, existence of a feasible $\boldsymbol{x}$ for (5.5) implies

$$g_{1j}(\boldsymbol{x}) \geq q_j, \quad q_j = \inf\{q \in \mathbb{R}^M : \Pr(q \geq Z_j) \geq 1 - \epsilon_j\}, \tag{5.8}$$

where q_j is the $1 - \epsilon_j$ quantile induced by the random variable $Z_j = g_{j2}(\boldsymbol{\xi})$. If the CDF $F_{Z_j} : \mathbb{R} \to [0, 1]$, $F_{Z_j}(q_j) = \Pr(q_j \geq Z_j)$ is continuous and strictly increasing, the $1 - \epsilon_j$ quantile q is defined by the inverse CDF $F_{Z_j}^{-1} : [0, 1] \to \mathbb{R}$ as $q = F_{Z_j}^{-1}(1 - \epsilon_j)$ [254, Section 4.3.2]. Therewith, (5.5) transforms to the equivalent deterministic constraint

$$g_{1j}(\boldsymbol{x}) \geq q. \tag{5.9}$$

Deterministic reformulations of (5.5) are also available for some non-separable functions. For example, with $g_j(\boldsymbol{x}, \boldsymbol{\xi}) = \sum_{i=1}^{p} a_{ij}(\boldsymbol{x})\xi_i - b_j$, the constraint

$$\Pr\left(\boldsymbol{a}_j^{\mathrm{T}}(\boldsymbol{x})\boldsymbol{\xi} \geq b_j\right) \geq 1 - \epsilon_j \tag{5.10}$$

defines a convex set if $[\boldsymbol{a}_j(\boldsymbol{x})]_i = a_{ij}(\boldsymbol{x})$ is affine in $\boldsymbol{x}$, $\boldsymbol{\xi}$ has a log-concave symmetric distribution, and $\epsilon_j \in [1/2, 1]$ [259, 260].[3] Furthermore, let $\boldsymbol{\xi} \sim \mathcal{N}(\boldsymbol{m}, \boldsymbol{C})$ and $\Phi : \mathbb{R} \to [0, 1]$ be the CDF of the normal distribution. Then, (5.5) becomes

$$\mathrm{p}_j(\boldsymbol{x}) = \Phi\left(\frac{\boldsymbol{a}_j(\boldsymbol{x})^{\mathrm{T}}\boldsymbol{m}_j - b_j}{\|\boldsymbol{C}_j^{1/2}\boldsymbol{a}_j(\boldsymbol{x})\|_2}\right) \geq 1 - \epsilon_j.$$

With the inverse CDF $q_j = \Phi^{-1}(1 - \epsilon_j)$, an equivalent deterministic formulation of (5.10) reads as

$$q_j \|\boldsymbol{C}^{1/2}\boldsymbol{a}_j(\boldsymbol{x})\|_2 \leq b_j - \boldsymbol{a}_j(\boldsymbol{x})^{\mathrm{T}}\boldsymbol{m}, \tag{5.11}$$

which is a SOC constraint if $\boldsymbol{a}_j : \mathbb{R}^p \to \mathbb{R}$ is affine in $\boldsymbol{x}$.

[3] The references assume $\boldsymbol{a}_j(\boldsymbol{x}) = \boldsymbol{x}$ for the Probabilistically constrained linear inequality (5.10).

Statements regarding convexity also exist for other non-separable stochastic functions. If g_j is jointly quasiconcave in $\boldsymbol{x}$ and $\boldsymbol{\xi}$, and $\boldsymbol{\xi}$ induces an α-concave probability measure $\Pr(\boldsymbol{\xi} \in \mathcal{A}_j(\boldsymbol{x}))$,[4] p_j will be quasiconcave [254, Theorem 4.39]. Hence, (5.5) has a convex representation for this case. Nevertheless, solving (5.6) generally requires repeated numerical integrations (cf. [254, Section 4.4.2]), which is intractable for many design scenarios. Therefore, state-of-the-art robust optimizations focus on conservative deterministic approximations for (5.5).

Another important non-separable function $g_j : \mathbb{C}^{NK} \times \mathbb{C}^N \to \mathbb{R}_+$ for chance constrained programming in wireless communication is

$$g_j(\boldsymbol{x}, \boldsymbol{\xi}) = \boldsymbol{\xi}^{\mathrm{H}} \boldsymbol{Q}(\boldsymbol{x})\boldsymbol{\xi} + \mathrm{Re}\left(\boldsymbol{a}^{\mathrm{H}}(\boldsymbol{x})\boldsymbol{\xi}\right) + b(\boldsymbol{x}), \tag{5.12}$$

where $\boldsymbol{Q}(\boldsymbol{x})$, $\boldsymbol{a}(\boldsymbol{x})$, and $b(\boldsymbol{x})$ are quadratic forms in $\boldsymbol{x} \in \mathbb{C}^{NK}$ and $\boldsymbol{\xi} \in \mathbb{C}^N$ is complex normal distributed, i.e., $\boldsymbol{\xi} \sim \mathcal{N}_{\mathbb{C}}(\mathbf{0}, \mathbf{I})$. The resulting chance constraint can be non-convex even if $g_j(\boldsymbol{x}, \boldsymbol{\xi}) \geq 0$ has a convex reformulation in $\boldsymbol{x}$. Moreover, computing $\mathrm{p}_j(\boldsymbol{x})$ from (5.5) requires a numerical integration [261, 262]. This makes (5.6) intractable for state-of-the-art convex programming solvers. To overcome this issue, several *conservative* approximations were proposed in the last years [35, 36, 44, 47, 51, 138, 169, 263, 264].[5] We distinguish two categories for the beamformer design in this work:

- *deterministic uncertainty models* for the unknown $\boldsymbol{\xi}$ and
- *probability bounds using moments* of $g_j(\boldsymbol{x}, \boldsymbol{\xi})$ or a related function.

Uncertainty models restrict $\boldsymbol{\xi}$ to reside in a pre-defined set $\mathcal{U}$. The probability of the joint event $\mathcal{A}_j(\boldsymbol{x}) \cap \mathcal{B}_j$, with $\mathcal{B}_j = \{\boldsymbol{\xi} \in \mathcal{U}\}$, is then an inner restriction for the chance constraint (5.5), i.e.,

$$\Pr(\mathcal{A}_j) \geq \Pr(\mathcal{A}_j(\boldsymbol{x}) \cap \mathcal{B}_j) = \Pr(\mathcal{A}_j(\boldsymbol{x})|\mathcal{B}_j)\Pr(\mathcal{B}_j) \geq 1 - \epsilon_j.$$

Setting $\Pr(\mathcal{B}_j) = 1 - \epsilon_j$ and $\Pr(\mathcal{A}_j(\boldsymbol{x})|\mathcal{B}_j) = 1$ imposes the uncertainty constraint

$$g_j(\boldsymbol{x}, \boldsymbol{\xi}) \geq 0, \quad \forall \boldsymbol{\xi} \in \mathcal{U}. \tag{5.13}$$

The degree of freedom is in the choice of $\mathcal{U}$. A tractable set for $\boldsymbol{\xi} \sim \mathcal{N}_{\mathbb{C}}(\mathbf{0}, \mathbf{I})$ is the convex ball $\mathcal{U} = \{\|\boldsymbol{\xi}\|_2 \leq d\}$, for example. Therewith, (5.6) becomes a quadratically constrained problem with norm bounded uncertainty [258, Chapter 6], [265]. This

[4] Generalized concavity theory is an utmost important tool for analyzing probability constraints, because the CDF of a random variable with an α-concave PDF is a quasiconcave function under mild restrictions on the concavity parameter α. The theory for this result and further analysis on concave probability measures for stochastic programming is provided by [254, Chapter 4].

[5] Wang [36] provides a summary on chance constrained programming for downlink optimization.

uncertainty can be removed via the S-lemma [103] or, if $g_j(\boldsymbol{x}, \boldsymbol{\xi}) \geq 0$ has a SOC reformulation, by the LMI constraint of [266, 267].[6]

Moment based bounds are obtained via Markov's and related inequalities. Markov's inequality bounds the probability of $f(\boldsymbol{x}, \boldsymbol{\xi}) \geq a \in \mathbb{R}_+$ exceeding as [211, Proposition 5.4]

$$\Pr\left(f(\boldsymbol{x}, \boldsymbol{\xi}) \geq a\right) \leq \frac{\mathrm{E}[f(\boldsymbol{x}, \boldsymbol{\xi})]}{a}, \qquad f(\boldsymbol{x}, \boldsymbol{\xi}) \in \mathbb{R}_+ \tag{5.14}$$

when $f : \mathbb{C}^{NK} \times \mathbb{C}^N \to \mathbb{R}_+$ and $a \in \mathbb{R}_+$ are from a reformulation of $g_j(\boldsymbol{x}, \boldsymbol{\xi}) \geq 0$. Bechar [172] derived two alternative "Bernstein-type" deviation bounds for (5.12):[7]

$$\Pr\left(g_j(\boldsymbol{x}, \boldsymbol{\xi}) - \mathrm{E}[g_j(\boldsymbol{x}, \boldsymbol{\xi})] \leq -\sqrt{s}\sqrt{2\|\boldsymbol{Q}(\boldsymbol{x})\|_\mathrm{F}^2 + \|\boldsymbol{a}(\boldsymbol{x})\|_2^2} - s\lambda_{\min}^+(\boldsymbol{Q}(\boldsymbol{x}))\right) \leq \mathrm{e}^{-s}, \tag{5.15}$$

$$\Pr\left(g_j(\boldsymbol{x}, \boldsymbol{\xi}) - \mathrm{E}[g_j(\boldsymbol{x}, \boldsymbol{\xi})] \geq \sqrt{s}\sqrt{2\|\boldsymbol{Q}(\boldsymbol{x})\|_\mathrm{F}^2 + \|\boldsymbol{a}(\boldsymbol{x})\|_2^2} + s\lambda_{\max}^+(\boldsymbol{Q}(\boldsymbol{x}))\right) \leq \mathrm{e}^{-s}, \tag{5.16}$$

where $\lambda_{\min}^+(\boldsymbol{A}) = \min\{\lambda \in \mathbb{R}_+ : \lambda\mathbf{I} + \boldsymbol{A} \succeq \mathbf{0}\}$ and $\lambda_{\max}^+(\boldsymbol{A}) = \min\{\lambda \in \mathbb{R}_+ : -\lambda\mathbf{I} + \boldsymbol{A} \preceq \mathbf{0}\}$. The name stems from its connection to Bernstein's inequality [269, Proposition 2.9]. These bounds allow to replace the probability inequality (5.5) by an ergodic constraint see Sects. 5.6.2 and 5.6.3.

An alternative approximation by Ben-Tal et al. [270, Section 10.4.2] results from the chance constraint of a linearly perturbed LMI if the inequality $g_j(\boldsymbol{x}, \boldsymbol{\xi}) \geq 0$ can be recast as a SOC in $\boldsymbol{\xi}$. The Vysochanskij–Petunin's inequality employed by [44] is for the special case where $\boldsymbol{a} = \mathbf{0}$, $\boldsymbol{Q}(\boldsymbol{x}) \in \mathcal{S}_+^q$, and $b(\boldsymbol{x}) < 0$ for (5.12).

5.1.2 Chance Constraints in Downlink Communication

Separable stochastic functions are found in PtP communication scenarios [1, Section 5.4.1]. For example, let

$$r = \log_2(1 + |h|^2\,\mathrm{SNR}) \tag{5.17}$$

be the rate of a scalar PtP channel, where $h \in \mathbb{C}$ denotes the random scalar channel. The constraint $\Pr(r \geq \rho) \geq 1 - \varepsilon$ has the deterministic counterpart [30, Section III]

$$\rho \leq \varrho = \log_2(1 + F_Z^{-1}(1 - \varepsilon)\,\mathrm{SNR}), \tag{5.18}$$

[6] Lorentz positive maps (LPMs) are linear mappings of vectors from a SOC into vectors of another SOC.

[7] A related bound for positive semidefinite $\boldsymbol{Q}(\boldsymbol{x}) \in \mathcal{S}_+$ is found in [268].

where F_Z^{-1} is the inverse CDF induced by $Z = |h|^2$. If $h \sim \mathcal{N}_{\mathbb{C}}(1, \sigma_\xi^2)$, Z is non-central chi-square distributed, i.e., $2\sigma_\xi^{-2} Z \sim \mathcal{X}_2^2(2\sigma_\xi^{-2})$, and $q = F_Z^{-1}(1-\varepsilon)$ is obtained via numerical integration [271, Section 29.8]. Its value enhances the noise power in (5.18), while the logarithmic expression remains.[8] Such a noise enhancement also occurs in the multi-user downlink with multiplicative fading (see Sect. 5.2).

Such deterministic formulations are unavailable for the vector BC with additive channel errors, where the probability constraints for the rate of receiver k read as

$$\Pr\left(\log_2\left(1 + \frac{|\boldsymbol{h}_k^{\mathrm{H}} \boldsymbol{t}_k|^2}{\sum_{i \neq k} |\boldsymbol{h}_k^{\mathrm{H}} \boldsymbol{t}_i|^2 + 1}\right) \geq \rho_k\right) \geq 1 - \epsilon_k. \tag{5.19}$$

The inner stochastic inequality remains non-separable even with pre-fixed ZF beamformers and only a power allocation that scales the beamformers. Such a power allocation for the multi-user downlink was also considered by [44] using an MMSE approximation for the rate and Vysochanskij–Petunin's inequality for approximating the chance constraints by deterministic requirements. Instead, Sect. 5.4 presents a power allocation for the QoS and RB optimizations based on the numerical probability computation from Sect. 5.3. This power allocation serves as a post-processing for the beamformer design with approximations for the probability constraints. Its solution satisfies all constraints $\varrho_j(\boldsymbol{t}) \geq 1 - \epsilon_j$ with equality for a sum power limitation. As a consequence, the proposed optimum by any conservative QoS optimization is reduced and the proposed rate of the RB problem is increased. The power allocation also provides a feasibility test for given beamformers using the fixed point map from [92] (see Sect. 5.4).

Utilizing the spatial multiplexing advances of the multi-antenna transmitter still requires a conservative beamformer design because repeated numerical integrations are often intractable. To this end, the previous chance constraint approximations were applied in literature [35, 36, 47, 104, 105, 138, 168–170]. Vorobyov et al. [168] have formulated a probability constraint for the useful signal power, i.e.,

$$\Pr\left(\mathrm{Re}(\boldsymbol{h}_k^{\mathrm{H}} \boldsymbol{t}_k) \geq \sqrt{\mathrm{S}_k}\right) \geq 1 - \epsilon_k, \tag{5.20}$$

and neglected the uncertainty for the sample mean of the total received signal power. For a Gaussian channel, the probability constraint matches (5.10). Thus, (5.20) features the SOC reformulation

$$\Phi^{-1}(1 - \epsilon_k) \|\boldsymbol{C}_k^{1/2} \boldsymbol{t}_k\|_2 \leq 2\left(\sqrt{\mathrm{S}_k} - \boldsymbol{t}_k^{\mathrm{H}} \bar{\boldsymbol{h}}_k\right). \tag{5.21}$$

The relation of (5.21) to robust uncertainty beamforming is analyzed by [35, 104].[9]

---

[8] A similar ϵ-outage rate expression (5.18) is also valid for vector PtP channels and additive fading, e.g., see [1, Section 5.4.2] for the multi-antenna receiver and [201] for the multi-antenna transmitter case.

[9] Chalise et al. [263] alternatively modeled the sample mean of the SINR with (5.10).

Uncertainty reformulations provide tractable approximations for the chance constraints of (5.3) and (5.4) if reliability is crucial. Shenouda and Davidson [47, 138] have exploited the deterministic approximation of a LMI reformulation for SINR requirements that is affine in the zero-mean Gaussian channel errors [270, Chapter 10]. This approximation can be too restrictive for the beamformer design [36, 144]. A somewhat improved performance provides the uncertainty formulation by [36, 105, 265], where the random channels are restricted to reside in a sphere around their estimates. The convex reformulation via the S-Lemma and an SDR has inspired further research on uncertainty constraints, e.g., [13, 105, 272]. The approximation steps for this reformulation are briefly recast in Sect. 5.5.

The approximation with the spherical shaped uncertainty region is still inaccurate. This motivates alternative uncertainty bounds. For example, Sect. 5.5 restricts the channel error that is orthogonal to the mean. The resulting bound promises performance gains for QoS and RB optimizations and increases the feasibility range for the QoS problem in comparison to the spherical uncertainty bound, while keeping the probability computation independent of the beamformer design.

Further gains in approximation accuracy were obtained from Wang et al. [36] by bounding the outage probability (5.19) with (5.15). This Bernstein-type inequality reformulation is recast in Sect. 5.6. The resulting constraint is again non-convex in the beamformers, but has a convex formulation after the SDR of $\boldsymbol{Q}_i = \boldsymbol{t}_i \boldsymbol{t}_i^{\mathrm{H}}, i = 1, \ldots, K$. Alternative direct beamformer optimizations are developed in Sect. 5.6.3. For example, Markov's bound relates the chance-constrained optimizations with the average MSE optimizations from Chapter 4. MSE optimizations are also obtained with the second Bernstein-type inequality (5.16). The RB performance of this scheme is a bit smaller than for the first Bernstein-type inequality approximation, but it outperforms the uncertainty approximations if the number of receivers is sufficiently large (see Sect. 5.7.3).

Remark 5.1 The simple form for the PtP ϵ-outage rate also motivated an outage analysis for other systems, e.g., interference channels [93, 273, 274], relay channels [30, 275, 276], and cognitive radio [277], that are beyond the scope of this work.

5.2 Multiplicative Fading Example

The multiplicative channel error model $\boldsymbol{h}_k = (1+\xi_k)\bar{\boldsymbol{h}}_k$ also features a deterministic expression for the ϵ-outage rate (5.2), which is like the perfect CSI rate, namely

$$\varrho_k(\boldsymbol{t}) = \log_2\left(1 + \frac{|\bar{\boldsymbol{h}}_k^{\mathrm{H}} \boldsymbol{t}_k|^2}{q_k + \sum_{i\neq k} |\bar{\boldsymbol{h}}_k^{\mathrm{H}} \boldsymbol{t}_i|^2}\right), \tag{5.22}$$

but with the channel's characteristics $\bar{\boldsymbol{h}}_k$ and an enhanced noise variance $q_k \in \mathbb{R}_+$. This result follows via separating the stochastic constraint $r_k(\boldsymbol{t}, \boldsymbol{h}_k) \geq \varrho$ into functions of $\boldsymbol{t}$ and $1 + \xi_k$ (cf. Sect. 5.1). In particular, the constraint[10]

$$\log_2\left(1 + \frac{|\bar{\boldsymbol{h}}_k^{\mathrm{H}}\boldsymbol{t}_k|^2}{\zeta_k + \sum_{i \neq k}|\bar{\boldsymbol{h}}_k^{\mathrm{H}}\boldsymbol{t}_i|^2}\right) \geq \varrho \tag{5.23}$$

can be separated into the random variable $\zeta_k = (1 + \xi_k)^{-2} \in \mathbb{R}_+$ and

$$g_{1k}(\boldsymbol{t}) = (2^{\varrho} - 1)^{-1}|\bar{\boldsymbol{h}}_k^{\mathrm{H}}\boldsymbol{t}_k|^2 - \sum_{i \neq k}|\bar{\boldsymbol{h}}_k^{\mathrm{H}}\boldsymbol{t}_i|^2 \geq \zeta_k. \tag{5.24}$$

Therewith, the probability (5.1) is the CDF $F_{\zeta_k} : \mathbb{R}_+ \to [0, 1]$ of ζ_k at $g_{k1}(\boldsymbol{t})$, i.e.,

$$\Pr\left(r_k(\boldsymbol{t}, (1 + \xi_k)\bar{\boldsymbol{h}}_k) \geq \varrho\right) = \Pr\left(g_{k1}(\boldsymbol{t}) \geq \zeta_k\right) = F_{\zeta_k}\left(g_{k1}(\boldsymbol{t})\right), \tag{5.25}$$

such that the chance constraint (5.19) has the deterministic counterpart [cf. (5.9)]

$$g_{k1}(\boldsymbol{t}) \geq q_k = F_{\zeta_k}^{-1}(1 - \epsilon_k).$$

Reformulating this constraint via the inverse operations that led to (5.24), we obtain the same data-rate constraint as (5.23), but with $q_k \in \mathbb{R}_+$ replacing ζ_k. Obviously, the maximum ϱ satisfying this constraint is the ϵ-outage rate (5.22).

Two examples are for a log-dB-normal distributed ζ_k and the Gaussian error model. For the log-dB-normal distribution model, the CDF of $\zeta_{k,\mathrm{dB}} = 10\log_{10}(\zeta_k)$, with $\ln(\zeta_{k,\mathrm{dB}}) \sim \mathcal{N}(\mu_k, \sigma^2_{\zeta_{k,\mathrm{dB}}})$, is

$$F_{\zeta_{k,\mathrm{dB}}}(z) = \Phi\left(\frac{\ln(z) - \mu_k}{\sigma_{\zeta_{k,\mathrm{dB}}}}\right),$$

where $\Phi : \mathbb{R} \to [0, 1]$ is the CDF of the normal distribution (A.55). The corresponding quantile function reads as [cf. (A.57)]

$$F_{\zeta_k}^{-1} = \exp\left(10^{-1}\ln(10)\exp\left(\mu_k + \sigma_{\zeta_{k,\mathrm{dB}}}\Phi^{-1}(x)\right)\right).$$

For the Gaussian model, the quantile function is found with the CDF and the inverse CDF of ζ_k^{-1}, respectively, as [cf. Lemma A.8 in Sect. A.8]

$$F_{\zeta_k}(g_{k1}(\boldsymbol{t})) = 1 - F_{\zeta_k^{-1}}(1/g_{k1}(\boldsymbol{t})), \tag{5.26}$$

[10] Here, $\Pr(\xi_k = -1) = 0$ and the noise variance is without loss of generality set to $\sigma_k^2 = 1$.

$$F_{\zeta_k}^{-1}(1-\epsilon_k) = 1/F_{\zeta_k^{-1}}^{-1}(\epsilon_k) = q_k. \tag{5.27}$$

For $\xi_k \sim \mathcal{N}_{\mathbb{C}}(0, \sigma_{\xi_k}^2)$, the normalized variable $\bar{\zeta}_k^{-1} = 2\sigma_{\xi_k}^{-2}\zeta_k^{-1}$ is distributed as $\bar{\zeta}_k^{-1} \sim \mathcal{X}_2^2(2\sigma_{\xi_k}^{-2})$. Therefore, the CDF of ζ_k^{-1} follows directly from the definition of the non-central chi-square distribution function for degree two (A.68):[11]

$$\begin{aligned} F_{\zeta_k^{-1}}(z) &= F_{\chi^2(2\sigma_{\xi_k}^{-2})}(2\sigma_{\xi_k}^{-2}z; 2) \\ &= 1 - \int_{\sqrt{2z\sigma_{\xi_k}^{-1}}}^{\infty} s\,\mathrm{e}^{-\frac{s^2}{2}-\sigma_{\xi_k}^{-2}} \sum_{n=0}^{\infty} \frac{(s^2\sigma_{\xi_k}^{-2}/2)^n}{(n!)^2}\,\mathrm{d}s. \end{aligned} \tag{5.28}$$

Fortunately, this CDF and its inverse has well-known approximations and computation algorithms (e.g., see [203, Section 26.4] and [279, Section 4]). For Rayleigh fading (2.19), where $2\sigma_{\xi_k}^{-2}\zeta_k^{-1} \sim \mathcal{X}_2^2$, the CDF (5.28) simplifies to

$$F_{\zeta_k^{-1}}(z) = F_{\chi^2}(2\sigma_{\xi_k}^{-2}z; 2) = 1 - \mathrm{e}^{-\sigma_{\xi_k}^{-2}z}, \tag{5.29}$$

where $F_{\chi^2}(x; 2)$ is the CDF of a (central) chi-square distributed random variable with degree two (see Sect. A.8). Therewith, the ϵ_k-quantile becomes

$$F_{\zeta_k^{-1}}^{-1}(\epsilon_k) = -\sigma_{\xi_k}^2 \ln(1-\epsilon_k). \tag{5.30}$$

Given the ϵ_k-outage rate (5.22), the QoS and RB optimization (5.3) and (5.4) are the same as for perfect CSI. Only the enhanced noise results in performance losses relative to perfect CSI, but the solutions are found via the same SOC or dual uplink formulation. Moreover, the feasible region $\mathcal{F}_{\mathrm{rate}}$ from (1.31) remains valid for (5.3). Hence, despite the robustness against the channel errors, solving (5.3) and (5.4) remains tractable for simulations with a hundred receivers K and antennas N.

Therefore, the multiplicative error approximation of Sect. 2.3 appears appealing also for chance constrained precoder designs with additive channel errors (2.1). However, the multiplicative fading approximation is opportunistic, that is, the optima for (5.3) and (5.4) bound the QoS and RB optima with additive channel errors from below and above, respectively. Furthermore, feasibility of (5.3) for the multiplicative fading approximation does not imply feasibility for the problem with additive channel errors. Whether the candidate $\boldsymbol{t}$ is feasible for (5.3) with additive fading has to be detected separately, e.g., via the next detailed numerical integration.

[11] The basis for this integral representation is the relation of the non-central chi-square distribution to the Marcum-Q function [278]. An alternative series expansion in terms of central chi-square distributions in Sect. A.8 follows from [203, Equation 26.4.25].

5.3 Outage Probability Computation for Additive Fading

While the outage probability for a multi-antenna PtP channel has a closed-form representation (cf. [280]), the probability computation for the intended multi-user scenario with adaptive beamforming requires a numerical evaluation. To this end, the probability of (5.1) is recast as

$$\Pr\big(Q(\boldsymbol{h}_k;\boldsymbol{B}_k)\geq\sigma_k^2\big)=1-\Pr\big(Q(\boldsymbol{h}_k;\boldsymbol{B}_k)\leq\sigma_k^2\big), \tag{5.31}$$

where the matrix $\boldsymbol{B}_k$ is a substitute for the indefinite quadratic form[12]

$$\boldsymbol{B}_k=\frac{1}{2^{\varrho}-1}\boldsymbol{t}_k\boldsymbol{t}_k^{\mathrm{H}}-\sum_{i\neq k}\boldsymbol{t}_i\boldsymbol{t}_i^{\mathrm{H}} \tag{5.32}$$

and $Q:\mathbb{C}^N\times\mathcal{H}^N\to\mathbb{R}$ defines the quadratic mapping

$$Q(\boldsymbol{z};\boldsymbol{A})=\boldsymbol{z}^{\mathrm{H}}\boldsymbol{A}\boldsymbol{z} \tag{5.33}$$

of the Gaussian random channel $\boldsymbol{h}_k\sim\mathcal{N}_{\mathbb{C}}(\bar{\boldsymbol{h}}_k,\boldsymbol{C}_k)$. Equality within (5.31) holds due to continuity of the probabilities in σ_k^2 (cf. Sect. 5.4.1).

Closed-form expressions and integral representations for the CDF of (5.33), with $\boldsymbol{z}\sim\mathcal{N}_{\mathbb{C}}(\boldsymbol{m},\boldsymbol{C})$, are summarized by Mathai and Provost [281, Chapter 4]. Examples are the distribution function for a positive definite $\boldsymbol{A}\succ\boldsymbol{0}$ and an indefinite $\boldsymbol{A}\in\mathcal{H}^N$ for $\boldsymbol{m}=\boldsymbol{0}$ [281, Section 4.3]. Integral expressions for the CDF of an indefinite quadratic form with non-zero mean are also shown by [174, 261, 282, 283].[13] We use Imhof's integration formula [174] to compute the probability (5.31). Let $\boldsymbol{C}_k=\boldsymbol{\Theta}_k\boldsymbol{\Theta}_k^{\mathrm{H}}$, where $\boldsymbol{\Theta}_k\in\mathbb{C}^{N\times M}$ has rank $M\leq N$. Then, the quadratic form is equivalently distributed as

$$Q(\tilde{\boldsymbol{z}};\boldsymbol{\Lambda})=\tilde{\boldsymbol{z}}^{\mathrm{H}}\boldsymbol{\Lambda}\tilde{\boldsymbol{z}}\simeq\bar{\lambda}_1\chi_1^2-\sum_{j=2}^{N_r}\bar{\lambda}_j\chi_j^2, \tag{5.34}$$

where $\tilde{\boldsymbol{z}}=\sqrt{2}\boldsymbol{V}^{\mathrm{H}}\boldsymbol{\Theta}_k^{\dagger}\boldsymbol{z}$ and the unitary and diagonal matrices $\boldsymbol{V}\in\mathbb{C}^{M\times M}$ and $\boldsymbol{\Lambda}\in\mathbb{R}^M$, respectively, follow from the Eigenvalue decomposition (EVD) $2^{-1}\boldsymbol{\Theta}_k^{\mathrm{H}}\boldsymbol{B}_k\boldsymbol{\Theta}_k=\boldsymbol{V}\boldsymbol{\Lambda}\boldsymbol{V}^{\mathrm{H}}$. For the stochastic equivalence (5.34), we considered N_r non-zero and distinct eigenvalues with magnitude $\bar{\lambda}_j\in\mathbb{R}_+$, each with multiplicity r_j and $\sum_{j=1}^{N_r}r_j\leq M$. In particular, only one single eigenvalue is positive, e.g., $\bar{\lambda}_1$,

[12] The probability is positive only if $\boldsymbol{B}_k$ has a positive eigenvalue and otherwise it is zero.

[13] Imhof's method and references to integral representations are also found in [281, Chapter 4].

due to the structure of $\boldsymbol{B}_k$. This leads to the above difference of the independent random variables $\chi_j^2 \sim \mathcal{X}_{2r_j}^2(\nu_j)$ with

$$\nu_j = 2 \sum_{i:|\bar{\bar{\lambda}}_i|=\bar{\lambda}_j} |[\boldsymbol{V}^{\mathrm{H}}\boldsymbol{\Theta}_k^{\dagger}\bar{\boldsymbol{h}}_k]_j|^2. \tag{5.35}$$

Based on Gil-Pelaez's inversion theorem [284] for the characteristic function of (5.34), Imhof's result for this probability reads as

$$\Pr\left(\bar{\lambda}_1\chi_1^2 - \sum_{j=2}^{N_r}\bar{\lambda}_j\chi_j^2 > x\right) = 1 - F_j(x) = \frac{1}{2} - \frac{1}{\pi}\int_0^{\infty}\frac{\sin(\theta(u))}{u\delta(u)}\,du, \tag{5.36}$$

where the functions $\theta : \mathbb{R} \to \mathbb{R}$ and $\delta : \mathbb{R} \to \mathbb{R}$ are given by

$$\begin{aligned}\theta(u) &= \frac{1}{2}\left(2r_1\tan^{-1}(\bar{\lambda}_1 u) + \frac{\nu_1\bar{\lambda}_1 u}{1+\bar{\lambda}_1^2u^2}\right)\\ &\quad - \frac{1}{2}\sum_{j=2}^{N_r}\left(2r_j\tan^{-1}(\bar{\lambda}_j u) + \frac{\nu_j\bar{\lambda}_j u}{1+\bar{\lambda}_j^2u^2}\right) - \frac{ux}{2},\\ \delta(u) &= \prod_{j=1}^{N_r}(1+\bar{\lambda}_j^2u^2)^{r_j/2}\exp\left(\frac{1}{2}\sum_{j=1}^{N_r}\frac{\nu_j\bar{\lambda}_j^2u^2}{1+\bar{\lambda}_j^2u^2}\right).\end{aligned}$$

An upper bounded numerical integration of (5.36) converges towards the exact value with an acceptable small error [174]. This computation is sufficiently accurate. A quantitative comparison to other numerical computations is provided by [285].

5.4 Power Allocation and Feasibility for Fixed Beamforming

The next detailed power allocation applies a sequence of the above probability evaluations within a fixed point algorithm. It requires a set of given unit-norm beamformers $\boldsymbol{u}_i \in \mathbb{C}^N$ from

$$\boldsymbol{t}_i = \boldsymbol{u}_i\sqrt{p_i}, \quad i = 1, \ldots, K \tag{5.37}$$

and allocates the transmit powers $p_i \in \mathbb{R}_+$ to refine the QoS or RB optimization. An example for $\boldsymbol{u}_i$, $i = 1, \ldots, K$ are normed MRT beamformers. Alternatives are the solution for either of the conservative beamformer designs or the multiplicative error approximation for the additive channel model and the result from Sect. 5.2.[14] With fixed beamformers, the remaining power allocation has the probability requirements

$$\mathrm{p}_k(\boldsymbol{p},\sigma_k^2;\rho_k)=\Pr\Big(\frac{1}{2^{\rho_k}-1}\beta_{kk}p_k-\sum_{i\neq k}\beta_{ki}p_i\geq\sigma_k^2\Big)\geq 1-\epsilon_k, \tag{5.38}$$

where $\boldsymbol{p}=[p_1,\ldots,p_K]^{\mathrm{T}}$, ρ_k is the QoS target, and $\beta_{ki}=|\boldsymbol{h}_k^{\mathrm{H}}\boldsymbol{u}_i|^2$ for $i,k=1,\ldots,K$ are chi-square distributed with degree two and mutually correlated. Furthermore, the reformulated power limitation $\boldsymbol{p}\in\tilde{\mathcal{P}}$ must be valid, where

$$\tilde{\mathcal{P}}=\{\boldsymbol{p}\in\mathbb{R}_+^L:\tilde{\boldsymbol{A}}\boldsymbol{p}\leq\boldsymbol{P}\} \tag{5.39}$$

is the convex polytope restricting $\boldsymbol{p}$, $[\tilde{\boldsymbol{A}}]_{\ell,i}=\|\boldsymbol{A}_{i,\ell}^{1/2}\boldsymbol{u}_i\|_2^2$ and $\boldsymbol{P}=[P_1,\ldots,P_L]^{\mathrm{T}}$.

Therewith, the QoS based power optimization reads as

$$\min_{p\geq 0,\boldsymbol{p}}\quad p\quad \text{s.t.}\quad p^{-1}\boldsymbol{p}\in\tilde{\mathcal{P}},\quad \mathrm{p}_k(\boldsymbol{p},\sigma_k^2;\rho_k)\geq 1-\epsilon_k,\quad k=1,\ldots,K, \tag{5.40}$$

and the ϵ-outage rate balancing optimization becomes

$$\max_{\rho,\boldsymbol{p}}\quad \rho\quad \text{s.t.}\quad \boldsymbol{p}\in\tilde{\mathcal{P}},\quad \mathrm{p}_k(\boldsymbol{p},\sigma_k^2;\rho\rho_k)\geq 1-\epsilon_k,\quad k=1,\ldots,K. \tag{5.41}$$

An analysis of the constraints, a feasibility test for the QoS problem, and the fixed point solutions for (5.40) and (5.41) are detailed next.

5.4.1 Characteristic of the Chance Constraints

Each chance constraint (5.38) forms a Probabilistically constrained linear inequality (PCLI) in $\boldsymbol{p}$ that reads as

$$\mathrm{p}_k(\boldsymbol{p},\sigma_k^2;\rho_k)=\Pr\left(\boldsymbol{\beta}_k^{\mathrm{T}}\boldsymbol{M}_k\boldsymbol{p}\geq\sigma_k^2\right)\geq 1-\epsilon_k, \tag{5.42}$$

with $\boldsymbol{\beta}_k=[\beta_{k1},\ldots,\beta_{kK}]^{\mathrm{T}}$ and the diagonal matrix $\boldsymbol{M}_k=\mathbf{I}-(1-2^{-\rho_k})^{-1}\mathbf{e}_k\mathbf{e}_k^{\mathrm{T}}$. Some characteristic properties of PCLIs (5.42) have been studied by Henrion [260, 286], e.g., continuity and monotonicity of the probability function. The following discussion restricts to cases where $\boldsymbol{C}_k$ is full-rank, such that the support of β_{ki} is $\mathbb{R}_+$ and the support of $\boldsymbol{\beta}_k^{\mathrm{T}}\boldsymbol{M}_k\boldsymbol{p}$ is all of $\mathbb{R}$ if $p_k>0$ and $p_i>0$ for at least one $i\neq k$.

---

[14]Adaptive rate control mechanisms are another method to decrease the conservatism in communication systems in favor for performance gains if fast feedback of single bits is available.

The probability $\mathrm{p}_k(\boldsymbol{p}, \sigma_k^2; \rho_k)$ is continuous at each point $(\boldsymbol{p}, \sigma_k^2) \in \mathbb{R}_+^K \times \mathbb{R}_+$ with $\mathrm{p}_k(\boldsymbol{p}, \sigma_k^2; \rho_k) > 0$ (cf. [260, Proof of Theorem 2.1]). First, the mapping from the power vector $\boldsymbol{p} \in \mathbb{R}_+^K$ to the set

$$\mathcal{T}(\boldsymbol{p}, \sigma_k^2) = \{\boldsymbol{\xi} \in \mathbb{R}_+^K : \boldsymbol{\xi}^{\mathrm{T}} \boldsymbol{M}_k \boldsymbol{p} \geq \sigma_k^2\} \tag{5.43}$$

is continuous, because the inequality within $\mathcal{T}$ defines a closed half-space.[15] Second, the weighted sum of the non-central chi-squared random variables β_{ki} is absolutely continuous with respect to the Lebesgue measure. Therefore, the probability function

$$\mathrm{p}_k(\boldsymbol{p}, \sigma_k^2; \rho_k) = \Pr(\boldsymbol{\beta}_k \in \mathcal{T}(\boldsymbol{p}, \sigma_k^2))$$

is continuous in $\boldsymbol{p}$.[16] Furthermore, the mapping $\sigma_k^2 \mapsto \mathcal{T}(\boldsymbol{p}, \sigma_k^2)$ is obviously also continuous in σ_k^2 when the support of $\boldsymbol{\beta}_k$ is all $\mathbb{R}_+^K$.

Additionally, the probability function $\mathrm{p}_k(\boldsymbol{p}, \sigma_k^2; \rho_k)$ has the same monotonicity properties as the stochastic function $g(\boldsymbol{p}, \sigma_k^2, \boldsymbol{\beta}_k) = \boldsymbol{\beta}_k^{\mathrm{T}} \boldsymbol{M}_k \boldsymbol{p} - \sigma_k^2$. In other words, the probability (5.42) is well-behaved with respect to preserving the monotonicity.

Definition 5.1 The probability function $p : \mathbb{R}_+ \to [0, 1]$ induced by $\boldsymbol{\xi}$ and $\mathrm{p}(\boldsymbol{x}) = \Pr(g(\boldsymbol{\xi}, \boldsymbol{x}) \geq 0)$ with $g : \mathbb{R}_+^K \times \mathbb{R}_+^M \to \mathbb{R}$ is monotonically well-behaved if it preserves the monotonicity properties of $g(\boldsymbol{\xi}, \boldsymbol{x})$ with regard to $\boldsymbol{x}$ for $\mathrm{p}(\boldsymbol{x}) \in (0, 1)$.

This property follows for a large class of absolutely-continuous distributed $\xi_i \in \mathbb{R}_+$ and stochastic functions. It is obvious for separable functions $g(\boldsymbol{\xi}, \boldsymbol{x}) = g_1(\boldsymbol{x}) - g_2(\boldsymbol{\xi})$ with continuous $g_1 : \mathbb{R}_+^M \to \mathbb{R}_+$ and $g_2 : \mathbb{R}_+^K \to \mathbb{R}_+$ if the support of the random value $g_2(\boldsymbol{\xi})$ is all $\mathbb{R}_+$. Also affine functions $g(\boldsymbol{\xi}, \boldsymbol{x}) = \boldsymbol{\xi}^{\mathrm{T}} \boldsymbol{x} - c$ with $c \geq 0$ and stochastic independent entries of $\boldsymbol{\xi}$ result in a monotonically well-behaved probability [260].

Lemma 5.1 *Moreover, the probability* $\mathrm{p}_k(\boldsymbol{p}, \sigma_k^2; \rho_k)$ *is well-behaved in that it is*

(i) *monotonically increasing in* p_k *for fixed* p_i, $i \neq k$ *and* σ_k^2;
(ii) *monotonically decreasing in each* p_i, $i \neq k$ *for fixed* p_k *and* σ_k^2;
(iii) *monotonically decreasing in* σ_k^2, *for fixed* p_i, $i = 1, \ldots, K$

for all probability values $\mathrm{p}_k(\boldsymbol{p}, \sigma_k^2; \rho_k) > 0$.

---

[15] The set-valued mapping particularly satisfies $\lim_{n\to\infty} \mathcal{T}(\boldsymbol{p}^{(n)}, \sigma_k^2) \to \mathcal{T}(\boldsymbol{p}^\star, \sigma_k^2)$ if $\boldsymbol{p}^\star$ is the limit point of $\boldsymbol{p}^{(n)}$ for $n \to \infty$, where convergence is in the Kuratowski–Painlevé sense [287].

[16] Generally, continuity in probability means that $\lim_{n\to\infty} \Pr(\mathcal{A}_n) = \Pr(\mathcal{A}^*)$ for any increasing sequence of events $\mathcal{A}_n \subseteq \ldots \subseteq \mathcal{A}_1$ with convergence set $\mathcal{A}^*$ and $\lim_{n\to\infty} \Pr(\mathcal{B}_n) = \Pr(\mathcal{B}^*)$ for any decreasing sequence of events $\mathcal{B}_n \supset \ldots \supset \mathcal{B}_1$ with convergence set $\mathcal{B}^*$. If the sequences of sets $\mathcal{A}_n$ and $\mathcal{B}_n$ converge in the Kuratowski–Painlevé sense, this results in the above convergence behavior in probability [287].

Proof As an immediate consequence of the set definition $\mathcal{T}(\boldsymbol{p}, \sigma_k^2)$ from (5.43)

$$\mathcal{T}(\boldsymbol{p}, x) \subseteq \mathcal{T}(\boldsymbol{p}, y), \quad \forall \boldsymbol{p} \in \mathbb{R}_+^K, x, y \in \mathbb{R}_+ : y \leq x,$$

where the subset inequality is strict if $y < x$ and $\mathcal{T}(\boldsymbol{p}, x)$ is non-empty. This results in $\Pr(\boldsymbol{\beta}_k \in \mathcal{T}(\boldsymbol{p}, x)) \leq \Pr(\boldsymbol{\beta}_k \in \mathcal{T}(\boldsymbol{p}, y)$ since the support of $\boldsymbol{\beta}_k$ is $\mathbb{R}_+^K$. Similarly,

$$\mathcal{T}(\boldsymbol{p}, x) \subseteq \mathcal{T}(\boldsymbol{p}', x), \quad \forall x \in \mathbb{R}_+, \ \boldsymbol{p}, \boldsymbol{p}' \in \mathbb{R}_+^K : p_k \leq p_k', \ p_i = p_i', \forall i \neq k,$$

$$\mathcal{T}(\boldsymbol{p}, x) \subseteq \mathcal{T}(\boldsymbol{p}', x), \quad \forall s \in \mathbb{R}_+, \ \boldsymbol{p}, \boldsymbol{p}' \in \mathbb{R}_+^K : p_k = p_k', \ p_i \geq p_i', \forall i \neq k,$$

where the set inequality is strict only if $p_k < p_k'$ and $p_i > p_i'$, respectively. Properties (i) and (ii) hold since $\mathcal{T}(\boldsymbol{p}, x) \subseteq \mathcal{T}(\boldsymbol{p}', x)$ imposes $\Pr(\boldsymbol{\beta}_k \in \mathcal{T}(\boldsymbol{p}, \sigma_k^2)) \leq \Pr(\boldsymbol{\beta}_k \in \mathcal{T}(\boldsymbol{p}, \sigma_k^2))$ where the inequality is strict if $\mathcal{T}(\boldsymbol{p}, \sigma_k^2)$ is a strict subset of $\mathcal{T}(\boldsymbol{p}', \sigma_k^2)$ and the probability for $\boldsymbol{\beta}_k \in \mathcal{T}(\boldsymbol{p}', \sigma_k^2) \setminus \mathcal{T}(\boldsymbol{p}, \sigma_k^2)$ is non-zero. □

5.4.2 *Fixed Point Framework for Power Allocation*

Due to these monotonicity properties of the probability, the solutions $\boldsymbol{p}^\star$ for (5.40) and (5.41) satisfy all probability requirements with equality, that is,

$$\mathrm{p}_k(\boldsymbol{p}^\star, \sigma_k^2; \rho\rho_k) = 1 - \epsilon_k, \quad k = 1, \ldots, K,$$

where $\rho = 1$ for the QoS optimization. Moreover, $\boldsymbol{p}' = \alpha \boldsymbol{p}$ with $\alpha > 1$ satisfies the probability constraints if so does $\boldsymbol{p}$, because $\mathrm{p}_k(\alpha \boldsymbol{p}, \sigma_k^2; \rho\rho_k) = \mathrm{p}_k(\boldsymbol{p}, \alpha^{-1}\sigma_k^2; \rho\rho_k)$ increases for decreasing $\alpha^{-1}\sigma_k^2$ (cf. Lemma 5.1). These properties motivate fixed point algorithms to solve (5.40) and (5.41). We define $I_k(\cdot; \rho\rho_k) : \mathbb{R}_+^K \to \mathbb{R}_+$ as

$$I_k(\boldsymbol{p}; \rho_k) = \min \left\{ x \in \mathbb{R}_+^K : \mathrm{p}_k([p_1, \ldots, p_{k-1}, x, p_{k+1}, \ldots, p_K]^\mathrm{T}, \sigma_k^2; \rho_k) = 1 - \epsilon_k \right\},$$
$$k = 1, \ldots, K. \tag{5.44}$$

Proposition 5.1 *The map* $\boldsymbol{I}(\cdot; \boldsymbol{\rho}) : \mathbb{R}_+^K \to \mathbb{R}_+^K$ *with entries* $I_k(\boldsymbol{p}; \rho_k)$ *of* (5.44) *is a standard interference function according to* Definition 1.1.

Proof Elementwise *positivity* of $\boldsymbol{I}(\boldsymbol{p}; \boldsymbol{\rho})$ follows by (5.44) and the monotonicity properties of Lemma 5.1 for $\epsilon_k \in (0, 1)$. Due to (ii) of Lemma 5.1, the inequality

$$\Pr\left(\beta_{kk} x - \sum_{i \neq k} \beta_{ki} p_i' \geq \sigma_k^2\right) \geq \Pr\left(\beta_{kk} x - \sum_{i \neq k} \beta_{ki} p_i \geq \sigma_k^2\right)$$

holds for all $x > 0$ and $\boldsymbol{p} \geq \boldsymbol{p}' \geq \boldsymbol{0}$. Together with Property (i) from Lemma 5.1, this results in $I_k(\boldsymbol{p}; \boldsymbol{\rho}) \geq I_k(\boldsymbol{p}'; \boldsymbol{\rho})$. Equality holds if $p'_l = p_l$ for all l and strict inequality holds if $p'_l > p_l$ for at least one $l \neq k$. This proves *monotonicity* of $\boldsymbol{I}(\boldsymbol{p}; \boldsymbol{\rho})$.

Sublinearity follows in turn from the following inequality for $\alpha > 1$:

$$\begin{aligned}
I_k(\alpha \boldsymbol{p}; \rho_k) &= \min\left\{x \in \mathbb{R}_+ : \Pr\left(\beta_{k,k}\frac{x}{\alpha} - \sum_{i \neq k} \beta_{k,i} p_i \geq \frac{\sigma_k^2}{\alpha}\right) = 1 - \epsilon_k\right\} \\
&= \alpha \min\left\{z \in \mathbb{R}_+ : \Pr\left(\beta_{k,k} z - \sum_{i \neq k} \beta_{k,i} p_i \geq \frac{\sigma_k^2}{\alpha}\right) = 1 - \epsilon_k\right\} \\
&< \alpha \min\left\{z \in \mathbb{R}_+ : \Pr\left(\beta_{k,k} z - \sum_{i \neq k} \beta_{k,i} p_i \geq \sigma_k^2\right) = 1 - \epsilon_k\right\} \\
&= \alpha I_k(\boldsymbol{p}; \rho_k),
\end{aligned}$$

where the first and second equality follow from (5.44) and substituting $z = x/\alpha$, and the inequality is due to Property (iii) of Lemma 5.1. □

With the standard interference property, the fixed point iteration

$$\boldsymbol{p}^{(n+1)} = \boldsymbol{I}(\boldsymbol{p}^{(n)}; \boldsymbol{\rho}) \tag{5.45}$$

globally converges to the unique solution $\boldsymbol{p}^\star$ of the QoS optimization (5.40) if the target vector $\boldsymbol{\rho} \in \mathbb{R}_+^K$ is attainable. This power allocation is independent from the type of the power constraint (5.39), in contrast to the solution of the RB optimization (5.41). For (5.41), the power allocation follows at convergence of the normalized self-mapping

$$\boldsymbol{p}^{(n+1)} = \boldsymbol{I}(\boldsymbol{p}^{(n)}; \rho^{(n+1)} \boldsymbol{\rho}). \tag{5.46}$$

The update $\rho^{(n+1)}$ ensures that the power limitation (5.39) remains satisfied, i.e.,

$$\rho^{(n+1)} = \max\left\{\rho \in \mathbb{R}_+ : \tilde{\boldsymbol{A}} \boldsymbol{I}(\boldsymbol{p}^{(n)}; \rho \boldsymbol{\rho}) - \boldsymbol{P} \leq \boldsymbol{0}\right\}. \tag{5.47}$$

For the implementation, $\rho^{(n+1)}$ is the largest positive root of $F : \mathbb{R}_+ \to \mathbb{R}$ with

$$F(\rho) = \max_\ell \mathbf{e}_\ell^{\mathrm{T}} \tilde{\boldsymbol{A}} \boldsymbol{I}(\boldsymbol{p}^{(n)}; \rho \boldsymbol{\rho}) - P_\ell$$

and obtained via a line search over ρ, e.g., a bisection.

5.4.3 Feasibility Detection for QoS Optimization

To test whether the chance constraints in (5.40) are feasible for practical systems, one may exploit that (5.45) defines a monotonically increasing sequence when starting from $\boldsymbol{p}^{(0)} = \boldsymbol{0}$ [75, Lemma 2]. If the sequence converges before exceeding a practical threshold, the convergence point is the solution. Otherwise, infeasibility may be declared.

However, since the transmit power is unbounded for (5.40), the above test fails to detect feasibility close to the limit $p \to \infty$. Then, $p^{-1}\sigma_k^2 \to 0$ and (5.44) becomes

$$I_k(\boldsymbol{p}; \rho_k) = \min\left\{x : \Pr\left(\boldsymbol{\beta}_k^{\mathrm{T}} \boldsymbol{M}_k \boldsymbol{p} \geq 0\right) = 1 - \epsilon_k\right\}. \tag{5.48}$$

While positivity and monotonicity from Proposition 5.1 are still valid, $\boldsymbol{I}(\boldsymbol{p}; \boldsymbol{\rho})$ strictly scales with $\boldsymbol{p}$ now, i.e., $\alpha \boldsymbol{I}(\boldsymbol{p}; \boldsymbol{\rho}) = \boldsymbol{I}(\alpha \boldsymbol{p}; \boldsymbol{\rho})$ for $\alpha \in \mathbb{R}_+$. In other words, the interference function is *linear* according to [11, Definition 2.1] (cf. Sect. 1.3).

An iterative test whether the constraints $\mathrm{p}_k(\boldsymbol{p}, 0; \rho_k) \geq 1 - \epsilon_k, k = 1, \ldots, K$ are achievable by a $\boldsymbol{p} \in \mathbb{R}_+^K$ follows by balancing the SIRs $p_k / I_k(\boldsymbol{p}; \rho_k)$ as [92][17]

$$C^{-1} = \max_{\boldsymbol{1}^{\mathrm{T}}\boldsymbol{p}=1} \min_k \frac{p_k}{I_k(\boldsymbol{p}; \rho_k)}. \tag{5.49}$$

If $C < 1$, the probability constraints of (5.40) can be met with finite $\boldsymbol{p}$ and infinite power is required for $C = 1$. For $C > 1$, the probability requirements are infeasible. The optimizer for (5.49) is found via the normalized fixed point iteration [92]

$$\boldsymbol{p}^{(n+1)} = \frac{\boldsymbol{I}(\boldsymbol{p}^{(n)}; \boldsymbol{\rho})}{\boldsymbol{1}^{\mathrm{T}} \boldsymbol{I}(\boldsymbol{p}^{(n)}; \boldsymbol{\rho})}. \tag{5.50}$$

If the sequence $\boldsymbol{p}^{(n)}$ has converged to $\boldsymbol{p}^\star$, C is the eigenvalue for the mapping $\boldsymbol{I} : \mathbb{R}_+^K \to \mathbb{R}_+^K$, that is, $C\boldsymbol{p}^\star = \boldsymbol{I}(\boldsymbol{p}^\star; \boldsymbol{\rho})$. Convergence of the sequence $\boldsymbol{p}^{(n)}$ and uniqueness of the optimizer $\boldsymbol{p}^\star$ are ensured since $\boldsymbol{I}(\boldsymbol{p}; \boldsymbol{\rho})$ is *primitive*. In other words, the self-mapping strongly preserves the order of $\boldsymbol{p}' \geq \boldsymbol{p}$, $\boldsymbol{p}' \neq \boldsymbol{p}$ in that $\boldsymbol{I}^{(m)}(\boldsymbol{p}'; \boldsymbol{\rho}) > \boldsymbol{I}^{(m)}(\boldsymbol{p}; \boldsymbol{\rho})$ for some $m \in \mathbb{N}$, where $\boldsymbol{I}^{(m)}(\boldsymbol{x}; \boldsymbol{\rho}) = \boldsymbol{x}^{(m)}$ denotes the m-th element of the sequence $\boldsymbol{x}^{(n+1)} = \boldsymbol{I}(\boldsymbol{x}^{(n)}; \boldsymbol{\rho})$ with starting point $\boldsymbol{x}^{(0)} = \boldsymbol{x}$ (cf. [92, Lemma 2]). Here, the integer is $m = 2$ because $I_k(\boldsymbol{p}; \rho_k)$ is strictly increasing in all p_l with $l \neq k$.

Remark 5.2 The numerical integration and line search algorithms for evaluating $I_k(\boldsymbol{p}; \rho_k)$ result in an intractable computational complexity if the dimensions of the communication system, i.e., the number of receivers K and/or the number of antennas N, are large. Then, the power allocation may only provide a benchmark.

For the QoS optimization (5.40), the optimal power allocation requires feasible beamformers. State-of-the-art convex optimization solvers can detect feasibility and infeasibility for conservative approximations of the QoS problem. However, the

[17] Here, $\boldsymbol{p}$ is normalized to $\boldsymbol{1}^{\mathrm{T}}\boldsymbol{p} = 1$ due to the scale-invariance of $p_k / I_k(\boldsymbol{p}; \rho_k)$.

power allocation cannot extend the feasible range of the approximate QoS problem when these solvers detect infeasibility and fail to deliver a reasonable beamformer, even though (5.3) is feasible. This motivates a non-conservative approximation for the beamformer design, e.g., the rank-one channel approximation of Sect. 2.3 and the solution of Sect. 5.2, before applying the power allocation.

5.5 Robust Uncertainty Reformulations

If reliability is crucial for the beamformer design, robust uncertainty reformulations provide tractable approximations for the chance constraints (5.19). The uncertainty approximation of Wang et al. [36], with a spherical uncertainty region for the additive channel error, is commonly employed for SINR requirements (see Sect. 5.5.1). Other literature with this uncertainty model, e.g., [170, 288], consider the MSE metric as an alternative.

A drawback of the spherical uncertainty region is that it restricts also channel realizations that improve the SINR, e.g., where the errors are positively collinear to the channel mean. Other quadratic uncertainty regions better suit the properties for downlink beamforming. Section 5.5.2 introduces an uncertainty region that bounds the channel error perpendicular to the direction of the channel mean. This choice is a trade-off between the complexity for computing the probability that a random channel lies within the uncertainty region and solving the resulting optimization.

5.5.1 *Sphere Bounds for the Additive Channel Errors*

Uncertainty constraints for beamforming restrict the additive error $\boldsymbol{e}_k$ in (2.1) to reside in a pre-defined set $\mathcal{U}_k$ with probability $\Pr(\boldsymbol{e}_k \in \mathcal{U}_k)$. The usually imposed uncertainty set is a symmetric ball around the uncorrelated part of the channel error (cf. [36, 105]):

$$\mathcal{U}_k = \left\{\boldsymbol{e}_k \in \mathbb{C}^N : \|\boldsymbol{C}_k^{-1/2}\boldsymbol{e}_k\|_2^2 \le d_k\right\}. \tag{5.51}$$

This additional requirement results in the probability lower bound

$$\Pr\left(g_k(\boldsymbol{t},\boldsymbol{e}_k) \ge \sigma_k^2\right) \ge \Pr\left(g_k(\boldsymbol{t},\boldsymbol{e}_k) \ge \sigma_k^2 \middle| \boldsymbol{e}_k \in \mathcal{U}_k\right)\Pr(\boldsymbol{e}_k \in \mathcal{U}_k), \tag{5.52}$$

where the stochastic constraint function $g_k : \mathbb{C}^{NK} \times \mathbb{C}^N \to \mathbb{R}$ reads as

$$g_k(\boldsymbol{t},\boldsymbol{e}_k) = \boldsymbol{e}_k^{\mathrm{H}}\boldsymbol{B}_k\boldsymbol{e}_k + 2\,\mathrm{Re}(\boldsymbol{e}_k^{\mathrm{H}}\boldsymbol{B}_k\bar{\boldsymbol{h}}_k) + \bar{\boldsymbol{h}}_k^{\mathrm{H}}\boldsymbol{B}_k\bar{\boldsymbol{h}}_k, \tag{5.53}$$

with $\boldsymbol{B}_k = \frac{1}{2^{\rho_k}-1}\boldsymbol{t}_k\boldsymbol{t}_k^{\mathrm{H}} - \sum_{i\neq k}\boldsymbol{t}_i\boldsymbol{t}_i^{\mathrm{H}}$ according to (5.32).

The right-hand side approximation must be larger than $1 - \epsilon_k$ to assure the probability constraint. Fixing $\Pr(\boldsymbol{e}_k \in \mathcal{U}_k) = 1 - \epsilon_k$, the remaining probability of (5.52) must equal one. In other words, the constraint $g_k(\boldsymbol{t}, \boldsymbol{e}_k) \geq \sigma_k^2$ must be fulfilled for all $\boldsymbol{e}_k \in \mathcal{U}_k$. The corresponding sphere bound for (5.51) is

$$d_k = F_{\chi^2_{2N}}^{-1}(1 - \epsilon_k)/2. \tag{5.54}$$

It is half the $1-\epsilon_k$-quantile of the chi-square distributed scalar $2\|\boldsymbol{C}_k^{-1/2}\boldsymbol{e}_k\|_2^2$. Hence, the requirements that replace the chance constraints of (5.3) and (5.4) are

$$g_k(\boldsymbol{t}, \boldsymbol{e}_k) \geq \sigma_k^2, \quad \forall \boldsymbol{e}_k : \|\boldsymbol{C}_k^{-1/2}\boldsymbol{e}_k\|_2^2 \leq d_k, \quad k = 1, \ldots, K. \tag{5.55}$$

This type of uncertainty constraint is well studied (e.g., see [105, 169, 170, 289, 290]). Moreover, it features a LMI reformulation that is affine in $\boldsymbol{Q}_k = \boldsymbol{t}_k\boldsymbol{t}_k^{\mathrm{H}}$ (e.g., [291, 292]) via the S-Lemma (e.g., see [258, Section B.2]). This LMI formulation of the above uncertainty constraint reads as (cf. [105])

$$\boldsymbol{\Psi}_k(\rho, \boldsymbol{Q}_1, \ldots, \boldsymbol{Q}_K, \lambda_k) \succeq \mathbf{0}, \tag{5.56}$$

where $\lambda_k \in \mathbb{R}_+$ and the matrix $\boldsymbol{\Psi}_k \in \mathbb{C}^{N+1\times N+1}$ read as

$$\begin{aligned}\boldsymbol{\Psi}_k(\rho, \boldsymbol{Q}_1, \ldots, \boldsymbol{Q}_K, \lambda_k) = {} & \big[\bar{\boldsymbol{h}}_k, \boldsymbol{C}_k^{1/2}\big]^{\mathrm{H}}\boldsymbol{B}_k\big[\bar{\boldsymbol{h}}_k, \boldsymbol{C}_k^{1/2}\big] \\ & + \lambda_k \,\mathrm{bdiag}(-d_k, \mathbf{I}_N) - \mathrm{bdiag}(\sigma_k^2, \mathbf{0}),\end{aligned} \tag{5.57}$$

and the matrix $\boldsymbol{B}_k$ is now a substitute for

$$\boldsymbol{B}_k = \frac{1}{2^{\rho\rho_k} - 1}\boldsymbol{Q}_k - \sum_{i\neq k}\boldsymbol{Q}_i. \tag{5.58}$$

Due to the affine structure in $\boldsymbol{Q}_1, \ldots, \boldsymbol{Q}_K$, a SDR became popular for solving the resulting QoS and RB optimizations (see Sects. 5.5.3 and 5.5.4).

Recently, an alternative reformulation for (5.55) was provided by Huang et al. [105], which is convex in the beamformers. He bounded the desired signal power as

$$|\boldsymbol{h}_k^{\mathrm{H}}\boldsymbol{t}_k| \geq \mathrm{Re}(\boldsymbol{h}_k^{\mathrm{H}}\boldsymbol{t}_k) \tag{5.59}$$

in order to approximate the uncertainty constraint by the SOC form

$$\boldsymbol{\Theta}_k(\boldsymbol{t})\begin{bmatrix}1\\ \boldsymbol{\xi}_k\end{bmatrix} \in \mathcal{L}^{2K-1}, \quad \forall \boldsymbol{\xi}_k : \begin{bmatrix}1\\ \boldsymbol{\xi}_k\end{bmatrix} \in \mathcal{L}^{2N}. \tag{5.60}$$

Here, the uncertainty vector is given by

$$\boldsymbol{\xi}_k = \begin{bmatrix} d_k^{-1/2}\,\mathrm{Re}(\boldsymbol{C}_k^{-1/2}\boldsymbol{e}_k) \\ d_k^{-1/2}\,\mathrm{Im}(\boldsymbol{C}_k^{-1/2}\boldsymbol{e}_k)\end{bmatrix},$$

and $\boldsymbol{\Theta}_k : \mathbb{C}^{NK} \to \mathbb{R}^{2K\times 2K+1}$ is an affine real-valued mapping in $\boldsymbol{t}$, i.e.,

$$\boldsymbol{\Theta}_k(\boldsymbol{t}) = \begin{bmatrix} \vartheta_k^{-1/2}\operatorname{Re}(\boldsymbol{t}_k^{\mathrm{H}}\bar{\boldsymbol{h}}_k) & d_k^{1/2}\vartheta_k^{-1/2}\operatorname{Re}(\boldsymbol{t}_k^{\mathrm{H}}\boldsymbol{C}_k^{1/2}) & -d_k^{1/2}\vartheta_k^{-1/2}\operatorname{Im}(\boldsymbol{t}_k^{\mathrm{H}}\boldsymbol{C}_k^{1/2}) \\ \operatorname{Re}(\boldsymbol{T}_{\setminus\{k\}}^{\mathrm{H}}\bar{\boldsymbol{h}}_k) & d_k^{1/2}\operatorname{Re}(\boldsymbol{T}_{\setminus\{k\}}^{\mathrm{H}}\boldsymbol{C}_k^{-1/2}) & -d_k^{1/2}\operatorname{Im}(\boldsymbol{T}_{\setminus\{k\}}^{\mathrm{H}}\boldsymbol{C}_k^{-1/2}) \\ \operatorname{Im}(\boldsymbol{T}_{\setminus\{k\}}^{\mathrm{H}}\bar{\boldsymbol{h}}_k) & d_k^{1/2}\operatorname{Im}(\boldsymbol{T}_{\setminus\{k\}}^{\mathrm{H}}\boldsymbol{C}_k^{-1/2}) & d_k^{1/2}\operatorname{Re}(\boldsymbol{T}_{\setminus\{k\}}^{\mathrm{H}}\boldsymbol{C}_k^{-1/2}) \\ \sigma_k & \boldsymbol{0}^{\mathrm{T}} & \boldsymbol{0}^{\mathrm{T}} \end{bmatrix},$$

where $\vartheta_k = 2^{\rho\rho_k} - 1$. In particular, (5.60) is a linear mapping of vectors from a SOC—a Lorentz cone—into vectors that also reside within a SOC. Such transformations are called Lorentz positive maps (LPMs) [266, 267]. They feature an LMI reformulation that is linear in $\boldsymbol{\Theta}(\boldsymbol{t})$ and, thus, also in $\boldsymbol{t}$ [105, Section III]. Therefore, these constraints are tractable by standard SDP solvers, e.g., SDPT3 [98].

5.5.2 *Quadratic Bounds for the Orthogonal Channel Errors*

The above SDR approximation for a quadratic constraint with a quadratic uncertainty restriction also applies for bounding the subspace error model [cf. (2.25)]

$$\boldsymbol{h}_k = (1+\xi_k)\bar{\boldsymbol{h}}_k + \boldsymbol{e}_k, \quad k = 1, \ldots, K. \tag{5.61}$$

Here, the multiplicative and additive Gaussian channel errors are

$$\boldsymbol{e}_k = \boldsymbol{C}_k^{1/2}\boldsymbol{V}_{N-1,k}\hat{\boldsymbol{e}}_{N-1,k}, \quad \hat{\boldsymbol{e}}_{N-1,k} \sim \mathcal{N}_{\mathbb{C}}(\boldsymbol{0}, \mathbf{I}),$$

$$\xi_k = \|\boldsymbol{C}_k^{-1/2}\bar{\boldsymbol{h}}_k\|_2^{-1}\hat{e}_{1,k}, \quad \hat{e}_{1,k} \sim \mathcal{N}_{\mathbb{C}}(0, 1),$$

where the subunitary matrix $\boldsymbol{V}_{N-1,k} \in \mathbb{C}^{N\times N-1}$ ensures that the additive error is orthogonal to the expected channel characteristics, i.e., $\boldsymbol{V}_{N-1,k}^{\mathrm{H}}\boldsymbol{C}_k^{-1/2}\bar{\boldsymbol{h}}_k = \boldsymbol{0}$, and $\hat{\boldsymbol{e}}_{N-1,k}$ and $\hat{e}_{1,k}$ are mutually independent.[18] Exploiting (5.61), we introduce the new uncertainty bound

$$\mathcal{U}_k = \left\{(\xi_k, \boldsymbol{e}_k) \in \mathbb{C}\times\mathbb{C}^N : \|\boldsymbol{C}_k^{-1/2}\boldsymbol{e}_k\|_2^2 \le a_k|1+\xi_k|^2\|\boldsymbol{C}_k^{-1/2}\bar{\boldsymbol{h}}_k\|_2^2 - b_k\right\}, \tag{5.62}$$

where we bound the additive channel error, which is orthogonal to the channel mean, dependent on the changing gain of the channel mean. The two design parameters $a_k, b_k \in \mathbb{R}_+$ are for meeting the probability requirement

$$\Pr((\xi_k, \boldsymbol{e}_k) \in \mathcal{U}_k) = 1 - \epsilon_k. \tag{5.63}$$

Their combination provides an additional degree of freedom for the approximation.

[18] This model is obtained when approximating the Gaussian additive channel error as in Sect. 2.3.

For the simulations, the parameterization is based on b_k, $k = 1, \dots, K$, while a_k is computed via a bisection, that repeatedly evaluates the probability of (5.63). This probability is a special case for (5.36), namely:

$$\Pr((\xi_k, \boldsymbol{e}_k) \in \mathcal{U}_k) = \Pr(a_k \chi_1^2 - \chi_2^2 > 2b_k)$$

where the random variables are $\chi_1^2 \sim \mathcal{X}_{2(N-1)}^2$ and $\chi_2^2 \sim \mathcal{X}_2^2(\nu_k)$, with

$$\nu_k = 2\|\boldsymbol{C}_k^{-1/2}\bar{\boldsymbol{h}}_k\|_2^2.$$

A direct integral form for the uncertainty probability reads as [cf. (A.73)]

$$\Pr(\chi_2^2 - a_k\chi_1^2 \leq -2b_k) = \int_0^\infty \frac{1}{2}\mathrm{e}^{-\frac{s+\nu_k}{2}} \Big(\sum_{j=0}^{\infty} \frac{(\nu_k s/4)^j}{(j!)^2}\Big) F_{\chi^2}(-2b_k + a_k s; 2(N-1))\,\mathrm{d}s.$$

The relation of the three parameters a_k, b_k, and ν_k can be tabulated with a predefined accuracy, which further simplifies the root search with respect to a_k.

Inserting (5.61) into (5.53), the quadratic constraint function becomes

$$g_k(\boldsymbol{t}, 1+\xi_k, \boldsymbol{e}_k) = |1+\xi_k|^2\bar{\boldsymbol{h}}_k^{\mathrm{H}}\boldsymbol{B}_k\bar{\boldsymbol{h}}_k + 2\,\mathrm{Re}\left((1+\xi_k)\bar{\boldsymbol{h}}_k^{\mathrm{H}}\boldsymbol{B}_k\boldsymbol{e}_k\right) + \boldsymbol{e}_k^{\mathrm{H}}\boldsymbol{B}_k\boldsymbol{e}_k. \quad (5.64)$$

An equivalent formulation for the requirement $g_k(\boldsymbol{t}, 1+\xi_k, \boldsymbol{e}_k) \geq \sigma_k^2$ reads as

$$\begin{bmatrix} 1 \\ 1+\xi_k \\ \hat{\boldsymbol{e}}_{N-1,k} \end{bmatrix}^{\mathrm{H}} \begin{bmatrix} -\sigma_k^2 & 0 & 0 \\ 0 & \bar{\boldsymbol{h}}_k^{\mathrm{H}}\boldsymbol{B}_k\bar{\boldsymbol{h}}_k & \bar{\boldsymbol{h}}_k^{\mathrm{H}}\boldsymbol{B}_k\boldsymbol{C}_k^{1/2}\boldsymbol{V}_{N-1,k} \\ 0 & \boldsymbol{V}_{N-1,k}^{\mathrm{H}}\boldsymbol{C}_k^{\mathrm{H}/2}\boldsymbol{B}_k\bar{\boldsymbol{h}}_k^{\mathrm{H}} & \boldsymbol{V}_{N-1,k}^{\mathrm{H}}\boldsymbol{C}_k^{\mathrm{H}/2}\boldsymbol{B}_k\boldsymbol{C}_k^{1/2}\boldsymbol{V}_{N-1,k} \end{bmatrix} \begin{bmatrix} 1 \\ 1+\xi_k \\ \hat{\boldsymbol{e}}_{N-1,k} \end{bmatrix} \geq 0, \quad (5.65)$$

while the quadratic uncertainty set reads equivalently as

$$\begin{bmatrix} 1 \\ 1+\xi_k \\ \hat{\boldsymbol{e}}_{N-1,k} \end{bmatrix}^{\mathrm{H}} \begin{bmatrix} -b_k & 0 & 0 \\ 0 & a_k\bar{\boldsymbol{h}}_k^{\mathrm{H}}\boldsymbol{C}_k^{-1}\bar{\boldsymbol{h}}_k & 0 \\ 0 & 0 & -\mathbf{I}_{N-1} \end{bmatrix} \begin{bmatrix} 1 \\ 1+\xi_k \\ \hat{\boldsymbol{e}}_{N-1,k} \end{bmatrix} \geq 0. \quad (5.66)$$

According to the S-Lemma [258, Section B.2], (5.65) implies (5.66) if there is a $\lambda_k \geq \sigma_k^2 b_k^{-1}$ satisfying (5.56), but with $\boldsymbol{\Psi}_k \in \mathbb{C}^{N\times N}$ given by

$$\begin{aligned}\boldsymbol{\Psi}_k(\rho, \boldsymbol{Q}_1, \dots, \boldsymbol{Q}_K, \lambda_k) = {} & \left[\bar{\boldsymbol{h}}_k, \boldsymbol{C}_k^{1/2}\boldsymbol{V}_{N-1,k}\right]^{\mathrm{H}}\boldsymbol{B}_k\left[\bar{\boldsymbol{h}}_k, \boldsymbol{C}_k^{1/2}\boldsymbol{V}_{N-1,k}\right] \\ & + \lambda_k\,\mathrm{bdiag}(-a_k\bar{\boldsymbol{h}}_k^{\mathrm{H}}\boldsymbol{C}_k^{-1}\bar{\boldsymbol{h}}_k, \mathbf{I}_{N-1}).\end{aligned} \quad (5.67)$$

Hence, (5.56) defines a convex constraint in the transmit covariance matrices.[19]

[19] A reformulation into the LPM (5.60) misses for the orthogonal channel uncertainty region.

Remark 5.3 (Asymptotic SNR Limit)
For $\sigma_k^2 \to 0$, the LMI from the orthogonal channel error bound (5.67) is less restrictive than that from the spherical uncertainty bound (5.57) if $d_k \geq a_k \bar{\boldsymbol{h}}_k^{\mathrm{H}} \boldsymbol{C}_k^{-1} \bar{\boldsymbol{h}}_k$. The offset parameter is $b_k = 0$ for this case, which also minimizes the parameter a_k.

5.5.3 QoS Optimization with Uncertainty Constraints

QoS and balancing optimizations with spherical uncertainty constraints are well studied [105, 169, 170, 289]. An alternating beamformer design and worst-case channel search locally converge to a candidate solution [169]. The common SDR approach exploits convexity of (5.57) in $\boldsymbol{Q}_k = \boldsymbol{t}_k \boldsymbol{t}_k^{\mathrm{H}}$, $k = 1, \ldots, K$, by relaxing $\operatorname{rank}\{\boldsymbol{Q}_k\} = 1$ (e.g., [36, 291, 292]). With this substitution and relaxation, the power constraint set (1.33) becomes

$$\mathcal{Q} = \left\{ \boldsymbol{Q}_i \in \mathcal{H}_+^N,\ i = 1, \ldots, K \,\middle|\, \sum_{i=1}^{K} \operatorname{tr}(\boldsymbol{Q}_i \boldsymbol{A}_{i,\ell}) \leq P_\ell,\ \ell = 1, \ldots, L \right\}. \tag{5.68}$$

It is a compact convex set of K semidefiniteness and L linear inequality constraints.
Hence, the relaxed QoS optimization reads as

$$\begin{aligned} \min_{p \in \mathbb{R}_+, \boldsymbol{Q}_i, \boldsymbol{\lambda} \in \mathbb{R}_+^K} p \quad \text{s.t.} \quad & (p^{-1} \boldsymbol{Q}_1, \ldots, p^{-1} \boldsymbol{Q}_K) \in \mathcal{Q}, \\ & \boldsymbol{\Psi}_k(1, \boldsymbol{Q}_1, \ldots, \boldsymbol{Q}_K, \lambda_k) \succeq \boldsymbol{0}, \quad k = 1, \ldots, K, \end{aligned} \tag{5.69}$$

where $\boldsymbol{\Psi}_k$ is affine in $\boldsymbol{Q}_k$, $k = 1, \ldots, K$ and given by (5.57) for the spherical uncertainty approximation. The solution is obtained with interior-point solvers, e.g., the disciplined convex programming toolbox CVX [99] with the solver SDPT3 [98].

The same convex optimization is valid for the orthogonal error bound, but with $\boldsymbol{\Psi}_k$ given by (5.67). Additionally, the optimization includes the parameters a_k and b_k that are connected via (5.63) for $k = 1, \ldots, K$. This joint optimization is non-convex due to the multiplication of λ_k and a_k in (5.67). To overcome this issue, we search the appropriate parameter combinations a_k and b_k in an outer optimization:

$$\min_{\boldsymbol{b}} p(\boldsymbol{b}) \quad \text{s.t.} \quad \boldsymbol{b} = [b_1, \ldots, b_K]^{\mathrm{T}} \in \mathbb{R}_+^K. \tag{5.70}$$

The objective is the optimum of (5.69) with fixed b_k and a_k for $k = 1, \ldots, K$. It is quasiconvex in one particular $b_k \in \mathbb{R}_+$ for given b_j, $j \neq k$. In particular, the objective first decreases when increasing b_k from zero until the minimum is found. Further increasing b_k, the objective increases until the inner QoS problem becomes infeasible. Due to this behavior, the solution $\boldsymbol{b} \in \mathbb{R}_+^K$ is found via a nested golden section search for $K = 2$ users. Unfortunately, this search becomes computationally demanding for more than two users. For similar statistical properties of the random channel, an approximation of this solution is based on a one-dimensional dependency of all b_k, $k = 1, \ldots, K$. For example, we used the convex combination

$$b_k = \alpha\sigma_k^2 + (1 - \alpha)\|\boldsymbol{C}_k^{-1/2}\bar{\boldsymbol{h}}_k\|_2^2, \quad k = 1, \ldots, K \tag{5.71}$$

to incorporate the gains of the channels' mean values and bound $\sigma_k^2 b_k^{-1} \geq 1$. Therewith, the outer optimization with respect to α is a line search. Its solution is obtained via a golden section search [227, Appendix C.3] if the objective remains quasiconvex. Section 5.6 shows a performance comparison to a nested golden section search for obtaining the optimal parameters b_k, $k = 1, \ldots, K$.

5.5.4 Balancing Optimization with Uncertainty Constraints

The RB optimization for the spherical uncertainty approximation reads as

$$\begin{aligned}\min_{\rho\geq 0, \boldsymbol{Q}_i, \boldsymbol{\lambda}\geq \boldsymbol{0}} \quad \rho \quad \text{s.t.} \quad & (\boldsymbol{Q}_1, \ldots, \boldsymbol{Q}_K) \in \mathcal{Q}, \\ & \boldsymbol{\Psi}_k(\rho, \boldsymbol{Q}_1, \ldots, \boldsymbol{Q}_K, \lambda_k) \succeq \boldsymbol{0}, \quad k = 1, \ldots, K.\end{aligned} \tag{5.72}$$

The constraint set is again convex in $\boldsymbol{Q}_k$, $k = 1, \ldots, K$, but it is non-convex in the balancing factor $\rho \in [0, \rho^{\text{UB}})$. An upper bound ρ^{UB} is the minimum of the single user transmit cases, i.e., with $\boldsymbol{Q}_i = \boldsymbol{0}$ and $\rho_i = 0$ for $i \neq k$ if k ist the index of the intended receiver's rate.[20]

With these bounds, a bisection search over ρ finds the solution for (5.72). A convex feasibility test is performed in each bisection step. The test checks whether a feasible tuple $(\boldsymbol{Q}_1, \ldots, \boldsymbol{Q}_K)$ exists that satisfies the constraints for the candidate ρ'. The solution for (5.69) with $\boldsymbol{\Psi}_k(\rho', \boldsymbol{Q}_1, \ldots, \boldsymbol{Q}_k, \lambda_k)$ can serve as such a test. If $p < 1$ is its optimum, ρ' is a lower bound for the optimum of (5.72) and it is an upper bound for $p > 1$. The bisection procedure is continued until the remaining bounds differ by $\rho^{\text{UB}} - \rho^{\text{LB}} < 10^{-4}$, where we declare ρ^{LB} as the optimal value.

For the orthogonal error uncertainty, the RB optimization (5.72) is additionally over the parameters b_k, $k = 1, \ldots, K$, which we find with the outer maximization

$$\max_{\boldsymbol{b}} \rho(\boldsymbol{b}) \quad \text{s.t.} \quad \boldsymbol{b} = [b_1, \ldots, b_K]^{\text{T}} \in \mathbb{R}_+^K, \tag{5.73}$$

where $\rho(\boldsymbol{b})$ is the optimum of (5.72). The objective of this problem is quasiconcave in $b_k \in \mathbb{R}_+$ for fixed b_j, $j \neq k$. This would allow a solution via a nested golden section search, but the inner problem is even more complex than for the QoS

[20]This single user rate maximization problem has an equivalent convex reformulation.

optimization. To overcome this complexity and analyze the average performance of the balancing optimization with the orthogonal error uncertainty, simulations are restricted to the convex combination (5.71) for $\boldsymbol{b} \in \mathbb{R}_+^K$. The joint decision parameter α is again computed via the golden section method [227, Appendix C.3], but with the aim to maximize ρ. Since this strategy incorporates a repeated evaluation of (5.72), it increases the complexity by the number of repetitions compared to RB with the spherical uncertainty constraint.

5.5.5 *Beamformer Reconstruction and Direct Beamformer Optimization*

The SDR for the above QoS and RB optimization requires a beamformer reconstruction from the solution $\boldsymbol{Q}_k$, $k = 1, \ldots, K$. If the covariance matrices are rank-one, $\boldsymbol{t}_k$ is the dominant eigenvector of $\boldsymbol{Q}_k$ scaled by the square root of the corresponding eigenvalue. For the spherical uncertainty bound, the rank-one property is likely met with a sum power constraint [36], especially, when $N = 2$ [291, Lemma 2] or the uncertainty radius (5.54) is sufficiently small. If (5.68) describes per-antenna constraints, the rank of the solutions can be larger than one [13, Remark 5]. Then, a Gaussian randomization [12, 107] finds candidate beamformers.

One can overcome this burden with the second uncertainty constraint reformulation in Sect. 5.5.1. With the convex dual formulation for the cone of LPMs [267], the QoS optimization with spherical uncertainty constraints becomes a convex optimization in the beamformers [105]. However, finding its solution via an interior-point solver is computationally more complex than the solution of (5.69). The reason is the dimension of the LMI constraints that describe the LPMs. Therefore, this approach is only tractable for small systems, e.g., [105] considered $N = 5$ antennas and $K = 3$ users. Furthermore, the optimum of the QoS optimization with LPMs generally upper bounds the optimum of (5.69) due to the additional approximation step (5.59). Similarly, the optimum of the related RB optimization with LPMs lower bounds the maximum of (5.72). The small difference is depicted in Sect. 5.7.2.

5.6 Tractable Bounds with Concentration Inequalities

Other approximations for the probability constraints of (5.3) and (5.4), i.e.,

$$\Pr(r_k(\boldsymbol{t}, \boldsymbol{h}_k) \geq \rho \rho_k) \geq 1 - \epsilon_k, \tag{5.74}$$

follow from *concentration inequalities* that bound the outage probability by an average measure of the stochastic function. In particular, Markov's inequality (5.14) and the Bernstein-type inequalities (5.15) and (5.16) provide bounds that guarantee the probability requirement. Section 5.6.1 shows an application of Markov's inequality that connects the outage rate optimization with the average MSE optimization from Chap. 4. Then follows a review of Wang et al.'s [36] application of the first Bernstein-type inequality for an SINR formulation of (5.74). The second Bernstein-type inequality also bounds an MSE approximation of (5.74).

5.6.1 Markov's Inequality Based MSE Approximation

While Markov's inequality is unsuitable for directly lower bounding the left-hand side probability of (5.74), the inequality let us restrict the MMSE reformulation as

$$\Pr(\mathrm{MMSE}_k(\boldsymbol{t}, \boldsymbol{h}_k) \leq 2^{-\rho\rho_k}) \geq 1 - \frac{\mathrm{E}[\mathrm{MMSE}_k(\boldsymbol{t}, \boldsymbol{h}_k)]}{2^{-\rho\rho_k}} \geq 1 - \epsilon_k.$$

This inner restriction for (5.74) results in the average MSE requirement

$$\mathrm{AMSE}_k(\boldsymbol{t}) = \mathrm{E}[\mathrm{MMSE}_k(\boldsymbol{t}, \boldsymbol{h}_k)] \leq \epsilon_k 2^{-\rho\rho_k}. \tag{5.75}$$

Replacing the probability constraints in (5.3) by these MMSE requirements, we end up with the approximate average QoS optimization from Sect. 4.3, i.e.,

$$\min_{p \geq 0,\, \boldsymbol{t}} \; p \quad \text{s.t.} \quad p^{-1}\boldsymbol{t} \in \mathcal{P}, \quad \mathrm{AMSE}_k(\boldsymbol{t}) \leq \varepsilon_k(\rho), \quad k = 1, \ldots, K, \tag{5.76}$$

but with MSE targets $\varepsilon_k(\rho) = \epsilon_k 2^{-\rho\rho_k}$, $k = 1, \ldots, K$. These targets are smaller than ϵ_k, which results in an infeasible QoS problem if the outage limits are low.

The resulting balancing optimization with the average MSE approximation reads as (4.109), but with the above MSE targets. A local solution of this problem is found by the provided ACS in Sect. 4.5. However, the corresponding objective is far below the actual optimum of (5.4), because the ϵ_k's scale down the targets. The balancing optimization can even become infeasible if ϵ_k is too small.

5.6.2 Bernstein-Type Inequality Bound for the SINR

The looseness of the previous deterministic approximations for (5.74), e.g., using either the spherical uncertainty approximation or the formulation of [47, 138], motivated Wang et al. [36] to study two further approximations of

$$\Pr(g_k(\boldsymbol{t}, \boldsymbol{e}_k) \geq \sigma_k^2) \geq 1 - \epsilon_k \tag{5.77}$$

with $g_k(\boldsymbol{t}, \boldsymbol{e}_k)$ from (5.53). One of these deterministic approximations employs the first Bernstein-type inequality (5.15) from [172] (see Sect. 5.1). It bounds the error

$$z = g_k(\boldsymbol{t}, \boldsymbol{e}_k) - \mathrm{E}[g_k(\boldsymbol{t}, \boldsymbol{e}_k)]$$

between the value $g_k(\boldsymbol{t}, \boldsymbol{e}_k)$ for a realization of the random error $\boldsymbol{e}_k$ and the mean

$$\mathrm{E}[g_k(\boldsymbol{t}, \boldsymbol{e}_k)] = \mathrm{tr}(\tilde{\boldsymbol{B}}_k) + \bar{\boldsymbol{h}}_k^{\mathrm{H}} \boldsymbol{B}_k \bar{\boldsymbol{h}}_k$$

$\boldsymbol{B}_k$ is from (5.32), by the following probability inequality [36, Lemma 1]:

$$\Pr\left(z > -\sqrt{2s}\sqrt{\|\tilde{\boldsymbol{B}}_k\|_{\mathrm{F}}^2 + 2\|\boldsymbol{C}_k^{\mathrm{H}/2}\boldsymbol{B}_k\bar{\boldsymbol{h}}_k\|_2^2} - s\lambda_{\min}^{+}(\tilde{\boldsymbol{B}}_k)\right) \geq 1 - \mathrm{e}^{-s}, \tag{5.78}$$

where $\tilde{\boldsymbol{B}}_k = \boldsymbol{C}_k^{\mathrm{H}/2}\boldsymbol{B}_k\boldsymbol{C}_k^{1/2}$ and $\lambda_{\min}^{+}(\tilde{\boldsymbol{B}}_k) = \min\{\lambda \in \mathbb{R}_+ : \lambda\mathbf{I} + \tilde{\boldsymbol{B}}_k \succeq \mathbf{0}\}$ is either minus the smallest non-positive eigenvalue of $\tilde{\boldsymbol{B}}_k$ or zero. Therefore, the chance constraint (5.77) is ensured when the right-hand side of (5.78) equals $1 - \epsilon_k$, that is, $s = -\ln(\epsilon_k)$. This imposes the deterministic inequality

$$\mathrm{E}[g_k(\boldsymbol{t}, \boldsymbol{e}_k)] - \sqrt{2s}\sqrt{\|\tilde{\boldsymbol{B}}_k\|_{\mathrm{F}}^2 + 2\|\boldsymbol{C}_k^{\mathrm{H}/2}\boldsymbol{B}_k\bar{\boldsymbol{h}}_k\|_2^2} - s\lambda_{\min}^{+}(\tilde{\boldsymbol{B}}_k) \geq \sigma_k^2. \tag{5.79}$$

An equivalent convex formulation with respect to the transmit covariance matrix $\boldsymbol{Q}_k = \boldsymbol{t}_k\boldsymbol{t}_k^{\mathrm{H}}$, $k = 1, \ldots, K$ and with $\boldsymbol{B}_k$ from (5.58) reads as [36, Table 1]

$$\begin{aligned}
&\mathrm{tr}(\tilde{\boldsymbol{B}}_k) + \bar{\boldsymbol{h}}_k^{\mathrm{H}}\boldsymbol{B}_k\bar{\boldsymbol{h}}_k - \sqrt{-2\ln(\epsilon_k)}x_k + \ln(\epsilon_k)y_k \geq \sigma_k^2, \\
&\left\|\mathrm{vec}\left(\boldsymbol{C}_k^{\mathrm{H}/2}\boldsymbol{B}_k\left[\sqrt{2}\boldsymbol{C}_k^{1/2}, \bar{\boldsymbol{h}}_k\right]\right)\right\|_2 \leq x_k, \\
&y_k\mathbf{I} + \tilde{\boldsymbol{B}}_k \succeq \mathbf{0}, \quad y_k \geq 0.
\end{aligned} \tag{5.80}$$

A solution for the QoS optimization with this constraint set is via a SDR of the rank-one covariance matrices $\boldsymbol{Q}_k = \boldsymbol{t}_k\boldsymbol{t}_k^{\mathrm{H}}$, $k = 1, \ldots, K$ and using interior-point solvers for the remaining SDP. The corresponding balancing solution is again via an additional bisection over ρ, because $\boldsymbol{B}_k$ from (5.58) is non-convex in this balancing level and, thus, also (5.80). The optimization procedure is the same as for the uncertainty constraints (see Sect. 5.5.4).

The results from [36] suspect that this Bernstein-type inequality approximation for the chance constraints is less conservative than the sphere bounded uncertainty approximation from Sect. 5.5. The simulations of Sect. 5.7 support this statement, that is, the QoS and RB results with (5.80) outperform the results based on (5.55).

5.6.3 Bernstein-Type Inequality Bound for the MSE

The second Bernstein-type inequality (5.16) bounds the MSE formulation

$$\Pr\left(\mathrm{MMSE}_k(\boldsymbol{h}_k, \boldsymbol{t}) \leq \varepsilon_k(\rho)\right) \geq \Pr\left(\mathrm{MSE}_k(\boldsymbol{h}_k, \boldsymbol{t}, f_k) \leq \varepsilon_k(\rho)\right) \geq 1 - \epsilon_k, \tag{5.81}$$

with the MSE target $\varepsilon_k(\rho) = 2^{-\rho\rho_k}$. The inner approximation holds for any deterministic equalizer $f_k \in \mathbb{C}$. Since the MSE is quadratic in $\boldsymbol{e}_k$ in this case, the deviation $z = \mathrm{MSE}_k(\boldsymbol{h}_k, \boldsymbol{t}, f_k) - \mathrm{E}[\mathrm{MSE}_k(\boldsymbol{h}_k, \boldsymbol{t}, f_k)]$ is bounded in probability by

$$\Pr\left(z < \sqrt{2s}\sqrt{\||f_k|^2 \tilde{\boldsymbol{Q}}_k\|_\mathrm{F}^2 + 2\|\boldsymbol{C}_k^{\mathrm{H}/2}\boldsymbol{Q}\bar{\boldsymbol{h}}_k|f_k|^2 - f_k^* \boldsymbol{C}_k^{\mathrm{H}/2}\boldsymbol{t}_k\|_2^2} + s\lambda_{\max}^+(|f_k|^2 \tilde{\boldsymbol{Q}}_k)\right)$$
$$\geq 1 - \mathrm{e}^{-s},$$

with $\tilde{\boldsymbol{Q}}_k = \boldsymbol{C}_k^{\mathrm{H}/2}\boldsymbol{Q}\boldsymbol{C}_k^{1/2}$ and its dominant eigenvalue $\lambda_{\max}^+(\tilde{\boldsymbol{Q}}) \in \mathbb{R}_+$. As a consequence, the following requirement ensures (5.81) if $s = -\ln(\epsilon_k)$:

$$\mathrm{E}[\mathrm{MSE}_k(\boldsymbol{h}_k, \boldsymbol{t}, f_k)] + \sqrt{2s}\sqrt{\||f_k|^2 \tilde{\boldsymbol{Q}}_k\|_\mathrm{F}^2 + 2\|\boldsymbol{C}_k^{\mathrm{H}/2}\boldsymbol{Q}\bar{\boldsymbol{h}}_k|f_k|^2 - f_k^* \boldsymbol{C}_k^{\mathrm{H}/2}\boldsymbol{t}_k\|_2^2}$$
$$+ |f_k|^2 s\lambda_{\max}^+(\tilde{\boldsymbol{Q}}_k) \leq \varepsilon_k(\rho). \tag{5.82}$$

This constraint has a convex formulation in either of the decision variables $\boldsymbol{t}$ and f_k if the other variable is fixed. In other words, the left-hand side of (5.82) is a biconvex function. First, the metric $\mathrm{AMSE}_k(\boldsymbol{t}, f_k) = \mathrm{E}[\mathrm{MSE}_k(\boldsymbol{h}, \boldsymbol{t}, f_k)]$, i.e.,

$$\mathrm{AMSE}_k(\boldsymbol{t}, f_k) = 1 - 2\,\mathrm{Re}\left(f_k^* \bar{\boldsymbol{h}}_k^\mathrm{H} \boldsymbol{t}_k\right) + \sum_{i=1}^{K} |f_k|^2 \boldsymbol{t}_i^\mathrm{H} \boldsymbol{R}_k \boldsymbol{t}_i + |f_k|^2 \sigma_k^2 \tag{5.83}$$

with $\boldsymbol{R}_k = \bar{\boldsymbol{h}}_k \bar{\boldsymbol{h}}_k^\mathrm{H} + \boldsymbol{C}_k$, is a biconvex function in f_k and $\boldsymbol{t}$. Second, the remaining summands have a convex representation if either f_k or $\boldsymbol{t}$ is fixed. If the beamformers $\boldsymbol{t}_k, k = 1, \ldots, K$ are fixed, the equivalent convex constraints for f_k are

$$\mathrm{AMSE}_k(\boldsymbol{t}, f_k) + \sqrt{-2\ln(\epsilon_k)}x_k - \ln(\epsilon_k)|f_k|^2 y_k \leq \varepsilon_k(\rho) \tag{5.84a}$$
$$\left\|\mathrm{vec}\left(\boldsymbol{C}_k^{\mathrm{H}/2}\left[\phi_k \boldsymbol{Q}\boldsymbol{C}_k^{1/2}, \sqrt{2}\phi_k \boldsymbol{Q}\bar{\boldsymbol{h}}_k - \sqrt{2} f_k^* \boldsymbol{t}_k\right]\right)\right\|_2 \leq x_k, \tag{5.84b}$$
$$|f_k|^2 \leq \phi_k, \tag{5.84c}$$

where $x_k, \phi_k \in \mathbb{R}_+$ are slack variables for the optimization of f_k, and $y_k = \lambda_{\max}^+(\tilde{\boldsymbol{Q}}_k)$ is fixed. For fixed f_k, the scalar $y_k \in \mathbb{R}_+$ imposes the constraint

$$-y_k \mathbf{I} + \boldsymbol{C}_k^{\mathrm{H}/2}\boldsymbol{Q}\boldsymbol{C}_k^{1/2} \preceq \mathbf{0} \tag{5.84d}$$

and the transmit covariance matrix $\boldsymbol{Q} \in \mathcal{H}_+^N$ replaces $\boldsymbol{T}\boldsymbol{T}^\mathrm{H}$, $\boldsymbol{T} = [\boldsymbol{t}_1, \ldots, \boldsymbol{t}_K]$. The relaxed version of this substitution, i.e., $\boldsymbol{T}\boldsymbol{T}^\mathrm{H} \preceq \boldsymbol{Q}$, reformulates to

$$\begin{bmatrix} \boldsymbol{Q} & \boldsymbol{T} \\ \boldsymbol{T}^\mathrm{H} & \mathbf{I} \end{bmatrix} \succeq \mathbf{0} \tag{5.84e}$$

when applying Schur's complement [293]. Hence, the precoder $\boldsymbol{t}$ is limited by two quadratic inequality constraints, (5.84a) and (5.84b), one LMI constraint per receiver (5.84d), and the LMI constraint (5.84e).

Due to the biconvex structure of (5.84), an ACS algorithm finds a local solution for the approximate QoS problem, i.e., with (5.84) replacing (5.81):

$$\min_{f,t} \; p \quad \text{s.t.} \quad p^{-1}t \in \mathcal{P}, \quad (5.84) \quad k = 1, \ldots, K, \tag{5.85}$$

The ACS updates the equalizers f_k and the precoder t as follows (cf. Sect. 4.3):

1. Given t, the QoS constraints restrict the equalizer f_k only by (5.84a)–(5.84c). Therefore, the solution equalizer minimizes the left-hand side of (5.84a), i.e.,

$$f_k^\star = \arg\min_{f_k} \; \text{AMSE}_k(t, f_k) + \sqrt{-2\ln(\epsilon_k)x_k} - \ln(\epsilon_k)|f_k|^2 y_k \tag{5.86}$$
$$\text{s.t.} \; \left\| \operatorname{vec}\left(C_k^{\mathrm{H}/2}\left[\phi_k Q C_k^{1/2}, \sqrt{2}\phi_k Q\bar{h}_k - \sqrt{2} f_k^* t_k\right]\right)\right\|_2 \le x_k, \; |f_k|^2 \le \phi_k.$$

   The result is obtained via a convex SOCP solver and generally differs in the magnitude and phase from a standard MMSE equalizer.
2. Given these equalizers, an SDP solver updates the beamformers $t_k, k = 1, \ldots, K$ for fixed equalizers $f_k, k = 1, \ldots, K$.

These steps are repeated until convergence. Convergence is ensured if starting from an initial feasible precoder t, because $p \in \mathbb{R}_+$ decreases in each iteration and the updates of the beamformers and the equalizers are unique [246, Theorem 4.9].

Finding an appropriate starting precoder for (5.85) and a criteria to test feasibility for given targets ρ_k, $k = 1, \ldots, K$ is a difficult task itself. A sufficient feasibility condition follows for ZF beamforming $t_k^{\mathrm{ZF}} = u_k\sqrt{p}$ with unit-norm $u_k \in \mathbb{R}^N$, e.g., $u_k = \alpha_k \bar{H}(\bar{H}^{\mathrm{H}}\bar{H})^{-1}\mathbf{e}_k$, $k = 1, \ldots, K$, and the substitutes $f_k = v_k p_k^{-1/2}$ with $v_k \in \mathbb{C}$. Therewith, the approximated chance-constraint (5.82) reads as

$$g_k(\boldsymbol{p}, v_k) \le 2^{-\rho\rho_k} \tag{5.87}$$

where the left-hand side function $g_k : \mathbb{R}_+^K \times \mathbb{C} \to \mathbb{R}_+$ is now defined as

$$g_k(\boldsymbol{p}, v_k) = 1 - 2\,\mathrm{Re}\left(v_k^* \bar{h}_k^{\mathrm{H}} u_k\right)$$
$$+ \frac{|v_k|^2}{p_k}\left(\sum_{i=1}^{K} p_i u_i^{\mathrm{H}} R_k u_i + \sigma_k^2 + \sqrt{-2\ln(\epsilon_k)x_k} - \ln(\epsilon_k)y_k\right),$$

and the slack variables x_k and y_k are given by

$$x_k = \sqrt{\|C_k^{\mathrm{H}/2} U P U^{\mathrm{H}} C^{1/2}\|_{\mathrm{F}}^2 + p_k^2 |v_k|^{-2} 2\|C_k^{\mathrm{H}/2} u_k\|_2^2 |u_k^{\mathrm{H}} \bar{h}_k v_k - 1|^2}$$
$$y_k = \lambda_{\max}^+(C_k^{\mathrm{H}/2} U P U^{\mathrm{H}} C^{1/2}),$$

with $P = \mathrm{diag}(p_1, \ldots, p_K)$ and the beamforming matrix $U = [u_1, \ldots, u_K]$.

Feasibility is again detected via generalized SIR balancing (cf. Sect. 5.4.3), but with the standard interference function $\boldsymbol{I}(\boldsymbol{p};\, \boldsymbol{\rho})$ and its entries

$$I_k(\boldsymbol{p};\, \rho\rho_k) = 2^{\rho\rho_k} \min_{v_k} p_k g_k(\boldsymbol{p}, v_k), \quad k = 1, \ldots, K. \tag{5.88}$$

The function $p_k g_k(\boldsymbol{p}, v_k)$ satisfies the three conditions of Definition 1.1, that is, it is positive, sublinear in a scaling of $\boldsymbol{p}$, and increases with each entry of $\boldsymbol{p}$ for fixed v_k. The minimization with respect to v_k preserves these properties (see Sect. A.2). In the infinite SNR limit, i.e., $\sigma_k^2 = 0$, $I_k(\boldsymbol{p};\, \rho\rho_k)$ is a linear interference function according to Sect. 1.3. Then, the generalized SIR balancing from Sect. 5.4.3 serves as a feasibility test. Literally, the targets $\rho\rho_k$, $k = 1, \ldots, K$ are achievable if the eigenvalue defined by $C\boldsymbol{p}^\star = \boldsymbol{I}(\boldsymbol{p}^\star;\, \boldsymbol{\rho})$ is $C > 1$. The solution $\boldsymbol{p}^\star$ is found via the converging normalized fixed point iteration (5.50).

The feasibility test is a special form of the corresponding RB problem:

$$\max_{\boldsymbol{f},\boldsymbol{t}} \rho \quad \text{s.t.} \quad \boldsymbol{t} \in \mathcal{P}, \quad (5.84), \quad k = 1, \ldots, K. \tag{5.89}$$

This problem is also solved via an ACS that switches between the equalizer and beamformer update. While the equalizer update remains the same, the beamformer update becomes a quasiconvex problem, because the targets $\varepsilon_k(\rho) = 2^{-\rho\rho_k}$, $k = 1, \ldots, K$ are convex in the additional decision variable $\rho \in \mathbb{R}_+$.

To overcome this difficulty, we approximate the target function $\varepsilon_k(\rho)$ by $\varepsilon\varepsilon_k{}'$ for the precoder update, where $\varepsilon_k{}' = \varepsilon_k(\rho')$. Due to linearity in $\varepsilon \in \mathbb{R}_+$, the minimum for each ACS iteration is found via a conic optimization. Therewith, the target update reads as

$$\rho^{(n+1)} = \max\left\{\rho \in \mathbb{R}_+ : \varepsilon\varepsilon_k{}' \leq 2^{-\rho\rho_k}\right\}.$$

This reduces the number of beamformer optimizations per ACS iteration to one, compared to perfectly solving the precoder update. The equalizer update between the beamformer updates partly compensates the target approximation, such that the number of additional ACS iterations remains small.

5.6.4 Related MSE Based Chance Constraint Approximation

The above MSE based approximation relates to the work by Vučić and Boche [44] when restricting to ZF beamforming and $f_k = p_k^{-1/2}(\boldsymbol{u}_k^{\mathrm{H}}\bar{\boldsymbol{h}}_k)^{-1}$, $k = 1, \ldots, K$. Then, the MSE simplifies to a quadratic form in the channel errors, i.e.,

$$\mathrm{MSE}_k(\boldsymbol{h}_k, \boldsymbol{t}^{\mathrm{ZF}}, p_k^{-1/2}(\boldsymbol{u}_k^{\mathrm{H}}\bar{\boldsymbol{h}}_k)^{-1}) = p_k^{-1}|\bar{\boldsymbol{h}}_k^{\mathrm{H}}\boldsymbol{u}_k|^{-2}(\boldsymbol{e}_k^{\mathrm{H}}\boldsymbol{U}\boldsymbol{P}\boldsymbol{U}^{\mathrm{H}}\boldsymbol{e}_k + \sigma_k^2). \tag{5.90}$$

With this MSE, the inequality (5.87) becomes $g_k(\boldsymbol{p}, |\bar{\boldsymbol{h}}_k^{\mathrm{H}}\boldsymbol{u}_k|^{-1}) \leq 2^{-\rho\rho_k}$, where

$$g_k(\boldsymbol{p}, |\bar{\boldsymbol{h}}_k^{\mathrm{H}}\boldsymbol{u}_k|^{-1}) = \frac{1}{p_k|\bar{\boldsymbol{h}}_k^{\mathrm{H}}\boldsymbol{u}_k|^2}\Big(\mathrm{tr}\big(\tilde{\boldsymbol{Q}}_k\big) + \sigma_k^2 + \sqrt{-2\ln(\epsilon_k)}\|\tilde{\boldsymbol{Q}}_k\|_{\mathrm{F}} - \ln(\epsilon_k)\lambda_{\max}^{+}\big(\tilde{\boldsymbol{Q}}_k\big)\Big) \tag{5.91}$$

and $\tilde{\boldsymbol{Q}}_k = \boldsymbol{C}_k^{\mathrm{H}/2}\boldsymbol{U}\boldsymbol{P}\boldsymbol{U}^{\mathrm{H}}\boldsymbol{C}_k^{1/2}$. The power allocation for the QoS and RB optimizations with this constraint approximation follows via normalized fixed point iterations using the interference function $\boldsymbol{I}(\boldsymbol{p}; \rho\boldsymbol{\rho})$, with the entries

$$I_k(\boldsymbol{p}; \rho\rho_k) = 2^{\rho\rho_k} p_k g_k(\boldsymbol{p}, |\bar{\boldsymbol{h}}_k^{\mathrm{H}}\boldsymbol{u}_k|^{-1}), \quad k = 1, \ldots, K.$$

The approximation [44, Equation (11)] results in a similar result, but with

$$g_k(\boldsymbol{p}, |\bar{\boldsymbol{h}}_k^{\mathrm{H}}\boldsymbol{u}_k|^{-1}) = \frac{1}{p_k|\bar{\boldsymbol{h}}_k^{\mathrm{H}}\boldsymbol{u}_k|^2}\Big(\mathrm{tr}\big(\tilde{\boldsymbol{Q}}_k\big) + \sigma_k^2 + 1/f(1-\epsilon_k)\|\tilde{\boldsymbol{Q}}_k\|_{\mathrm{F}}\Big), \tag{5.92}$$

where $f(x) = \sqrt{(1-x)/(x-5/9)}$ for $x \geq 5/6$ and $f(x) = \sqrt{(4/3-x)/x}$ for $x < 5/6$. Since $M^{-1/2}\|\tilde{\boldsymbol{Q}}_k\|_{\mathrm{F}} \leq \lambda_{\max}^{+}(\tilde{\boldsymbol{Q}}_k)$ for $M = \min(N, K)$, (5.92) ensures a tighter probability approximation than (5.91) if $1/f(1-\epsilon_k) \leq \sqrt{-1\ln(\epsilon_k)} \leq \sqrt{-1\ln(\epsilon_k)} - M^{-1/2}\ln(\epsilon_k)$, that is, for $\epsilon_k \leq 0.07055$. Compared to the constraint functions of (5.87) and (5.82), the gain in approximation accuracy for the probability constraint is for the cost of a fixed equalizer and precoder. A numerical comparison in Sect. 5.7.3 shows that this loss of flexibility results in a performance loss if $\min(N, K) \geq 4$ and ϵ_k is small.

5.7 Numerical Results for Chance-Constrained Optimization

The simulation results for chance-constrained beamforming are for randomly drawn channel mean realizations from a standard complex Gaussian distribution and a scaled identity covariance matrix for the channel errors. They show a performance comparison for RB and QoS optimization with the above approximations for the chance constraints and quantify the limits of QoS feasibility.

Section 5.7.1 discusses the effect of the post-processing power allocation for different beamformer designs. It depicts the performance of the non-conservative beamforming based on the rank-one channel approximation and state-of-the-art conservative approximations: the SOCP reformulation from [138], the spherical uncertainty approximation, and the approximation with the SINR based Bernstein-type inequality. The comparison shows that the power allocation partly compensates for the approximate beamforming. It reduces the proposed transmit power of the conservative QoS designs and ensures reliable data transmission for the non-conservative approximation approach.

The difference of the orthogonal bound to the standard spherical bound for the uncertainty approximation is analyzed in Sect. 5.7.2 via the minimal transmit power for QoS optimization. Numerical results for chance-constrained RB are then presented in Sect. 5.7.3. Therein, the focus is on the Bernstein-type inequality approximation with the MSE metric. Dependent on the system dimensions, this approximation outperforms either of the uncertainty approximations, while the SINR based Bernstein-type inequality approximation benchmarks these results.

5.7.1 Post-Processing Power Allocation for QoS Optimization

Simulations with the post-processing power allocation were performed for a scenario with $K = N = 3$ antennas and receivers, $\sigma_k^2 = 0.1$, and a sum power restriction for $\mathcal{P}$. The channel means are drawn from a standard complex Gaussian distribution and the error covariance matrices are fixed to $\boldsymbol{C}_k = \frac{0.01}{N}\mathbf{I}_N$. Moreover, the outage rate is restricted to $\epsilon_k = 0.1, k = 1, \ldots, K$ and the relative rate targets are $\rho_1 = \rho_3 = 0.1$ and $\rho_2 = 0.2$. These targets have been successively increased with the scaling factor $\rho \in \mathbb{R}_+$ to measure the resulting transmit power in dB over the imposed data rate (see Fig. 5.1).

Figure 5.1a particularly shows the minimal transmit power p over ρ for an exemplary channel mean realization. The plot additionally includes the ZF beamforming

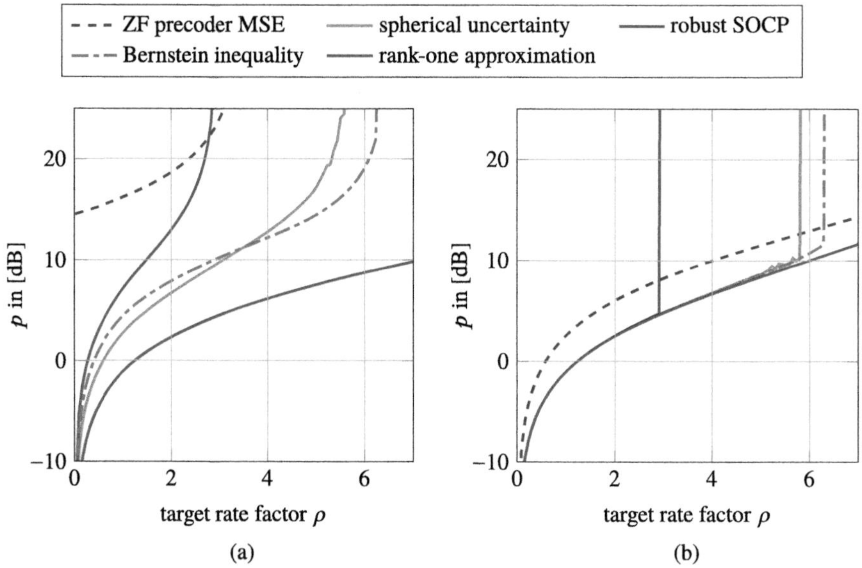

Fig. 5.1 Minimal transmit power p for the QoS optimization by the conservative and non-conservative designs with and without the post-processing power allocation. (**a**) Proposed transmit power. (**b**) After power allocation

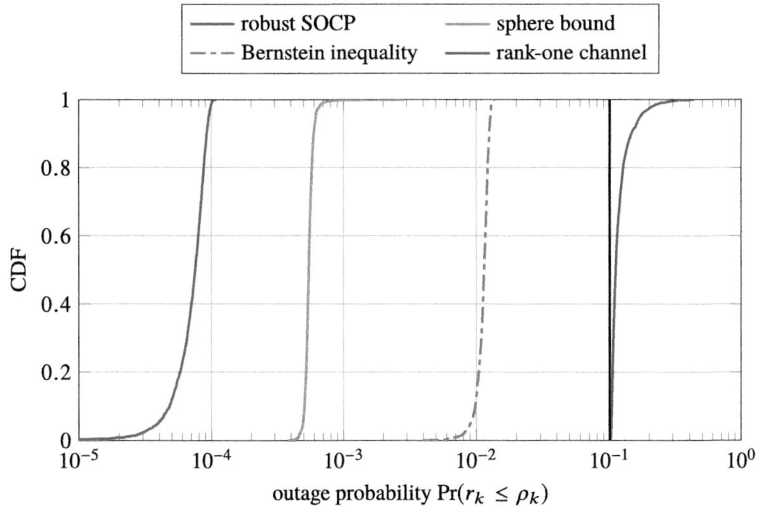

Fig. 5.2 Empirical CDF of the actual outage probability after the approximated chance-constrained QoS optimization with $\epsilon_k = 0.1$, a sum power limitation, and Gaussian channel fading, i.e., $\boldsymbol{h}_k \sim \mathcal{N}_{\mathbb{C}}(\bar{\boldsymbol{h}}_k, \boldsymbol{C}_k)$ with $\boldsymbol{C}_k = \frac{0.01}{N}\mathbf{I}$, $k = 1, \ldots, 3$

and MSE based QoS optimization form [44] with targets $\varepsilon_k = 2^{-\rho\rho_k}$. The proposed optima for the conservative designs lie considerably above that for the rank-one channel approximation, which serves as a lower bound for the actual optimum.

After the additional power allocation, all adaptive beamforming approaches achieve similar p within their feasible range. Only ZF beamforming falls behind this performance. Above the feasibility threshold, which is $\rho = 2.8$ for the robust SOCP approximation, $\rho = 5.2$ for the spherical uncertainty approximation, and $\rho = 6.3$ for the Bernstein-type inequality approximation, the employed interior-point solvers fail to provide a reasonable precoder. No post-processing is possible in this case. In contrast, the rank-one channel approximation delivers reasonable beamformers over the whole range for ρ. Its result is close to that after the post-processing for $\rho \leq 3$. For $\rho > 3$, the beamformers are still adequate, so that the power allocation meets the outage requirements. In this way, a wider range for ρ can be supported than for the conservative beamformer designs.

The power gain of the additional power allocation is about 3dB for the spherical uncertainty approximation, 5dB for the Bernstein's inequality approximation, and about 7dB for the robust SOCP formulation at $\rho = 1$. The gain is even larger, when ρ is close to the infeasibility bound of these conservative designs. The power loss for reaching the required reliability with the beamformers of the rank-one channel approximation is small for $\rho \leq 3$, but it increases with ρ (see Fig. 5.1). In other words, the larger the rate targets, the less accurate is the rank-one approximation.

Figure 5.2 depicts the CDF curves of the outage probabilities for the QoS optimization results with the above conservative approximations and the rank-one channel approximation. To create the curves, 500 channel mean realizations were drawn at random. For each channel mean realization, the beamformer designs were performed with $\rho = 2$ and the outage probabilities were computed with Imhof's method. The plot restricts to the cases, where the conservative designs are feasible.

The outage probabilities of the conservative designs are below 0.015 for this case. This promises considerable performance gains for the post-processing power allocation. In contrast, the rank-one channel approximation results in beamforming that violates the outage requirements. However, the outage limit $\epsilon_k = 0.1$ is met after the additional power allocation. In other words, the rank-one approximation requires an additional power allocation to meet the outage limits.

5.7.2 QoS Optimization with Channel Uncertainty Constraints

The performance comparison for the uncertainty approximations of the chance constraints is for $K = 2, 3, 4$ users and $N = 4, 6, 8$ transmit antennas, respectively. Here, the following Rician fading model is employed for the analysis:

$$\boldsymbol{h}_k = \sqrt{\frac{\kappa}{1+\kappa}}\bar{\boldsymbol{z}}_k + \sqrt{\frac{1}{1+\kappa}}\boldsymbol{C}_k^{1/2}\boldsymbol{z}_k$$

with $\boldsymbol{z}_k \sim \mathcal{N}_{\mathbb{C}}(\mathbf{0}, N^{-1}\mathbf{I}_N)$ and $\kappa = 5\,\text{dB}, 10\,\text{dB}, 15\,\text{dB}$. One hundred realizations for the channels' means $\bar{\boldsymbol{z}}_k$ are drawn at random from a standard complex Gaussian distribution for each setup. Furthermore, the noise variances are fixed at $\sigma_k^2 = 1$, the outage limits are $\epsilon_k = 0.1$ and the targets are $\rho_k = 1$, $k = 1, \ldots, K$.

Figure 5.3 shows the average minimal power p for the QoS optimization with $K = 2$ receivers and the approximated chance constraints. Besides the results for the uncertainty approximations, the plots depict the minimal transmit power for the first Bernstein-type inequality approximation and for perfect CSI. The average is over all channel realizations where both uncertainty approximations and the first Bernstein-type inequality approximation are feasible at $\rho = 2.0$. Here, this are exactly the 61 and 97 channel realizations for $\kappa = 5\,\text{dB}$ and $\kappa = 10\,\text{dB}$, respectively, where the spherical uncertainty approximation is feasible. For $\kappa = 15\,\text{dB}$, all the approximations have been feasible for all 100 channel mean realizations.

The base scenario is for $\kappa = 10\,\text{dB}$ in Fig. 5.3a. The proposed p for the spherical uncertainty approximation with the LPM and SDR reformulations are equal until close to the infeasibility bound, i.e., $\rho = 2.2$. Then, the LPM reformulation becomes infeasible for at least one channel scenario, while the SDR reformulation remains feasible for all 97 scenarios until $\rho = 2.6$. The reason is that the LPM formulation is more conservative than its SDR counterpart (see Sect. 5.5.1 and [105]).

The optimum decreases when using the orthogonal uncertainty model for the channel errors instead of the spherical uncertainty bound. It improves the approximation for the set of channel errors that satisfy the chance constraints. When

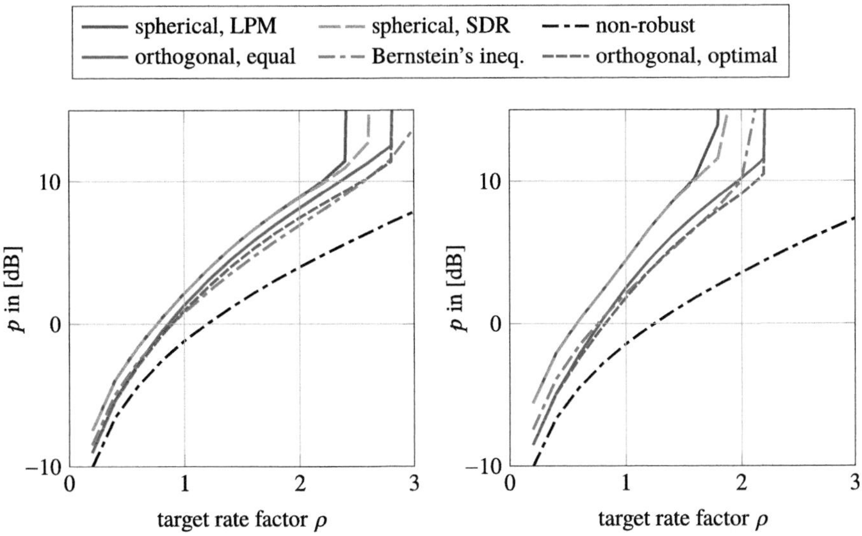

(a)

(b)

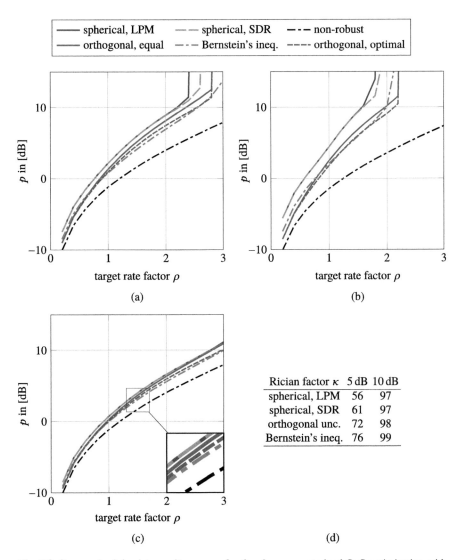

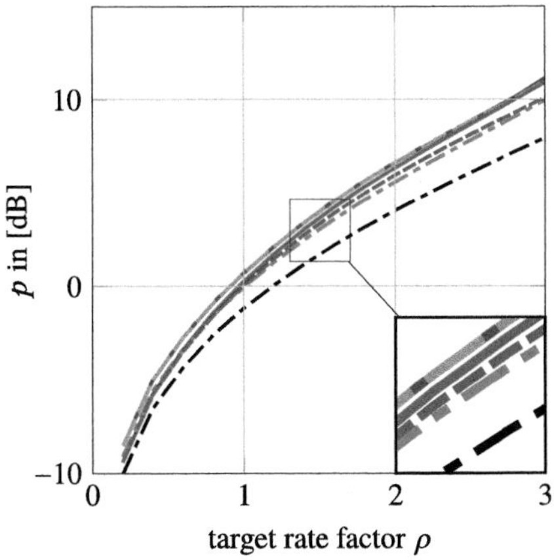

(c)

Rician factor κ	5 dB	10 dB
spherical, LPM	56	97
spherical, SDR	61	97
orthogonal unc.	72	98
Bernstein's ineq.	76	99

(d)

Fig. 5.3 Proposed minimal transmit power p for the chance constrained QoS optimization with $\epsilon_k = 0.1$ and $K = 2$ receivers. The plots are an average over the channel realizations where the spherical uncertainty approximation and the first Bernstein-type inequality approximation are feasible at $\rho = 2.0$. (**a**) $\kappa = 10\,\text{dB}$ and equal ρ_k. (**b**) $\kappa = 5\,\text{dB}$ and equal ρ_k. (**c**) $\kappa = 15\,\text{dB}$ and equal ρ_k. (**d**) # feasible scenarios for $\rho = 2$

the offset parameters for the orthogonal uncertainty regions are from the convex combination (5.71), i.e., with equal α for all $k = 1, \ldots, K$, the gain to the spherical uncertainty restriction is only about 1 dB at $\rho = 2$ for $\kappa = 10\,\text{dB}$. For $\kappa = 5\,\text{dB}$ (see Fig. 5.3b), this gain increases up to 3 dB at $\rho = 1.5$. A further reduction of the minimal transmit power p provides nested golden section search for the uncertainty parameters b_k, $k = 1, \ldots, K$. This even closes the performance gap between the uncertainty approximation and the Bernstein-type inequality approximation for the chance constraints. In other words, the orthogonal uncertainty restriction with optimized parameters reduces the power consumption by 2.5 dB and 4 dB for $\kappa = 10$ dB and $\kappa = 5$ dB, respectively, compared to the spherical uncertainty restriction (see Fig. 5.3a, b).

For accurate channel knowledge, i.e., $\kappa = 15$ dB, the performance differences between the robust QoS solutions are only within 1 dB in Fig. 5.3c. Then, the computationally less complex spherical uncertainty bound approximates the chance constraints sufficiently accurate. The improvements by the orthogonal uncertainty and Bernstein-type inequality approximations are only small.

Figure 5.4 presents the QoS solution for increased system dimensions. The left column shows the minimal transmit power and feasibility range for $K = 3$ receivers and $N = 6$ antennas while the right column presents the results for $K = 4$ and $N = 8$. The optimal parameter search for the orthogonal uncertainty bound is too complex for computing the results for these scenarios and the LPM formulation for the spherical uncertainty approximation becomes intractable for $K = 4$ and $N = 8$.

The order for the performance curves of the remaining approximations is the same as for $K = 2$.[21] The orthogonal uncertainty approximation outperforms the spherical one and achieves the Bernstein-type inequality approximation for $\rho < 0.8$. For $\rho > 0.8$, the transmit power gap between the orthogonal and the spherical uncertainty approximation is only small. The gap even vanishes for $\kappa = 10$ dB if $1 < \rho < 2$ and $K = 3$ or $1 < \rho < 2.4$ and $K = 4$. Within this range, the distance to the Bernstein-type inequality approximation and the perfect CSI curve increases with ρ. This difference to the $K = 2$ user scenario is to some extent due to the scaling of the channel covariance matrix with N^{-1} and the fact that $K \ll N$.

Besides the performance, the orthogonal uncertainty approximation also increases the feasible range of the spherical uncertainty approximation. On the one hand, it increases the worst-case achievable ρ by 0.2 in Fig. 5.4a–c. On the other hand, it also supports a larger number of channel scenarios at $\rho = 2.0$.

This improvement is by closer fulfilling the probability requirements $\epsilon_k = 0.1$. To show this, Fig. 5.5 depicts the empirical CDFs of the actual outage probabilities from the QoS optimization results for Rician fading with $\kappa = 5$ dB and $\rho_k = 1$, $k = 1, \ldots, K$. The CDF curves for $K = 2, 3, 4$ receivers stem from all feasible channel realizations of the chance constraint approximations with $\rho = 2.0$.

Figure 5.5a shows the CDFs for the $K = 2$ receiver case. While the non-robust optimization fails to fulfill the outage requirements, the spherical uncertainty approximation overfulfills the requirements by keeping the outage probability below 0.2%. The SDR method provides slightly higher outage values than the LPM formulation. In contrast, the orthogonal uncertainty approximation with either the

[21] Even though distinct rate targets do not change this relation between the orthogonal and the spherical uncertainty approximation for the chance constraints. Section 5.7.3 shows that the orthogonal uncertainty approximation outperforms the spherical bound also for $\rho_1 = 1.25$ and $\rho_2 = 0.75$ if $K = 2$ for example.

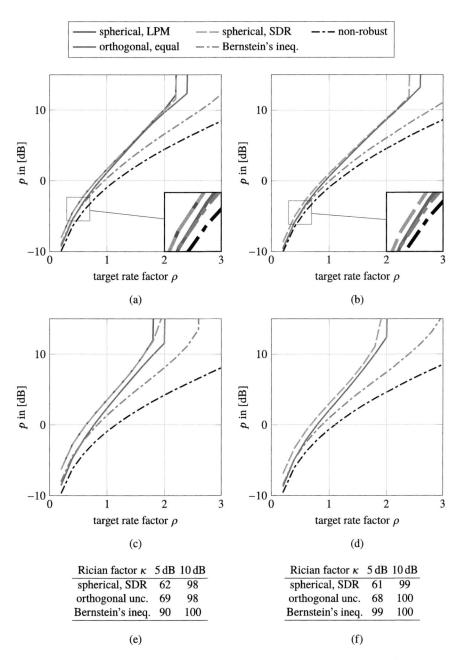

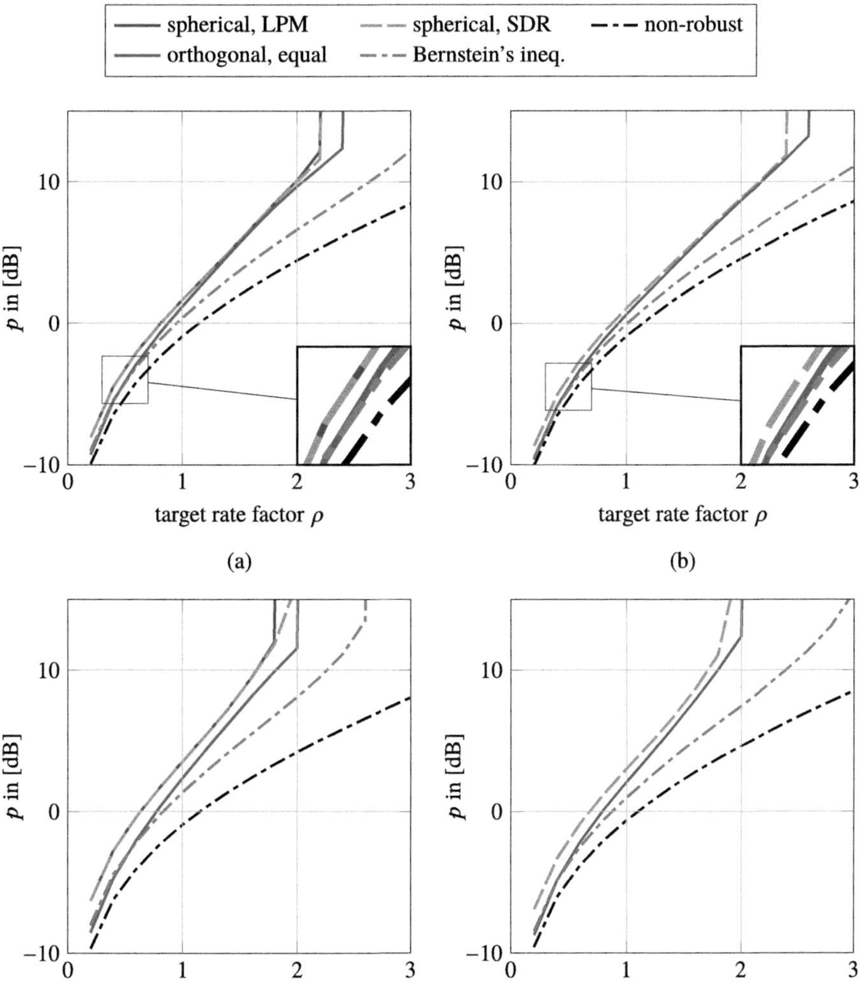

(c) (d)

Rician factor κ	5 dB	10 dB
spherical, SDR	62	98
orthogonal unc.	69	98
Bernstein's ineq.	90	100

(e)

Rician factor κ	5 dB	10 dB
spherical, SDR	61	99
orthogonal unc.	68	100
Bernstein's ineq.	99	100

(f)

Fig. 5.4 Minimum p of the QoS optimization with $\epsilon_k = 0.1$, $\rho_k = 1$, and either $K = 3$ or $K = 4$. The plots are an average for the channel mean realizations where both, the spherical uncertainty approximation and the Bernstein-type inequality approximation are feasible at $\rho = 2.0$. (**a**) $K = 3$ and $\kappa = 10\,\text{dB}$. (**b**) $K = 4$ and $\kappa = 10\,\text{dB}$. (**c**) $K = 3$ and $\kappa = 5\,\text{dB}$. (**d**) $K = 4$ and $\kappa = 5\,\text{dB}$. (**e**) # Feasible scenarios for $\rho = 2$, $K = 3$. (**f**) # Feasible scenarios for $\rho = 2$, $K = 4$

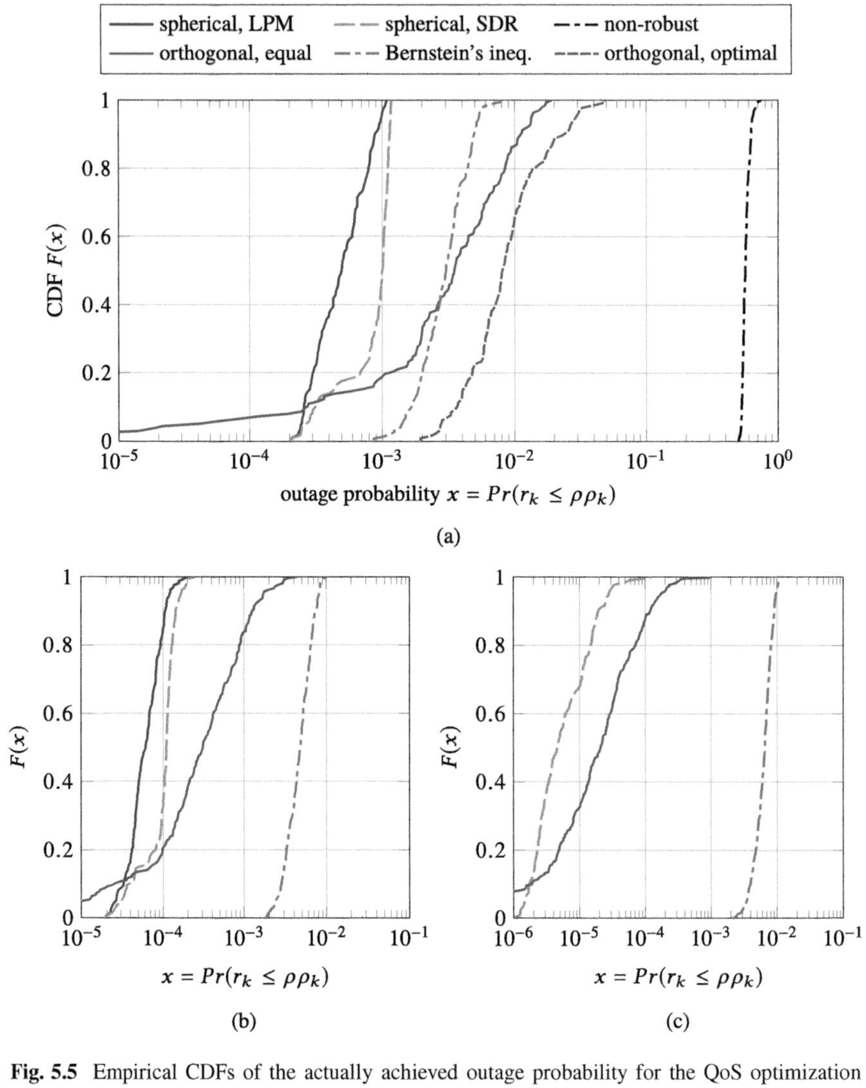

Fig. 5.5 Empirical CDFs of the actually achieved outage probability for the QoS optimization with $\epsilon_k = 0.1$ and $\rho_k = 1$. The plots are an average over all feasible scenarios for the target rate $\rho = 2.0$ and Rician fading with $\kappa = 5$ dB. (**a**) $K = 2$, $\kappa = 5$ dB, and $\rho = 2.0$. (**b**) $K = 3$, $\kappa = 5$ dB, and $\rho = 2.0$. (**c**) $K = 4$, $\kappa = 5$ dB, and $\rho = 2.0$

joint or optimal parameter search results in outage probabilities of up to 2% and 5%, respectively. This leads to the performance improvement compared to the spherical uncertainty approximation of the chance constraint.

Unfortunately, the outage probabilities of the joint parameter search vary with the channel realizations. This indicates the performance loss for jointly searching parameters of the uncertainty region. In contrast, the optimal search for b_k, $k =$

$1, \ldots, K$ results in a much smaller variation of the outage probabilities and closer matches the outage requirements than the joint search. It even surpasses the resulting outage probabilities of the Bernstein-type inequality approximation, which indicates a potential performance gain also with respect to this approximation (cf. Fig. 5.3b).

For the $K = 3$ and $K = 4$ receiver cases (see Fig. 5.5b, c, respectively), the probability of the orthogonal uncertainty approximation is closer to the curve of the spherical uncertainty approximation than to the curve of the first Bernstein-type inequality approximation. This loss in approximation accuracy is the reason for the relative performance decrease in comparison to the Bernstein-type inequality approximation when increasing the number of receivers and antennas. Enhanced parameter search strategies for the orthogonal uncertainty approximation, that overcome this behavior, remain a topic for further research.

5.7.3 Rate Balancing with Approximated Chance Constraints

The simulations for the chance-constrained RB problem are for the same setup and Rician channel model as for the QoS optimization. Again, 100 realizations for the channel means $\bar{\boldsymbol{h}}_k$, $k = 1, \ldots, K$ are drawn at random for the $K = 2$ receiver setup, while 25 realizations are drawn for the $K = 3$ and $K = 4$ user setups. However, the rate targets ρ_k are changed to the distinct values from Table 5.1.

Figure 5.6 shows the on average achieved performance of the RB optimization for the base setup, i.e., with $K = 2$ receivers, and with the chance constraint approximations from Sects. 5.5 and 5.6. Since the RB problem is always feasible, the average is over the optima of all 100 channel mean realizations. The parameters b_k, $k = 1, \ldots, K$ for the orthogonal uncertainty approximation are chosen with (5.71), because the optimal search is computationally too demanding for the average performance comparison. Additionally, the plots show the balancing results for the MSE based Bernstein-type inequlity approximation from Sect. 5.6.3.

While the perfect CSI curve grows unboundedly, the robust RB approaches saturate. The saturation level depends on the quality of the CSI. Figure 5.6a–c shows the resulting optima over P in dB for $\kappa = 5$, $\kappa = 10\,\text{dB}$, and $\kappa = 15\,\text{dB}$, respectively, while Fig. 5.6d depicts ρ over κ for $P = 25\,\text{dB}$. The first Bernstein-type inequality approximation has the highest saturation level of the approximations, i.e., $\rho \approx 2.5$ for $\kappa = 5\,\text{dB}$, $\rho \approx 4.2$ for $\kappa = 10\,\text{dB}$, and $\rho \approx 5.7$ for $\kappa = 15\,\text{dB}$. About the same values are achieved by the orthogonal uncertainty approximation.

With a saturation level of $\rho \approx 2.2$ for $\kappa = 5\,\text{dB}$, the spherical uncertainty approximation falls behind the previous two approximations. The proposed ρ for the

Table 5.1 Rate targets for chance-constrained QoS and RB optimization

# Users	Distinct targets
$K = 2$	$\rho_1 = 1.25$, $\rho_2 = 0.75$
$K = 3$	$\rho_1 = 1$, $\rho_2 = 0.5$, $\rho_3 = 1$
$K = 4$	$\rho_1 = 1.25$, $\rho_2 = 1$, $\rho_3 = 0.75$, $\rho_4 = 1$

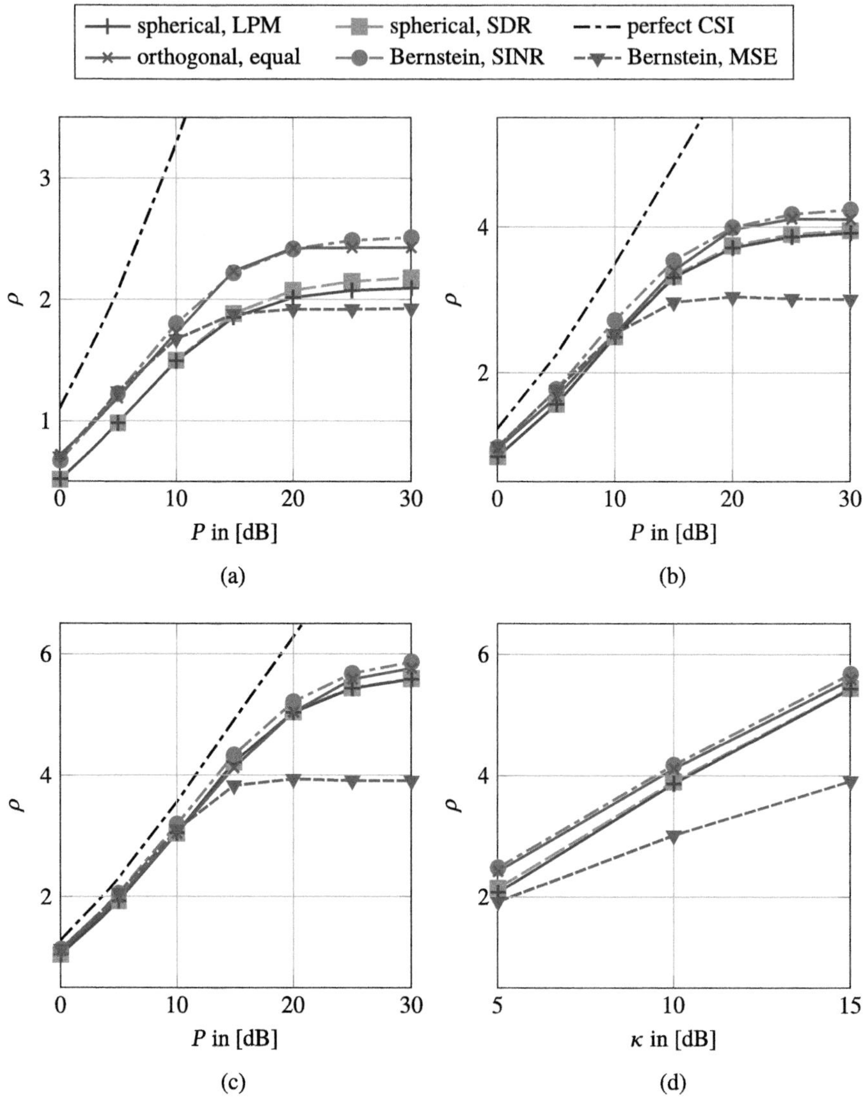

Fig. 5.6 Optimum ρ of the RB optimization for $K = 2$, $\epsilon_1 = \epsilon_2 = 0.1$, and target rates from Table 5.1. The plots show an average over 100 channel mean realizations. (**a**) $K = 2$ and $\kappa = 5$ dB. (**b**) $K = 2$ and $\kappa = 10$ dB. (**c**) $K = 2$ and $\kappa = 15$ dB. (**d**) $K = 2$ and $P = 25$ dB

LPM and SDR constraint reformulation are almost equal below $P = 15$ dB, but the saturation level of the LPM reformulation is smaller than that of the SDR approach because of the additional approximation (cf. Sect. 5.5.1). The accuracy of these uncertainty approximations increases with the quality of the CSI and becomes close to the first Bernstein-type inequality approximation for $\kappa = 15\,\text{dB}$. The spherical uncertainty suits for the chance-constrained RB optimization in this case.

The smallest saturation level for the $K = 2$ receiver case is provided by the second Bernstein-type inequality approximation, i.e., it is $\rho \approx 1.9$ for $\kappa = 5\,\text{dB}$ and increases only up to $\rho \approx 4$ for $\kappa = 15\,\text{dB}$. This scheme achieves the performance of the first Bernstein-type inequality approximation and the orthogonal uncertainty approximation for $P \leq 10\,\text{dB}$. In this low power regime, the spherical uncertainty approximation has a gap of about 0.2 to the other approximations for $\kappa = 5\,\text{dB}$.

The accuracy of the MSE based Bernstein-type inequality approximations improves when increasing the number of receivers and antennas to $K = 3$ and $N = 6$ or even to $K = 4$ and $N = 8$. The corresponding performance plots are shown in Fig. 5.7. The MSE based Bernstein-type inequality approximation outperforms both uncertainty approximations for low to medium transmit power P for $\kappa = 10\,\text{dB}$ (see Fig. 5.7b, d for the $K = 3$ and $K = 4$ receiver setups, respectively). For $\kappa = 5\,\text{dB}$, the channels' error variance has a serious impact on the performance. Then, the second Bernstein-type inequality approximation has even a higher saturation level than the orthogonal uncertainty approximation. However, there remains a performance gap of about 0.7 and 0.5 to the SINR based inequality approximation for $K = 3$ and $K = 4$ receivers, respectively (see Fig. 5.7a, c).

In contrast to the MSE based approximation, the relative performance of the orthogonal uncertainty approximation to that of the spherical uncertainty approximation decreases for $K = 4$. This shows that the joint parameter search is unsuitable for an increasing number of receivers. An enhanced parameter search, e.g., the optimal search, may reduce this relative performance loss.[22]

The achieved outage probability for $K = 2$ and $K = 4$ indicates this relative performance change of the orthogonal uncertainty and the MSE based Bernstein-type approximations to the other two approximations. Figure 5.8 depicts the empirical CDF of the outage probability for the solution beamformers from the approximate RB optimizations, for Rician fading with $\kappa = 5\,\text{dB}$. Clearly, the non-robust optimization fails to fulfill the outage probability requirements. In contrast, the spherical uncertainty approximation is far too conservative for achieving an adequate performance.[23]

For $K = 2$ receivers (see Fig. 5.8a), the outage probabilities for the orthogonal uncertainty approximation vary on a broad range. In less than 5% of the cases, the outage probabilities are smaller than for the spherical uncertainty approximation. However, for about 72% of the cases, the outage probabilities are even larger than that of the SINR based Bernstein-type inequality approximation. In these cases, it achieves the closest match for the outage requirements. This explains the good RB performance of the orthogonal uncertainty approximation for $K = 2$ receivers.

[22] Investigations on a tractable but close to optimal parameter search is an open topic for research.

[23] Again, the SDR method for the spherical uncertainty approximation provides higher outage values than the LPM formulation due to the additional approximation (cf. Sect. 5.5).

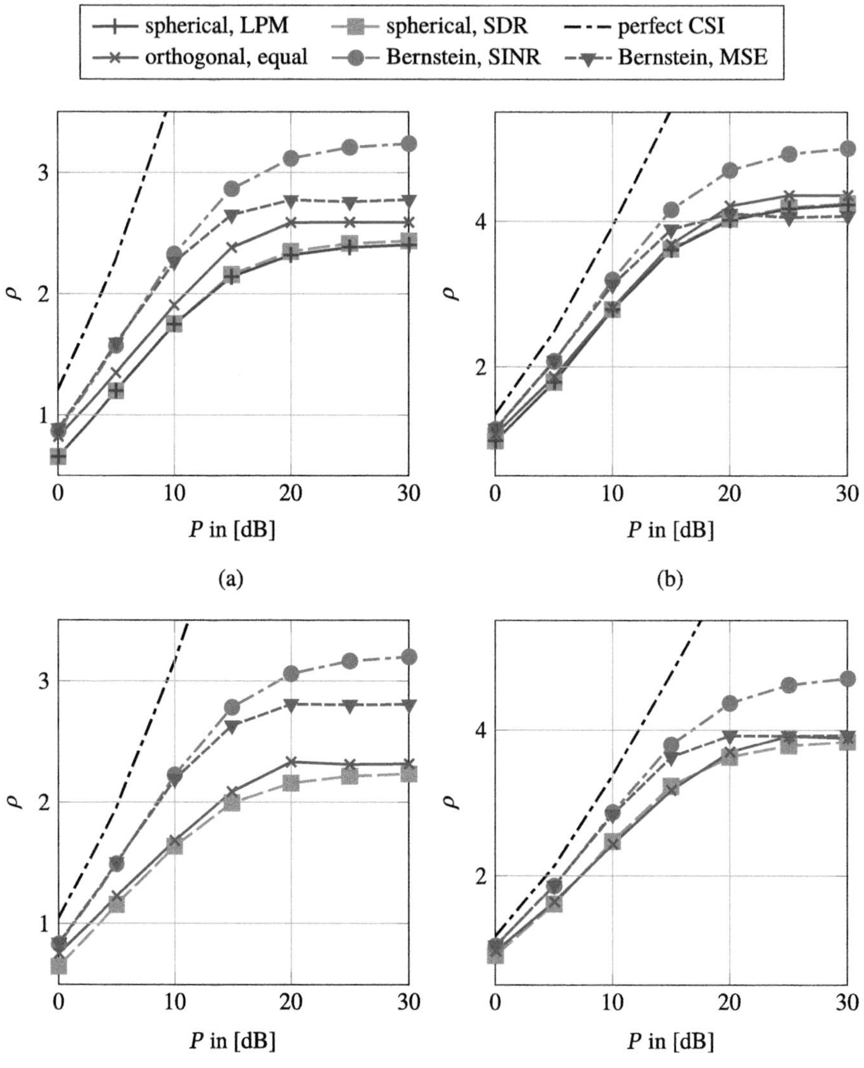

Fig. 5.7 Optimum ρ of the chance-constrained RB optimization for the $K = 3$ and $K = 4$ receiver scenarios, $\epsilon_k = 0.1$ for $k = 1, \ldots, K$, and the rate targets from Table 5.1. The plots show an average over 25 channel mean realizations. (**a**) $K = 3$ and $\kappa = 5$ dB. (**b**) $K = 3$ and $\kappa = 10$ dB. (**c**) $K = 4$ and $\kappa = 5$ dB. (**d**) $K = 4$ and $\kappa = 10$ dB

However, the approximation accuracy for the orthogonal uncertainty region falls behind both Bernstein-type inequality approximations when increasing the number of receivers to $K = 3$ for Fig. 5.8b or to $K = 4$ for Fig. 5.8c. Then, the orthogonal uncertainty approximation becomes overly conservative for many channel mean

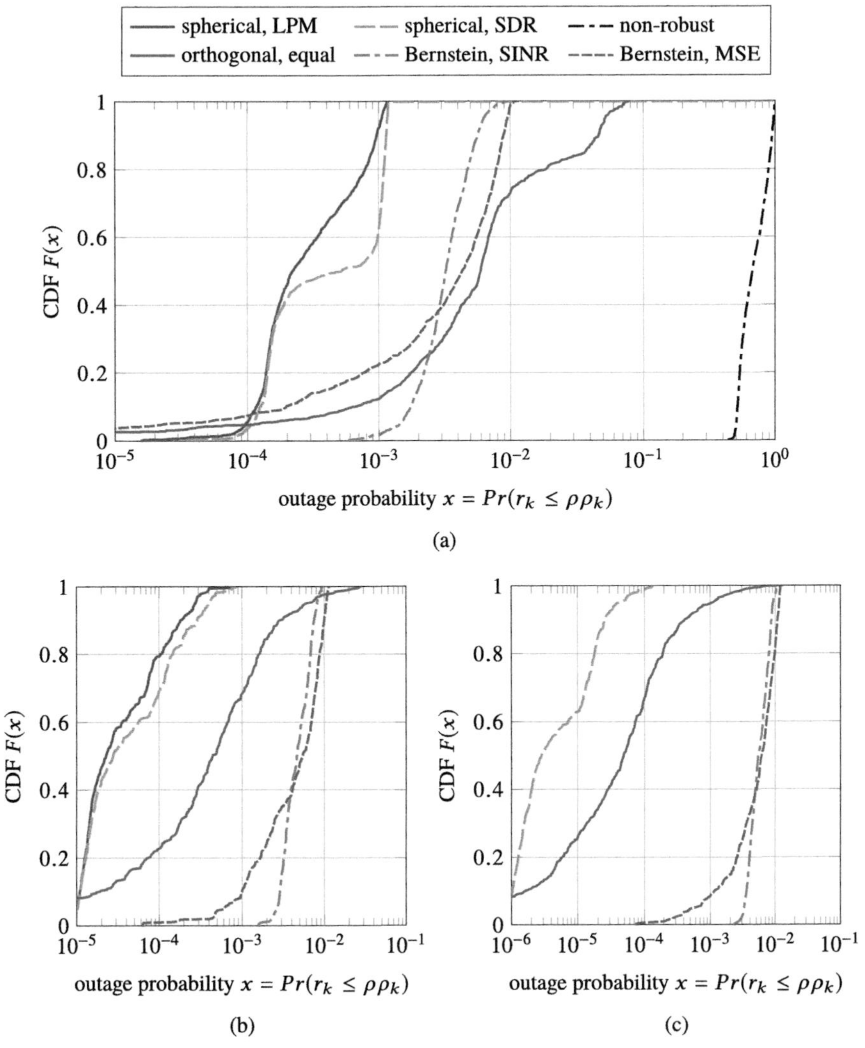

Fig. 5.8 Empirical CDF of the outage probability from the RB optimization results with $\epsilon_k = 0.1$ and the rate targets from Table 5.1. The plots use all channel realizations and all realizations of $P = 0\,\text{dB}, 5\,\text{dB}, \ldots, 30\,\text{dB}$ and the Rician fading with $\kappa = 5\,\text{dB}$. (**a**) $K = 2$ and $\kappa = 5\,\text{dB}$. (**b**) $K = 3$ and $\kappa = 5\,\text{dB}$. (**c**) $K = 4$ and $\kappa = 5\,\text{dB}$

realizations, similar to the spherical uncertainty approximation. This indicates the performance loss compared to the first Bernstein-type inequality approximation.

Both Bernstein-type inequality approximations keep their accuracy for the increased system dimensions. This explains the relative performance improvement of the MSE based approximation compared to the uncertainty approximations when changing the setup from $K = 2$ to $K = 4$ receivers. The SINR based Bernstein-type

Table 5.2 Percentage of rank-one RB solutions for the SDR based constraint approximations

# Users and antennas	Spherical, SDR	Bernstein, SINR	Orthogonal, equal
$K = 2, N = 4$	99.5%	96.0%	98.2%
$K = 3, N = 6$	97.3%	99.0%	80.6%
$K = 4, N = 8$	99.0%	99.0%	68.2%

The RB is for ρ_k from Table 5.1, $\epsilon_k = 0.1$, Rician fading with $\kappa = 5$ dB, and $P = 30$ dB. A covariance matrix $\boldsymbol{Q}_k$ is rank-one if the ratio of the two largest eigenvalues is $\lambda_{2,k}/\lambda_{1,k} \leq 0.01$

inequality approximation still results in a smaller outage probability variation than that for the MSE. For 40% of the channel mean realizations, the outage probability for the MSE based Bernstein-type inequality approximation is smaller than that of the SINR based approximation and it is slightly larger for the remaining 60%. This explains the small average performance loss of the MSE based approximation compared to the SINR based approximation for $\kappa = 5$ dB. However, this gap increases with the quality of the CSI and the transmit power P (see Fig. 5.7).

The advantage of the MSE based Bernstein-type inequality approximation is that it results in a convex constraint set for the beamformers, while the SINR based approximation requires a SDR and, thus, a post-processing beamformer reconstruction. In this context, Table 5.2 shows the percentage of rank-one solution transmit covariance matrices $\boldsymbol{Q}_k$, $k = 1, \ldots, K$ for the SDR based RB optimizations. These results are from the simulations to Fig. 5.7, particularly, for Rician fading with $\kappa = 5$ dB and $P = 30$ dB. A covariance matrix $\boldsymbol{Q}_k$ is denoted as rank-one if the ratio of the largest two eigenvalues $\lambda_{1,k} > \lambda_{2,k} \geq 0$ is smaller than $\lambda_{2,k}/\lambda_{1,k} \leq 0.01$.

The SINR based Bernstein-type inequality approximation and the spherical uncertainty approximation both result in rank-one transmit covariances in 99% of the simulated cases. Even a hundred percent are reached below saturation, i.e., for $P \leq 20$ dB. For the spherical uncertainty approximation, this is in concordance with [291], which states that the optimal transmit covariance matrices are rank-one if the limit for the sum transmit power is low. Hence, the advantage of a beamformer based design using the spherical uncertainty approximation and the LPM formulation or the second Bernstein-type inequality approximation instead of the SDR methods is only marginal for a sum power limitation. The number of solution transmit covariance matrices $\boldsymbol{Q}_k$, $k = 1, \ldots, K$ with $\text{rank}\{\boldsymbol{Q}_k\} > 1$, is apparently large for the orthogonal uncertainty approximation.

5.7.4 Comparison of the MSE Based Approximations

A performance comparison of the MSE based Bernstein-type inequality approximation with the Vysochanskij–Petunin's inequality approximation from [44] is shown in Fig. 5.9. Note that the latter approximation requires ZF beamforming and scaled Matched filter (MF) equalizers. The results are for $N = 4$ transmit antennas and Rician fading with $\kappa = 10$ dB. The maximum outage limits are $\epsilon_k = 0.1$,

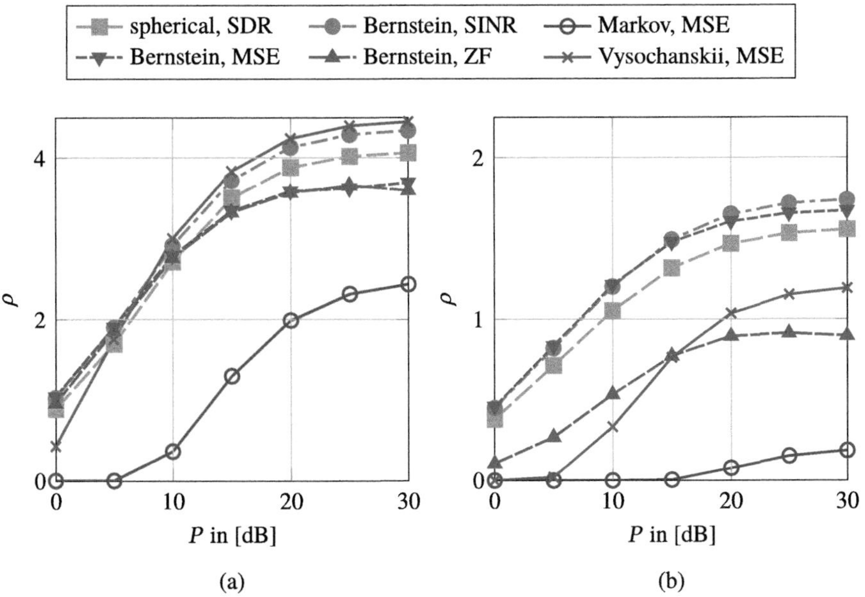

Fig. 5.9 Objective of the RB optimization for $K = 2$ and $K = 4$ receivers and $N = 4$, $\kappa = 10\,\text{dB}$, and $\rho_k = 1$. The plots show an average over 100 channel mean realizations. (**a**) $K = 2$, $N = 4$, $\epsilon_k = 0.10$. (**b**) $K = N = 4$ and $\epsilon_k = 0.10$

$k = 1, \ldots, K$, that is, the Vysochanskij–Petunin's inequality tighter approximates the probability constraints than the Bernstein-type inequality approximation with ZF beamforming and scaled MF equalizers (cf. Sect. 5.6.4). Furthermore, the MSE approximation via Markov's inequality is loose for these outage limits and even becomes infeasible for many channel realizations. In particular, the balanced rate ρ is set to zero for these cases.

For $K = 2$ receivers (see Fig. 5.9a), the Vysochanskij–Petunin's inequality approximation is also tighter than the Bernstein-type inequality approximation with optimized equalizers and either ZF or flexible beamforming for $P \geq 10\,\text{dB}$. It even outperforms the SINR based Bernstein-type inequality approximation that serves as a reference in this plot. First, ZF beamforming is close to optimal, because $N = 2K$ and the channel means are drawn at random from a standard complex Gaussian distribution. Second, the Bernstein-type inequality approximation is loose if $M = \min(N, K) = 2$ and, therefore, $\lambda^+_{\max}(\tilde{\boldsymbol{Q}}_k)$ is close to $\|\tilde{\boldsymbol{Q}}_k\|_\text{F}$ within (5.82).[24]

For $K = 4$ (see Fig. 5.9b), the approximation via the Vysochanskij–Petunin's inequality still outperforms that via the Bernstein-type inequality and ZF beamforming for $P \geq 15\,\text{dB}$. However, the strict ZF beamforming and scaled MF equalizers apparently degrade the performance in comparison to the other approximations,

[24]The dominant positive eigenvalue of $\boldsymbol{A} \in \mathcal{H}_+^N$ is bounded as $N^{-1/2}\|\boldsymbol{A}_k\|_\text{F} \leq \lambda^+_{\max}(\boldsymbol{A}_k) \leq \|\boldsymbol{A}\|_\text{F}$.

especially for $P < 15\,\text{dB}$. This requirement even results in an infeasible RB problem for 65% of the channel mean realizations at $P = 10\,\text{dB}$.[25] In contrast, the MSE based Bernstein-type inequality approximation improves with increasing M. For $\tilde{Q}_k \approx \alpha\mathbf{I}$, the inequality $M^{-1/2}\|\tilde{Q}_k\|_{\text{F}} \leq \lambda^{+}_{\text{max}}(\tilde{Q}_k)$ for the last summand of (5.82) becomes tight. This term vanishes asymptotically with increasing M. Therefore, the performance of the MSE approximation with flexible beamforming resides far above that from [44] for $K = N = 4$.

[25] When the RB problem becomes infeasible due to the approximation, ρ is set to zero.

Chapter 6
Applications in Satellite Communication

Research on SatCom nowadays aims at high data television broadcast and on-demand data transfer at a total rate ranging from several gigabit for mobile services up to one terabit for fixed terminals [294–297]. To pursue this goal, researchers have strengthened investigations to increase the number of spotbeams for Geostationary earth orbit (GEO) satellites, increase the frequency reuse, and use higher frequency bands. For example, the S-band (2–4 GHz) and Ka-band (20–30 GHz) technologies gained importance over the L-band (1–2 GHz) and the Ku-band (12–14 GHz) for mobile and fixed terminal applications, respectively, because it allows to realize smaller transmit and receive apertures [204]. The Ku-band and other bands above 10 GHz have also been investigated for mobile SatCom services [298, 299], e.g., for trains, cars, and boats, such that mobile and fixed terminals may be served simultaneously in these bands.

Adaptive physical layer beamforming and power adaptation are seen as a key technology in this context [53–58, 60]. The reason is that these techniques allow the reuse of the same resources, e.g., bandwidth and time, by limiting the inter-spotbeam interference. Here, the task is to find designs that can cope with the users' demands and take care on simultaneous transmission to fixed and mobile user terminals. Moreover, since the channels to the terminals inherently suffer from fading [59], the beamformer design must be robust to provide reliable data services. Section 6.1 depicts the imposed channel characteristics and fading models and the other sections show the applications of the previous beamforming strategies. The results are separate for the ergodic and outage constrained optimizations (see Sects. 6.3 and 6.5, respectively).

A. Gründinger, *Statistical Robust Beamforming for Broadcast Channels and Applications in Satellite Communication*, Foundations in Signal Processing, Communications and Networking 22, https://doi.org/10.1007/978-3-030-29578-3_6

The large number of receivers and frequency reuse one are additionally challenging. Interference from neighboring spotbeams becomes the ultimate limitation for downlink transmission in these setups [57].[1] To keep the computational complexity tractable and manage the resources, [300] has used adaptive phased antenna arrays and pre-fixed illumination patterns. In contrast, this work studies fully adaptive beamforming techniques to coordinate the interference in neighboring spotbeams.

6.1 Satellite Channel Characteristic

The satellite forward link—transmission from the satellite to the (fixed or mobile) terminals on the earth's surface—has the properties of a vector BC if the terminals are served simultaneously in the same frequency band (e.g., see [52, 53, 301, 302]). The transmit and receive signal model for the terminal (fixed or mobile) follows that of Sect. 1.1 when the satellite linearly precodes the independent data signals.[2] However, the structure and fading properties of the channels $\boldsymbol{h}_k$, $k = 1, \ldots, K$ distinguish from the standard terrestrial models. SatCom channels have an inherently strong LoS component, but they suffer from phase fading, shadowing, atmospheric attenuation, and scattering at obstacles close to the terminals. The limiting fading effect changes with the carrier frequency and also depends on the antenna characteristic of the terminals. Arapoglou et al. [59] summarized these fading models for SatCom. The well-accepted multi-spotbeam channel [52] and its properties is detailed next.

6.1.1 Multi-Spotbeam Model

The considered SatCom setup is for a GEO satellite that works in either S-band for mobile terminals or Ka-band for fixed terminals.[3] The satellite has a multi-spotbeam antenna system to cover its service area, e.g., Europe [53] (or the US [307]). A large number of spotbeams with narrow beamwidth divides this coverage area into a cell

[1]Interference becomes severe if the frequency reuse factor is one, i.e., the same bands are used in all cells, or the beamwidth is decreased below the usually assigned 3 dB area.

[2]For multiple frequency bands, the model results in parallel vector BCs for the distinct frequency bands. These parallel BCs can be combined to a MIMO BC model if the satellite and the terminals allow for carrier cooperation [303]. Carrier cooperation requires sufficiently different channels in the different bands to gain in performance over separately treating the different bands [304].

[3]Mobile and fixed terminal data services use the DVB-SH [305] and the DVB-S2 standards [306].

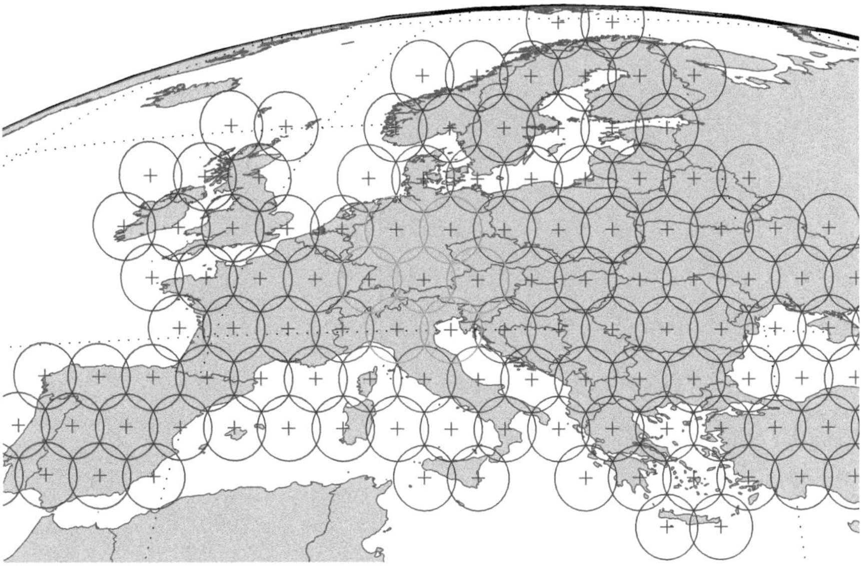

Fig. 6.1 Exemplary beam structure to cover Europe with 121 spotbeams. Clusters of 7 and 19 beams for the simulations, respectively, are highlighted in green and red

structure (see Fig. 6.1). Each spotbeam is created by the radiated power of a horn antenna. This power is focused by a circular reflector [308], [309, Section 9.8].[4] Each cell is bounded by the 3 dB loss of the experienced radiation gain at the earth's surface in comparison to the link-gain along the central path (cf. Fig. 6.2). Spotbeam diffractions due to the reflector aperture are neglected herein.

The link-gain along the central path of a spotbeam is given by the antenna gain G_{tx}, the operating transmit power [309, Chapter 5], the fading losses (see Sect. 6.1.2), the effective terminal gain G_{rx}, and the noise level at the terminals. Even though the gain characteristics offside the central path of a spotbeam depends on the antenna and reflector aperture [308], an undeformed tapered-aperture beam-structure model is commonly employed [309, Section 9.8]. Therewith, the gain $g_{i,k} = g(\theta_{i,k})$ between the ith spotbeam from the satellite to the kth user is a function of the angle $\theta_{i,k}$ between the beamcenter axis and the user's location (see Fig. 6.2) and the one-sided half-power beamwidth $\theta_{3\,\text{dB}}$ (see [301] and [57] for details).

The angular position of the users and the geometric structure of the antenna and reflector aperture also specify the error free phase shifts at the satellite, e.g., for clear sky conditions, idealized knowledge of the atmospheric effects, and fixed terminals. Let $\phi_{i,k}$ be the phase shift of the ith spotbeam to the kth terminal in the

[4]A spotbeam may also be created by an adaptive phased-array architecture.

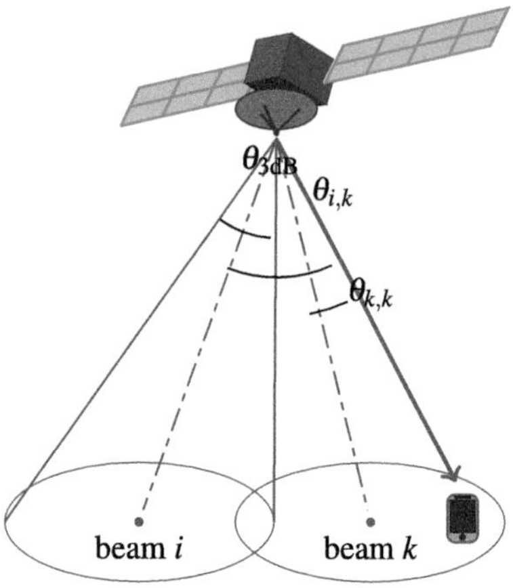

Fig. 6.2 Angular model for multi-spotbeam satellite communication

Table 6.1 Link budget parameters of the (mobile) terminal SatCom system (cf. [52])

Parameter	S-band	Ka-band
Satellite configuration	GEO	GEO
Beamwidth $\theta_{3\,\text{dB}}$ (in degree)	0.4	0.4
Frequency reuse	1	1
Antenna gains $G_{\text{tx}}/G_{\text{rx}}$	52 dBi/3 dBi	42 dBi/41.7 dBi
Approximate FSL	190 dB	210 dB
Noise power	−133 dBW	−143 dBW
Rician fading factor κ	5.10 dB	5.10 dB
Rain fading $m_{\text{rain},k}/\sigma^2_{\text{rain},k}$ [52]	−2.62 dB/1.63 dB, 3.26 dB	−2.62 dB/1.63 dB
Transmit power $P = \sum_{\ell=1}^{L} P_\ell$	0, 5, ..., 30 dB	0, 5, ..., 30 dB

kth spotbeam area. These phase shifts are small due to the narrow angular distance of the spotbeams within the coverage area. We approximate them by assuming that the antennas reside in a plane that is perpendicular to the pointing direction of the satellite. Therewith, the relative beamgain from the satellite's ith antenna element to the kth receiver is

$$b_{i,k} = \sqrt{G_{\text{rx}} G_{\text{tx}}} \sqrt{g_{i,k}(\theta_{i,k})} \mathrm{e}^{-\mathrm{j}\phi_{i,k}}, \quad i = 1, \ldots, N, \quad k = 1, \ldots, K. \tag{6.1}$$

Exemplary values for G_{rx}, G_{tx}, and $\theta_{3,\text{dB}}$ for a mobile terminal system that operates in the S-band and a fixed terminal system for the Ka-band are given in Table 6.1 (cf. [52]). The large satellite antenna gain is required to overcome the FSL between the satellite and the terminals on the earth's surface. It is proportional to the square of the carrier frequency and the distance, e.g., about 36, 000 km for the GEO satellite orbit. Its inverse is denoted as $g_{\text{FSL},k}$. The other entries are the parameters of the channel fading and the powers for the simulations within Sects. 6.3 and 6.5.

6.1.2 Fading Characteristics

The fading of the SatCom channels' stems from the atmospheric attenuation along the LoS path and the scattering in the vicinity of the terminal. The strength of either of these effects depends on the operating frequency and the employed antennas at the users (cf. [2]). For example, above 10 GHz and for terminals with reflector antennas, scattering can be neglected. The influence of the atmosphere and the scattering at small water drops [204], also known as *rain fading*, dominate in these bands. The detailed fading impairments have been summarized by Arapoglou et al. [310] and can be simulated as shown in the ITU-R Recommendations P.618 [205], P.840 [311], and P.676 [312]. The provided simulations focus on the slow-fading rain attenuation $a_{\mathrm{rain},k}$ and the fast-fading scattering at particles in the atmosphere.

The rain attenuation has a log-dB-normal distribution [313], i.e., $a_{\mathrm{rain},k}$ in dB is distributed as $\ln(a^{(\mathrm{dB})}_{\mathrm{rain},k}) \sim \mathcal{N}_{\mathbb{C}}(m_{\mathrm{rain},k}, \sigma^2_{\mathrm{rain},k})$, and is independent for terminals in adjacent spotbeams. Therewith, the channel to the kth terminal would read as

$$\boldsymbol{h}_k = \sqrt{\frac{g_{\mathrm{FSL},k}}{a_{\mathrm{rain},k}}} \mathrm{e}^{\mathrm{j}\psi_k} \boldsymbol{b}_k, \tag{6.2}$$

where the phase $\psi_k \in [0, 2\pi)$ is uniformly distributed and $\boldsymbol{b}_k = [b_{1,k}, \ldots, b_{N,k}]^{\mathrm{T}}$ comprises the beamgains to terminal k. The fast atmospheric multi-path fading effects add a Rayleigh, Rician, or Nakagami component to the channel model [59]. Herein, this additional component is modeled by Rician fading with factor $\kappa \in \mathbb{R}_+$ and $\tilde{z} \sim \mathcal{N}_{\mathbb{C}}(0, 1)$, that is, the channels read as

$$\boldsymbol{h}_k = \sqrt{\frac{g_{\mathrm{FSL},k}}{a_{\mathrm{rain},k}}} \mathrm{e}^{\mathrm{j}\psi_k} \left(\sqrt{\frac{\kappa}{\kappa+1}} + \sqrt{\frac{1}{\kappa+1}} \tilde{z}_k\right) \boldsymbol{b}_k. \tag{6.3}$$

For carrier frequencies below 10 GHz, rain fading is weak. The scattering and shadowing at obstacles in the terminals' vicinity dominates [2]:

- *Slow shadow fading* and blocking, e.g., due to trees and houses near the mobile terminals in rural environments, influences the LoS path.
- *Fast multi-path fading* for the non-LoS signal path follows from scattering and dispersion, e.g., in urban environments or if the elevation angle is low.

The blocking and shadow fading from obstacles close to mobiles may be modeled via a multi-state process [2]. A three-state Markov chain is suggested by [314] and the ITU Recommendation R.681 [315] to account for such environmental changes.

The herein employed S-band channel model from [52] (see also [57]) neglects such blocking and only accounts for the corresponding multi-path effects with a Rician fading model. Accordingly, the channel to the kth (mobile) terminal reads as

$$\boldsymbol{h}_k = \sqrt{\frac{g_{\mathrm{FSL},k}}{a_{\mathrm{rain},k}}} \mathrm{e}^{\mathrm{j}\psi_k} \left(\sqrt{\frac{\kappa}{\kappa+1}} \boldsymbol{b}_k + \sqrt{\frac{1}{\kappa+1}} \boldsymbol{B}_k \tilde{\boldsymbol{z}}_k\right), \tag{6.4}$$

where $\boldsymbol{B}_k = \mathrm{diag}(b_{1,k}, \ldots, b_{N,k})$, $\tilde{\boldsymbol{z}}_k \sim \mathcal{N}_{\mathbb{C}}(\boldsymbol{0}, \boldsymbol{C}_{\tilde{z}_k})$, and $\kappa = 5\,\mathrm{dB}, 10\,\mathrm{dB}, 15\,\mathrm{dB}$.

The beamgain characteristic fluctuates with the realization of $\tilde{z}_k$ in the neighborhood of $\boldsymbol{b}_k$. If the scattering at obstacles in the vicinity of the mobile terminal varies only the signal intensity, but originates in the same beamgain characteristic, the correlation matrix has rank-one, i.e., $\boldsymbol{C}_{\tilde{z}_k} = \mathbf{1}_{N\times N}$.[5] The channel is then equivalent to the multiplicative Rician fading model (6.3). The other extreme with no correlations for the entries in $\tilde{z}_k$, i.e., $\boldsymbol{C}_{\tilde{z}_k} = \mathbf{I}_N$, can be seen as the worst-case scattering environment with many far distant scattering obstacles. The channel has then an additive and multiplicative fading component. A compromise between these extreme covariance matrices is the exponential correlation model $[\boldsymbol{C}_{\tilde{z}_k}]_{i,j} = \phi_k^{|i-j|}$ with $\phi_k \in (0, 1]$, $i, j = 1, \ldots, N$, where the correlation is decaying with the distance to the diagonal entries. Then, the strongest correlation is to the 6 surrounding spotbeams of a terminal in cell k (cf. Fig. 6.1).

6.1.3 Channel Error Model

While the channel states $\boldsymbol{h}_k$ can be estimated for clear sky conditions and fixed terminals, the fast attenuation fluctuations from the atmosphere and the scattering at obstacles on the ground usually result in imperfect CSI. Pilot aided channel training in either the return link—from the mobile to the satellite—or the forward link, i.e., with channel estimation at the terminals and feedback to the gateway, reduces the channel uncertainty [316]. However, the gateway information of the terminals' channels is still outdated due to the long round trip times within SatCom [56, 64].

Then, the beamformer design has only access to the statistics of the fading channel via the feedback, e.g., it has access to an estimate $\bar{\boldsymbol{h}}_k$ and the covariance matrix $\boldsymbol{C}_k$. An appropriate channel error model reads as [cf. (6.2)]

$$\boldsymbol{h}_k = (1 + \xi_{\text{rain},k})(\bar{\boldsymbol{h}}_k + \boldsymbol{e}_k), \quad k = 1, \ldots, K. \tag{6.5}$$

Here, the inverse multiplicative attenuation factor $\zeta_{\text{rain},k} = (1+\xi_{\text{rain},k})^{-2}$ is modeled via a log-db-normal distribution to account for effects of the atmosphere on the path from the satellite to the earth. The unknown additive error shall again be circularly symmetric complex Gaussian. It results from a changing scattering, e.g., due to receiver mobility or particles in the atmosphere, between the training phase and the forward-link transmission based on the estimated channel information.

This model is equivalent to (6.2) if $\zeta_{\text{rain},k}$ equals the rain attenuation $a_{\text{rain},k}$, i.e., $\xi_{\text{rain},k} = a_{\text{rain},k}^{-1/2} - 1$, and the channel estimate and the covariance matrix are

[5] A rank-one covariance matrix is used in [141, 143] for a low-complex robust physical layer design. Beamformer designs for a full-rank channel covariance matrix are used in [63, 146].

$$\begin{aligned}\bar{\boldsymbol{h}}_k &= \sqrt{g_{\mathrm{FSL},k}\frac{\kappa}{\kappa+1}}\boldsymbol{b}_k,\\ \boldsymbol{C}_k &= \frac{g_{\mathrm{FSL},k}}{\kappa+1}\boldsymbol{B}_k\boldsymbol{C}_{\tilde{z}_k}\boldsymbol{B}_k^{\mathrm{H}},\end{aligned} \tag{6.6}$$

respectively. In contrast to the error model from Sect. 2.3, the multiplicative error $\xi_{\mathrm{rain},k}$ also affects $\boldsymbol{e}_k \sim \mathcal{N}_{\mathbb{C}}(\boldsymbol{0}, \boldsymbol{C}_k)$. If this additive error has a rank-one covariance matrix, e.g., for the Ka-band, the error model becomes [cf. (6.3)]

$$\boldsymbol{h}_k = (1+\xi_{\mathrm{rain},k})(1+\xi_k)\bar{\boldsymbol{h}}_k, \quad k = 1, \dots, K, \tag{6.7}$$

where $\xi_k \sim \mathcal{N}_{\mathbb{C}}(0, \kappa^{-1})$ results from the rank-one additive error covariance matrix.

6.2 Balancing Optimization for Satellite Communication

Since on-demand data and video services are the considered target for SatCom, the performance metric of interest is the spectral efficiency (cf. [57, 255, 317, 318]). Besides throughput maximization and rate matching, QoS and RB optimizations are undoubtedly the most important physical layer designs in SatCom [57]. Since SatCom channels are inherently subject to fading, the robust beamforming strategies from Chaps. 3 and 5 are appealing to avoid unwanted outages.

For dominating fast fading effects and long codes, the beamforming design is with respect to the ergodic rate (1.35). Then, the strategies of Chap. 3 are applied as long as the channel uncertainty is sufficiently well modeled by the rank-one covariance matrix model. When the ergodic rate is unavailable, e.g., because the additive fading does not feature a rank-one covariance matrix, the average MMSE balancing from Chap. 4 is applied instead. The MSE metric was already exploited to find efficient beamforming schemes for SatCom [56, 60–63]. In contrast to the standard precoding strategies within these works, herein the MMSE balancing problem is solved directly for characteristic power constraints.

If the channels are constant during a transmission block but changes between the training and transmission phase, the epsilon-outage rate (5.2)—the achievable rate under limited probability of an outage—is also the measure of interest for SatCom (e.g., see [64]). To ensure reliable access and data transmission, the outage limits ϵ_k are kept small for SatCom, e.g., $\epsilon_k \leq 0.1$. The thereof obtained outage constrained RB optimization also has to cope with the multiplicative attenuation error from (6.5). While this does not change the reformulation for rank-one additive fading (6.7), the deterministic approximations for a full-rank additive channel error cannot be applied directly. For this reason, the outage probabilities are separated with respect to the multiplicative attenuation uncertainty and the additive channel error from the Rician multi-path fading model. This separation provides an additional degree of freedom to trade-off the probability bounds between the two error types. For an optimized choice, the chance constraint approximations and the resulting optimizations are repeated for various such trade-off values (see Sect. 6.5).

Chance-constrained beamforming for SatCom has also been studied by Gharanjik et al. [64]. His channel model distinguishes from (6.5) in that the attenuation is a known component and the beamgains are subject to phase fading, that is, only the phases $\phi_{i,k}$ in (6.1) are unknown. The corresponding approximation for the chance constraints does not apply for the channel error model at hand.

Besides the channel statistics, both robust RB optimizations have to deal with multiple power constraints within $\mathcal{P}$, which are imposed by the satellites high-power RF amplifier chains and the available power budget (cf. Sect. 1.4):

- per-antenna constraints if each spotbeam has one RF amplifier chain,
- per-array constraints if small groups of antennas are served by the same resource,
- a total sum requirement to model the overall transmit power limits.

These conic quadratic constraints appropriately model the power consumption at the satellite for the provided beamformer designs and their performance comparison within SatCom. Alternatively, Zheng et al. [57] used generic non-linear functions to model the input-to-output relation of the satellite's high-power RF amplifiers, which operate near to their saturation. If the output power is a concave function of the input power (cf. [309, Section 9.2.1]), the linear model is a local approximation.

6.3 Ergodic Rate and Mean Square Error Optimization

To apply the ergodic balancing optimizations from Chaps. 3 and 4, the different error sources for SatCom have to be taken into account. If the atmospheric attenuation changes only slowly, $a_{\mathrm{rain},k}$ can perfectly be estimated, i.e., the pilot-based training of the channel is sufficient to set the error $\xi_{\mathrm{rain},k}$ to zero. Then, the error models (6.5) and (6.7) are equivalent to the additive and multiplicative error models (2.1) and (2.2), respectively. The ergodic RB strategies from Chap. 3 and the MSE balancing methods from Chap. 4 can directly be applied in this case.

When the atmospheric attenuation fluctuates too fast for an accurate estimation, the corresponding error cannot be neglected. This additional error source complicates the expectation evaluations for the ergodic rate and the average MSE. The joint expectation computations are overcome by bounding $R_k(\boldsymbol{t})$ in two steps. First, the ergodic rate is bounded with respect to the attenuation $\zeta_{\mathrm{rain},k} = |1 + \xi_{\mathrm{rain},k}|^{-2}$:

$$R_{\mathrm{rain},k}(\boldsymbol{t}) = \mathrm{E}_{\boldsymbol{e}_k}[r_k^{\mathrm{LB1}}(\boldsymbol{t}, \bar{\boldsymbol{h}}_k + \boldsymbol{e}_k)] \leq \mathrm{E}_{\boldsymbol{h}_k}[r_k(\boldsymbol{t}, \boldsymbol{h}_k)] = R_k(\boldsymbol{t}), \tag{6.8}$$

where the inequality is due to LB1 from Sect. 3.2.2. The rate bound r_k^{LB1} reads as

$$r_k^{\mathrm{LB1}}(\boldsymbol{t}, \bar{\boldsymbol{h}}_k + \boldsymbol{e}_k) = \log_2\left(1 + \frac{|(\bar{\boldsymbol{h}}_k + \boldsymbol{e}_k)^{\mathrm{H}}\boldsymbol{t}_k|^2}{\mathrm{E}[\zeta_{\mathrm{rain},k}]\sigma_k^2 + \sum_{i\neq k}|(\bar{\boldsymbol{h}}_k + \boldsymbol{e}_k)^{\mathrm{H}}\boldsymbol{t}_i|^2}\right), \tag{6.9}$$

where the noise is enhanced by $\mathrm{E}[\zeta_{\mathrm{rain},k}]$. This mean value is computed numerically for the simulations. For the rank-one channel error model from (6.7), the ergodic rate (6.9) simplifies to $r_k^{\mathrm{LB1}}(\boldsymbol{t}, (1+\xi_k)\bar{\boldsymbol{h}}_k)$.

Balancing the ratios $R_{\mathrm{rain},k}(\boldsymbol{t})/\rho_k$ instead of $R_k(\boldsymbol{t})/\rho_k$ for SatCom, i.e.,

$$\max_{\boldsymbol{t}} \min_{k} \frac{R_{\mathrm{rain},k}(\boldsymbol{t})}{\rho_k} \quad \text{s.t.} \quad \boldsymbol{t} \in \mathcal{P}, \tag{6.10}$$

results in a tight lower bound for the actual problem and enables the RB solution strategies from Chap. 3 for the rank-one channel covariance matrix case. If the covariance matrix $\boldsymbol{C}_k$ has a rank larger than one, the related MMSE formulation from Chap. 4 conservatively approximates this ergodic RB problem.

6.4 Results for Rate and Mean Square Error Balancing

Numerical simulations for the rate and MSE balancing optimizations are performed for the Ka-band and the S-band model in Table 6.1, respectively. While simulations with the full-rank covariance matrix scenario are only for a cluster of 7 neighboring spotbeams, the computations with the rank-one channels are also tractable for 19 and 121 spotbeams (e.g., see Fig. 6.1), even for per-antenna constraints.

6.4.1 Results for Rain and Rank-One Additive Channel Fading

For the simulations with the rank-one fading model (6.7), the target rates are $\rho_j = 2$ and $\rho_i = 1$ for terminals $j = 2k-1$ and $i = 2k$, $k = 1, \ldots, K/2$. The channels are for the Ka-band system (see Table 6.1) and for setups with either $N = K = 7$, $N = K = 19$, or $N = K = 121$ spotbeams and terminals. A sum power limitation is imposed for a performance comparison of the approximate RB optimization with either the ergodic rate bounds or the SCS strategy from Sects. 3.2 and 3.5.

For these approximation strategies, the optimum of (6.10) is computed for 100 channel realizations. Each of these realizations corresponds to a randomly drawn placement of the terminals within the spotbeams and independently drawn rain attenuation coefficients. The satellite approximates the variation in the rain attenuation using the rate bounds (6.9) and complete knowledge of the fading statistic.

Figure 6.3 depicts the average of the optima ρ for the channel realizations over the transmit power P in dB for the above configurations of the multi-spotbeam satellite system. The curves within the (sub-)figures are for the upper bound UB2 and the lower bounds LB2 and ALB for $\mathrm{E}[r_k^{\mathrm{LB1}}(\boldsymbol{t}, (1+\xi_k)\bar{\boldsymbol{h}}_k)]$ from Sect. 3.2.2, the ZF beamforming with perfect power allocation, and the SCS from Sect. 3.5. Figure 6.3a, b is for $K = N = 7$ spotbeams and terminals, Fig. 6.3c shows the

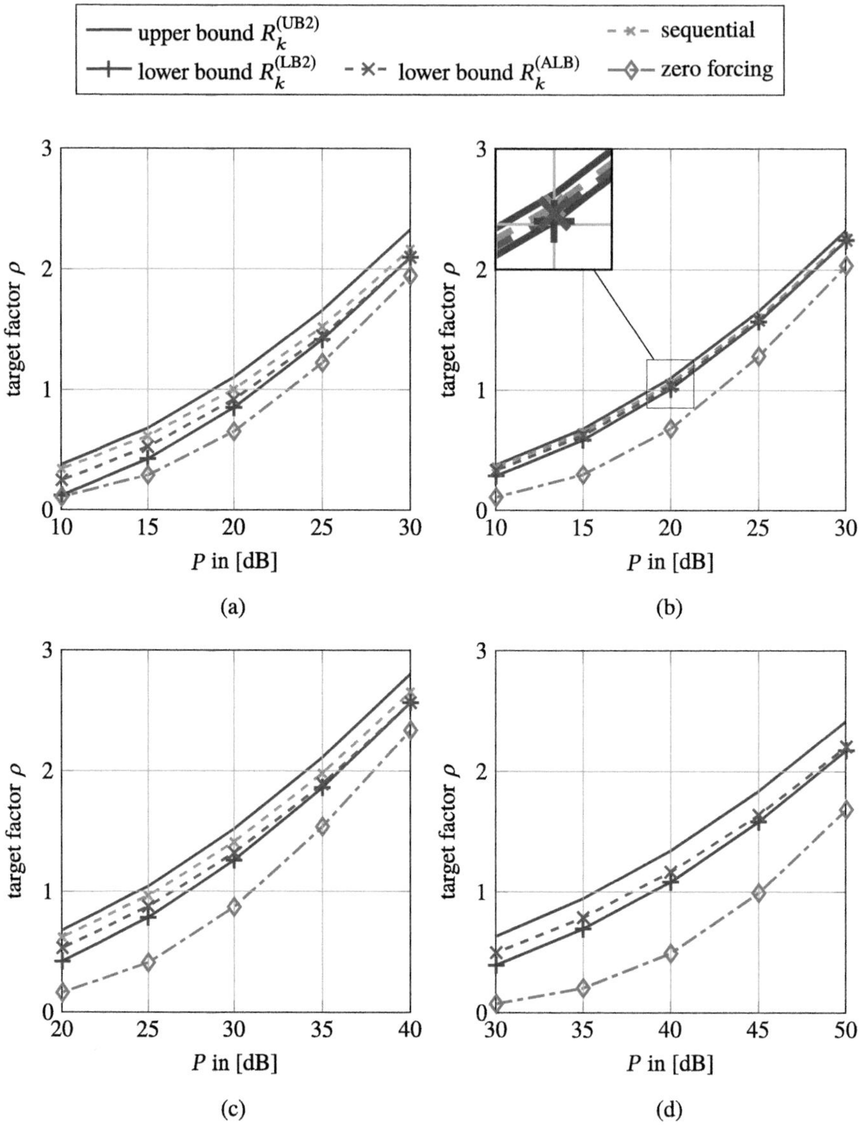

Fig. 6.3 Maximum balancing level ρ vs. sum transmit power P in dB for ergodic RB optimization in the multi-spotbeam SatCom setup in the Ka-band with either $K = 7$, $K = 19$, or $K = 121$ terminals with known rain attenuation and Rician fading. (**a**) $N = K = 7$, $\kappa = 5\,$dB. (**b**) $N = K = 7$, $\kappa = 10\,$dB. (**c**) $N = K = 19$, $\kappa = 5\,$dB. (**d**) $N = K = 121$, $\kappa = 5\,$dB

average ρ for $K = N = 19$, and Fig. 6.3d is for the $K = N = 121$ spotbeams covering Europe (cf. Fig. 6.1). The repeated RB optimization within the SCS method is too complex for $K = N = 121$, but the rate bounds are still tractable as only one single RB optimization has to be solved for each channel realization.

The curves for the SCS strategy lie between the upper and lower bounds as expected. Moreover, the ALB curves outperform the LB2 curves, because the ALB bound tighter approximates the average rate. The ZF strategy obviously falls behind the obtained performance of the lower bounds. The reason is that strict ZF leads to a remarkably decrease of the useful signal power for SatCom. Only for a transmit power over 50 dB for $K = N = 7$, ZF beamforming reaches the LB2 and ALB curves. This loss for strict interference avoidance increases with the system dimension. Then, the adaptive interference management with the ergodic rate bounds gains importance.

When the channel uncertainty is decreased by increasing κ from 5 dB to 10 dB, the distance between the UB2 curve and the lower bound curves becomes small. In contrast, the distance between the bounds remains almost constant when increasing the system dimension to either $N = K = 19$ or $N = K = 121$. This indicates that these bounds are useful also for very large spotbeam scenarios if the channel covariance matrices are rank-one. To reach a similar rate target range, the focus of the plots is shifted by 10 dB from the $N = K = 7$ case in Fig. 6.3a to the $N = K = 19$ case in Fig. 6.3c, and by another 10 dB for the $N = K = 121$ scenario in Fig. 6.3d. Since the inter-spotbeam interference increases with the system dimensions, a larger transmit power is required to achieve the same performance within the large multi-spotbeam SatCom scenario.

6.4.2 *Results for Rain and Full-Rank Additive Channel Fading*

Simulations for the full-rank channel covariance matrix model are based on the MSE. The MSE balancing optimizations are performed for the S-band SatCom system (see Table 6.1) with a cluster of $K = N = 7$ neighboring spotbeams and terminals (see Fig. 6.1). Furthermore, the per-spotbeam power limitations $P_\ell = P/L$ for $\ell, k = 1, \ldots, K$ are imposed and the frequency reuse is one, i.e., the satellite serves all 7 terminals simultaneously in the same frequency band. The mobile terminals are randomly placed within the 3 dB area of the SatCom beams. In this way, 100 different channel characteristics are created. Additionally, one terminal placement is drawn for the 121 spotbeams that covers Europe (cf. Fig. 6.1).

Figure 6.4 depicts the mean of the balanced MSEs versus P in dB for the 100 channel realizations. The perfect CSI curve provides a lower bound for the achievable MSEs per-user. The imperfect CSI curves use the MSE balancing method from Sect. 4.5, but for imperfect transmitter and receiver CSI to keep the complexity

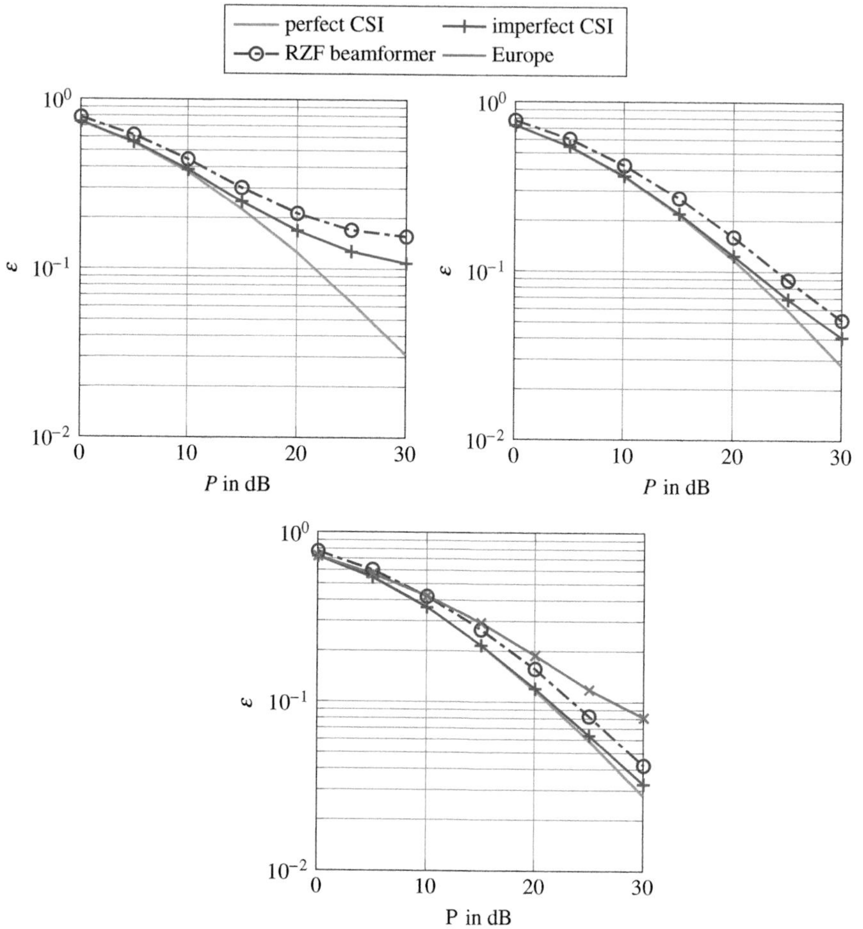

Fig. 6.4 Average MSE balancing results versus P for the S-band spotbeam SatCom model with $N = K = 7$ neighboring spotbeams or $N = K = 121$ spotbeams, targets $\varepsilon_k = 1$, and per-spotbeam power limits $P_\ell = P/N$, $\ell = 1, \ldots, N$. (**a**) $\kappa = 5\,\text{dB}$. (**b**) $\kappa = 10\,\text{dB}$. (**c**) $\kappa = 15\,\text{dB}$

for the expectation computations sufficiently small for the large SatCom system.[6] The RZF beamformer scheme from [52] is based on the channel estimates. A subsequent power allocation jointly minimizes the average MSEs.

Both these imperfect CSI curves are close to the perfect CSI curve for small P. For $\kappa = 5\,\text{dB}$ and $P \geq 20\,\text{dB}$, the imperfect CSI MSE curve flattens out and finally saturates (see Fig. 6.4a). The saturation level is about 0.1 for $\kappa = 5\,\text{dB}$ and strictly decreases with increasing quality of the channel information. Only the flattening

[6]With up to 121 spotbeams and simultaneously served terminals in one frequency band, multi-spotbeam SatCom systems are larger than common terrestrial systems, e.g., Wi-Fi and LTE.

behavior remains visible for $\kappa = 10\,\text{dB}$. For $\kappa = 15\,\text{dB}$, the imperfect CSI curve meets the perfect CSI curve also for $P = 25\,\text{dB}$. The multi-path scattering may therefore be neglected in this case.

The MSE balancing curves are strictly below the RZF beamforming curve. The loss of RZF beamforming is a consequence of the neglected SatCom specific beamgain characteristic for the regularization term. This characteristic deforms the channel mean and error covariance alike according to the power profile [cf. (6.4)]. The scaled identity regularization matrix results in a mismatched upscaling of the errors from neighboring spotbeams by the inverse matrix $(\bar{\boldsymbol{H}}\bar{\boldsymbol{H}}^{\mathrm{H}}+\alpha\mathbf{I})^{-1}$ if $\alpha \in \mathbb{R}_{+}$ is too large. The RZF scheme has therefore a 26% higher MSE per-user than the direct MSE balancing for the 7 spotbeam system with $\kappa = 15$ dB and $P = 20\,\text{dB}$.

In addition to the 7 cell MSE curves, Fig. 6.4c shows the balanced MSE for one exemplary channel realization with 121 spotbeams. This curve decreases apparently slower than the 7 spotbeam curves. The reason is the increased inter-spotbeam interference in this scenario. Users in boundary cells experience less interference on average than users from cells in the inner region, i.e., with 6 neighbor cells, and the ratio of inner to boundary cells is increased for the 121 cell scenario. In other words, the system becomes interference limited in the operating power regime.

6.4.3 Performance Limits with Per-Antenna Power Constraints

To sketch the effects of the various power constraints for SatCom, the optimization is simplified to a sum MSE minimization for $K = 2$ terminals and $N = 4, 8$ spotbeams that are placed in a row (cf. [255]). This leads to a simplified exponential power profile model that forms the channels' gain characteristic, which is then corrupted by the log-normal rain attenuation and Rician fading. In this setup, the transmitter prefers serving the kth user with the kth spotbeam antenna and the rain attenuation determines the terminals' link quality.

Figure 6.5 depicts the average of the minimized sum MSEs for 1000 rain attenuation realizations for the simplified system and either $N = 4$ antennas (see Fig. 6.5a) or $N = 8$ antennas (see Fig. 6.5b). The main three lines in each subfigure correspond to a sum power limitation P, per-array power constraints with $P_\ell = P/2$, and per-antenna constraints with $P_\ell = P/N$. Here, the jth per-array constraint joins the power for the amplifiers of antennas $i = 2j$ and $i = 2j + 1$, $j = 1, \ldots, N/2$. The minimum sum MSEs and optimal beamformers for the per-array and per-antenna constraint cases were computed with the ACS from Sect. 4.5.

With the above choices for the transmit power limits P_ℓ, the per-antenna constraints are stricter than the per-array constraints and the sum power constraint. For this reason, the average sum MSE increases with the number of power constraints.[7] In other words, the sum power limitation curve is a lower bound for the achievable

[7] The same holds for the computational complexity of the beamformer computation (cf. Sect. 4.7).

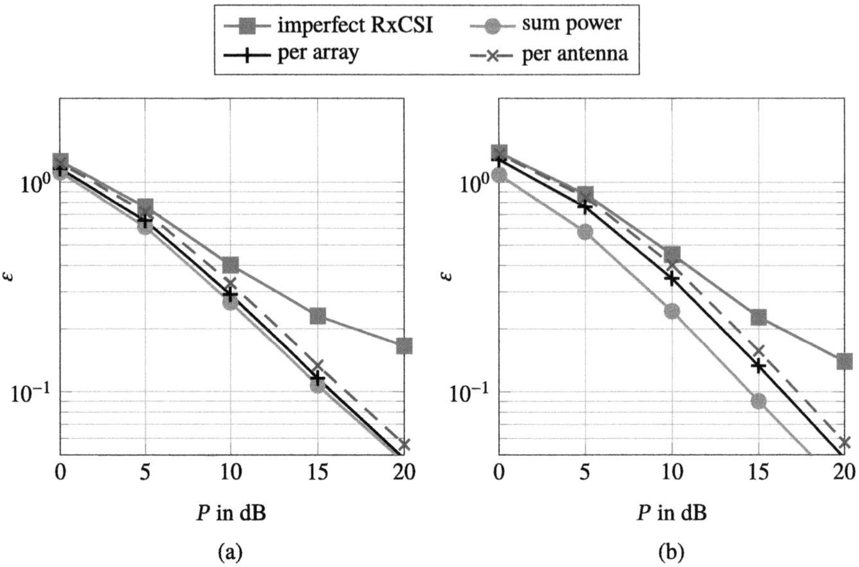

Fig. 6.5 Sum MSE versus P for a row of N spotbeam antennas serving $K = 2$ terminals. The transmission is subject to per-antenna constraints, per-array constraints, or a sum power constraint at the satellite and suffers from log-normal rain attenuation and Rician fading with $\kappa = 10\,\text{dB}$. (**a**) $N = 4$ antennas. (**b**) $N = 8$ antennas

sum MSEs with per-array or per-antenna constraints. An upper bound is obtained by assuming that the terminals and the satellite have the same imperfect CSI and per-antenna constraints.

The difference between the sum power and the per-array and per-antenna power constraint lines increases with N (for constant K). While the system still has full flexibility to allocate all power to the two main serving antennas for the sum power constraint, this flexibility is restricted by the per-antenna and per-array constraints.

Remark 6.1 The losses of the per-antenna power limitation compared to the sum power limitation are not as substantial for spotbeam scenarios with $N = K$. On the one hand, there is one terminal in each coverage cell. On the other hand, the deviation of the maximum link-gains $b_{k,k}$, $k = 1, \ldots, K$ in neighboring spotbeams is only about 3 dB. Therefore, the available power budget is restricted by serving all the terminals with their corresponding spotbeam antennas.

The sum MSE results imply only a small loss for per-antenna or per-array constraints instead of a single sum power constraint. Therefore, the computationally less complex sum power constraint provides a good benchmark for the achievable MSE performance for SatCom. However, the optimization with the sum power constraint fails to deliver a beamformer approximation for the per-antenna or per-array constraint cases. The sum constraint beamformers dramatically increase the dynamic range for the actual per-antenna/per-array powers $\hat{P}_\ell$. Figure 6.6 depicts the empirical CDF curves of the largest per-antenna/per-array power $\hat{P}_{\max} = \max_\ell \hat{P}_\ell$

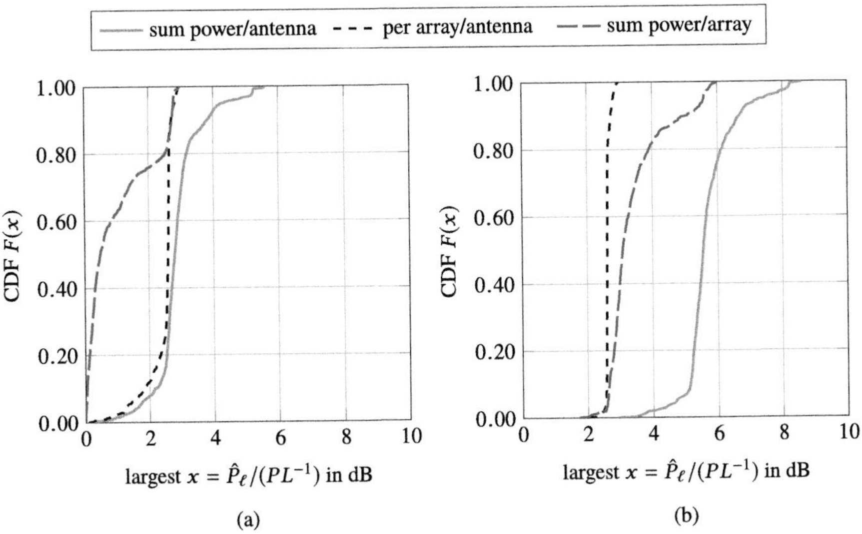

Fig. 6.6 Empirical CDF of the largest per-antenna/array power $\hat{P}_\ell$ normalized to P/L in dB for the sum MSE optimal beamformers with a sum power constraint or per-array power limits, rain fading, and Rician fading with $\kappa = 10\,\text{dB}$. (**a**) $N = 4$ antennas. (**b**) $N = 8$ antennas

when minimizing the average sum MSE with a sum power or per-array constraints. Here, $\hat{P}_{\max}$ is normalized such that the power limits $P_\ell = P/L$ represent 0 dB.

For $N = 4$ and a sum power constraint, a double of the per-antenna bound (3 dB) is surpassed in more than 20% of the channel realizations. This 20% bound increases to 6 dB for $N = 8$ due to two properties. First, the exponential channel characteristic leads preferably to a power profile with a large power for only the two antennas with the highest channel gains. Second, since the channels are subject to rain fading, their gains are likely to differ. This results in an unbalanced power distribution also for the main antennas. These properties also explain the increased sum MSE gap between the sum power and the per-antenna curves for $N = 8$ compared to $N = 4$ in Fig. 6.5.

If $N/2$ per-array constraints are imposed instead of per-antenna constraints, the relative power for one antenna stays below the 3 dB bound. Astonishingly, 2 dB is surpassed by 85% of the channel realizations for $N = 4$ and by 99% if $N = 8$. Hence, almost all the power is allocated to the main spotbeams for $K \leq N/2$.

Figure 6.6 also shows the maximum per-array power at the satellite if a sum power limitation was imposed for the optimization. While this maximum stays close to the actual per-array limits for 50% of the realizations if $N = 4$, almost all realizations have a maximum per-array power larger than 2.5 dB for $N = 8$.

Hence, the results confirm the above statement: The sum power limitation provides a good benchmark for the achievable performance when optimizing the multi-spotbeam SatCom forward link with respect to the MSE. However, solely including the sum power limit within $\mathcal{P}$, the optimization fails to predict an appropriate precoder for the transmit scenarios at hand.

6.5 Outage Constrained Rate Optimization

While Chap. 5 strictly distinguishes between the multiplicative and full-rank additive channel error models for the chance-constrained beamformer design, the chance constraints in SatCom are with respect to the combined multiplicative and additive error model (6.5). If the error model reduces to the joint multiplicative error model (6.7), the epsilon-outage rate still features the deterministic rate formulation (5.22). However, the enhanced noise variance $q_k \in \mathbb{R}_+$ is now the $1-\epsilon_k$ quantile of the product $z = \zeta_k \zeta_{\text{rain},k}$ of the independent random variables $\zeta_k = |1+\xi_k|^2$ and $\zeta_{\text{rain},k} = (1+\xi_{\text{rain},k})^2$ [cf. (6.7)]. Since $2\sigma_{\xi_k}^{-2}\zeta_k^{-1} \sim \mathcal{X}_2^2(2\sigma_{\xi_k}^{-2})$ for this product, the CDF of z reads as [cf. (A.72)]

$$F_z(t) = \int_0^\infty \frac{1}{2}\mathrm{e}^{-\frac{s}{2}+\sigma_{\xi_k}^{-2}} \left(\sum_{j=0}^{\infty} \frac{(\sigma_{\xi_k}^{-2}s/2)^j}{(j!)^2}\right) \Phi\left(\frac{\ln(10\log_{10}(s2^{-1}\sigma_{\xi_k}^2 t)) - \mu_{y_{\mathrm{dB}}}}{\sigma_{y_{\mathrm{dB}}}^2}\right) \mathrm{d}s. \tag{6.11}$$

Evaluating this CDF requires a numerical integration. The quantile $q_k = F_z^{-1}(1-\epsilon_k)$ is then obtained via a line search that repeatedly evaluates (6.11).

A joint approximation of the multiplicative attenuation uncertainty and the Gaussian additive channel error is unavailable if the Rician fading is not rank-one. An appropriate uncertainty region for the joint additive and multiplicative error model is unknown and the Bernstein-type inequalities (5.15) and (5.16) are invalid for non-Gaussian random vectors. Furthermore, accurately computing the transmit success probability requires a nested numerical integration for additive Gaussian and multiplicative errors. Such an integral representation reads as

$$\Pr\left(g_k(\boldsymbol{t},\boldsymbol{e}_k) \geq \sigma_k \zeta_{\text{rain},k}\right) = \int_0^\infty f_{\zeta_{\text{rain},k}}(z) \Pr\left(g_k(\boldsymbol{t},\boldsymbol{e}_k) \geq \sigma_k^2 \zeta_{\text{rain},k} \middle| \zeta_{\text{rain},k} = z\right) \mathrm{d}z, \tag{6.12}$$

where $g_k(\boldsymbol{t},\boldsymbol{e}_k)$ is given by (5.53) and $\zeta_{\text{rain},k}$ follows a log-dB-normal distribution similar to the rain attenuation, i.e., with the PDF from (A.70). The inner conditional probability of (6.12) is computed with Imhof's method [174] for all required values z to numerically evaluate the outer integral with a quadrature integration rule and a truncation from above. Since the PDF (A.70) and the inner probability (6.12) are exponentially decreasing in z, the truncation error remains small.

For a conservative approximation of the chance constraints we separate the outage probability with respect to the multiplicative and additive channel errors. In particular, we separate the optimization variables and the additive error $\boldsymbol{e}_k$ from the attenuation $\zeta_{\text{rain},k}$ as within the integration of (6.12). If some upper bound for the effective noise $\tilde{\sigma}_k^2 = \sigma_k^2 \zeta_{\text{rain},k}$ were available, the remaining uncertainty would only be due to the additive error $\boldsymbol{e}_k$.[8] Let $a_k \in \mathbb{R}_+$ denote such a bound. Then, (6.12) can be approximated as

$$\Pr\left(g_k(\boldsymbol{t}, \boldsymbol{e}_k) \geq \tilde{\sigma}_k^2\right) > \Pr\left(g_k(\boldsymbol{t}, \boldsymbol{e}_k) \geq a_k \middle| a_k \geq \tilde{\sigma}_k^2\right) \cdot \Pr\left(\tilde{\sigma}_k^2 \leq a_k\right), \qquad (6.13)$$

where the right-hand side is separated also in probability in the attenuation uncertainty and the additive channel errors. Furthermore, fixing $\alpha_k = \Pr(\tilde{\sigma}_k^2 \leq a_k)$ within its domain, i.e., $\alpha_k \in [1 - \epsilon_k, 1]$, it remains the chance constraint[9]

$$\Pr\left(g_k(\boldsymbol{t}, \boldsymbol{e}_k) \geq a_k(\alpha_k)\right) \geq (1 - \epsilon_k)\alpha_k^{-1} \qquad (6.14)$$

for the additive fading. Here, $a_k(\alpha_k)$ is the α_k-quantile of the random noise $\tilde{\sigma}_k^2$, i.e.,

$$a_k(\alpha_k) = \sigma_k^2 F^{-1}_{\zeta_{\text{rain},k}}(\alpha_k).$$

Approximation (6.14) is conservative, but improves by an appropriate choice for α_k. Therefore, the RB optimization additionally includes $\boldsymbol{\alpha} = [\alpha_1, \ldots, \alpha_K]^{\mathrm{T}}$ as a decision variable when replacing the initial chance constraints by (6.14)

$$\begin{aligned} \max_{\boldsymbol{\alpha}, \boldsymbol{t}, \rho} \rho \quad \text{s.t.} \quad & \boldsymbol{t} \in \mathcal{P}, \quad \Pr\big(g_k(\boldsymbol{t}, \boldsymbol{e}_k) \geq a_k(\alpha_k)\big) \geq (1 - \epsilon_k)\alpha_k^{-1}, \\ & 1 - \epsilon_k \leq \alpha_k \leq 1, \qquad k = 1, \ldots, K. \end{aligned} \qquad (6.15)$$

This chance-constrained optimization is obviously more complex than that for only additive channel errors (cf. Chap. 5). However, the clear separation of the attenuation and the additive fading allows to divide (6.15) into an inner beamformer optimization for fixed $\boldsymbol{\alpha}$ and an outer decision about these prior probabilities.

6.5.1 Conservative Inner Optimization

With fixed probability values α_k, $k = 1, \ldots, K$, the inner RB optimization becomes

$$\max_{\boldsymbol{t}, \rho} \rho \quad \text{s.t.} \quad \boldsymbol{t} \in \mathcal{P}, \quad \Pr\left(g_k(\boldsymbol{t}, \boldsymbol{e}_k) \geq a_k(\alpha_k)\right) \geq (1 - \epsilon_k)\alpha_k^{-1}, \quad k = 1, \ldots, K. \qquad (6.16)$$

Its probability constraints feature the deterministic approximations of Sects. 5.5 and 5.6, i.e., based on an uncertainty region for the additive channel error or using the Bernstein-type inequalities, respectively. Therefore, either of the corresponding balancing approaches may be employed for lower bounding the optimum of (6.16).

[8] Vice versa, if one was aware of $\boldsymbol{e}_k$, the probability would be the CDF of $\sigma_k \zeta_{\text{rain},k}$ at $g_k(\boldsymbol{t}, \boldsymbol{e}_k)$.

[9] The conditioning is removed due to independence of $\zeta_{\text{rain},k}$ and $\boldsymbol{e}_k$.

However, further requirements shrink the number of suitable approaches. First, a global solution of the approximated problem is required to uniquely identify whether a candidate $\boldsymbol{\alpha}'$ increases the balancing level ρ compared to $\boldsymbol{\alpha}$ or not. Second, the computational complexity for finding this solution has to remain small when repeatedly evaluating the inner beamformer design for the outer search of $\boldsymbol{\alpha}$. These properties make the spherical uncertainty and the Bernstein-type inequality approximation of [36] attractive for (6.16).

With the spherical uncertainty restriction (5.51) for the additive error $\boldsymbol{e}_k \in \mathcal{U}_k$, the conservative constraint approximation follows directly the steps in Sect. 5.5.1, but with $a_k(\alpha_k)$ replacing the noise variance. Furthermore, the probability for $\boldsymbol{e}_k \in \mathcal{U}_k$ is fixed to $\Pr(\boldsymbol{e}_k \in \mathcal{U}_k) = (1-\epsilon_k)\alpha_k^{-1}$ to assure the kth probability constraint of (6.16). Therewith, the squared sphere radius for the uncertainty region (5.51) becomes

$$d_k(\alpha_k) = \frac{1}{2} F^{-1}_{\chi^2_{2N}}\left(\frac{1-\epsilon_k}{\alpha_k}\right). \tag{6.17}$$

This value increases with decreasing α_k, while $a_k(\alpha_k)$ decreases with decreasing α_k. The probabilities α_k, $k = 1, \ldots, K$, thus, trade the channel uncertainty off against the noise power. The inner uncertainty constrained RB optimization then reads as

$$\max_{\rho,\boldsymbol{t}} \rho \quad \text{s.t.} \quad \boldsymbol{t} \in \mathcal{P},\ g_k(\boldsymbol{t}, \boldsymbol{e}_k) \geq a_k(\alpha_k),\quad \forall \boldsymbol{e}_k{:}\|\boldsymbol{C}_k^{-1/2}\boldsymbol{e}_k\|_2^2 \leq d_k(\alpha_k),\ k{=}1, \ldots, K.$$

Applying the SDR for $\boldsymbol{Q}_k = \boldsymbol{t}_k\boldsymbol{t}_k^{\mathrm{H}}$, i.e., dropping $\mathrm{rank}\{\boldsymbol{Q}_k\} = 1$, and the S-Lemma as in Sect. 5.5.1, the LMIs that replace the uncertainty constraints read as

$$\begin{aligned}\boldsymbol{\Psi}_k(\rho, \boldsymbol{Q}_1, \ldots, \boldsymbol{Q}_K, \lambda_k) = {} & \big[\bar{\boldsymbol{h}}_k, \boldsymbol{C}_k^{1/2}\big]^{\mathrm{H}} \boldsymbol{B}_k \big[\bar{\boldsymbol{h}}_k, \boldsymbol{C}_k^{1/2}\big] \\ & + \lambda_k\, \mathrm{bdiag}(-d_k(\alpha_k), \mathbf{I}_N) \\ & - \mathrm{bdiag}(a_k(\alpha_k), \boldsymbol{0}) \succeq \boldsymbol{0}.\end{aligned} \tag{6.18}$$

The induced trade-off is now between the constant and the variable portion that both degrade the useful signal power, i.e., the upper-left most entry of the first summand.

The alternative approximation via the Bernstein-type inequality and the above SDR results in the following set of convex requirements (cf. Sect. 5.6.2):

$$\begin{aligned}&\mathrm{tr}\left(\boldsymbol{C}_k^{\mathrm{H}/2}\boldsymbol{B}_k\boldsymbol{C}_k^{1/2}\right) + \bar{\boldsymbol{h}}_k^{\mathrm{H}}\boldsymbol{B}_k\bar{\boldsymbol{h}}_k - \sqrt{-2\ln(\beta_k(\alpha_k))}x_k + \ln(\beta_k(\alpha_k))y_k \geq a_k(\alpha_k), \\ &\left\| \mathrm{vec}\left(\boldsymbol{C}_k^{\mathrm{H}/2}\boldsymbol{B}_k\big[\boldsymbol{C}_k^{1/2}, \sqrt{2}\bar{\boldsymbol{h}}_k\big]\right)\right\|_2 \leq x_k, \\ &y_k\mathbf{I} + \boldsymbol{C}_k^{\mathrm{H}/2}\boldsymbol{B}_k\boldsymbol{C}_k^{1/2} \succeq \boldsymbol{0}, \quad y_k \geq 0,\end{aligned} \tag{6.19}$$

with $\boldsymbol{B}_k = \frac{1}{2^{\rho\rho_k}-1}\boldsymbol{Q}_k - \sum_{i\neq k}\boldsymbol{Q}_i$ from (5.57) and $\beta_k(\alpha_k) = 1 - \frac{1-\epsilon_k}{\alpha_k}$. Here, the trade-off due to the choice of α_k is between the scale of the effective noise variance and the influence of x_k and y_k, which bound the uncertainty of the useful signal and interference powers. Decreasing α_k increases $-\ln(\beta_k(\alpha_k))$ and, thus, also the induced losses by these terms.

Therefore, the inner optimization with either of these approximations reads as

$$\max_{\rho, Q_1,\ldots,Q_K, y, x} \rho \quad \text{s.t.} \quad (Q_1, \ldots, Q_k) \in \mathcal{Q}, \quad (6.18) \text{ or } (6.19) \quad k = 1, \ldots, K. \tag{6.20}$$

For both types of constraints, (6.20) is monotonic but non-convex in ρ and convex in all the other variables. The optimum and the corresponding transmit covariance matrices $Q_1, \ldots, Q_K$ are, therefore, obtained via a bisection search over ρ, as detailed in Sect. 5.5.4. Since the Bernstein-type inequality approximation is usually less conservative than the uncertainty approximation (cf. Sect. 5.7.3), the solution of (6.20) with (6.19) will outperform the results with (6.18).

6.5.2 Outer Optimization of Priors

Let $\rho(\boldsymbol{\alpha})$ denote the optimum of (6.20). Then, the outer optimization reads as

$$\max_{\boldsymbol{\alpha}} \rho(\boldsymbol{\alpha}) \quad \text{s.t.} \quad \mathbf{1} - \boldsymbol{\epsilon} \leq \boldsymbol{\alpha} \leq \mathbf{1}. \tag{6.21}$$

It represents the maximization of a lower bound for the achievable data rates.

Two properties are deduced from the influence of α_k on the kth constraint of (6.20). First, the objective is zero if any entry of $\boldsymbol{\alpha}$ approaches either of its bounds, that is, $\rho(\boldsymbol{\alpha}) = 0$ if $\alpha_i = 1$ or $\alpha_i = 1 - \epsilon_k$ for at least one $i = 1, \ldots, K$. For $\alpha_i = 1$, the effective noise power becomes $\lim_{\alpha_i \to 1} a_i(\alpha_i) \to \infty$. For $\alpha_i = 1 - \epsilon_i$, the squared radius of the uncertainty bound (6.18) is $\lim_{\alpha_i \to 1-\epsilon_i} d_i(\alpha_i) \to \infty$ and $\lim_{\alpha_i \to 1-\epsilon_i} \ln(1 - \frac{1-\epsilon_i}{\alpha_i}) \to \infty$ for the Bernstein-type inequality approximation (6.19). Both these cases result in zero rate, while $\rho(\boldsymbol{\alpha})$ is non-zero between these extreme choices for α_k, $k = 1, \ldots, K$.[10]

Second, the objective is quasiconcave in α_k if the other entries α_i, $i \neq k$ are fixed. Due to the trade-off property that the choice of α_k represents for (6.18) and (6.19), the kth receiver's rate is quasiconcave in α_k for fixed transmit covariance matrices. It increases with α_k from $1 - \epsilon_k$ until some α'_k and decreases when further increasing α_k. If the rate at α'_k is larger than $\rho(\boldsymbol{\alpha})$, the solution of (6.20) will also satisfy $\rho(\boldsymbol{\alpha}') \geq \rho(\boldsymbol{\alpha})$ for $\alpha'_i = \alpha_i$, $i \neq k$. This monotonicity implies that also $\rho(\boldsymbol{\alpha})$ is quasiconcave in $\alpha_k \in [1 - \epsilon_k, 1]$.

Due to these properties, a nested line search based on the golden-section method finds the solution of (6.21). Computationally less complex is a sequential update of the α_k's in each iteration. Within the kth step of the ith iteration, the maximum

[10]This holds if all α_k are sufficiently large, e.g., $\|C_k^{-1/2}\bar{h}_k\|_2^2 > d_k(\alpha_k)$ must be satisfied for (6.18).

$$\rho^{(i,k)} = \max_{\alpha_k} \rho\left([\boldsymbol{\alpha}_{\underline{k}}^{(i+1),\mathrm{T}}, \alpha_k, \boldsymbol{\alpha}_{\bar{k}}^{(i),\mathrm{T}}]^{\mathrm{T}}\right) \quad \text{s.t.} \quad 1 - \epsilon_k \leq \alpha_k \leq 1 \tag{6.22}$$

is found with the golden section search from [227, Appendix C.3], where the auxiliary vectors in the (i, k)th iteration are $\boldsymbol{\alpha}_{\underline{k}} = [\alpha_1, \ldots, \alpha_{k-1}]^{\mathrm{T}}$ and $\boldsymbol{\alpha}_{\bar{k}} = [\alpha_{k+1}, \ldots, \alpha_K]^{\mathrm{T}}$. The sequence $\{\rho^{(i,k)}\}_{i,k}$ is non-decreasing when starting from an initial feasible $\boldsymbol{\alpha}^{(0)}$. Since the sequence is moreover bounded above by the power limitations and the approximated chance-constraints, this sequential update rule converges in the objective. Convergence in the decision variables $\boldsymbol{\alpha}^{(i)}$ is expected as well because $\rho(\boldsymbol{\alpha})$ is a continuous function.

The computational costs for this sequential search increase more than linearly with the number of terminals K. This disadvantage is overcome by choosing equal $\alpha_k, k = 1, \ldots, K$ i.e., $\boldsymbol{\alpha} = \alpha_0 \mathbf{1}$, and only optimize $\alpha_0 \in [1 - \min\{\epsilon_k\}, 1]$:

$$\max_{\alpha_0} \rho(\alpha_0 \mathbf{1}) \quad \text{s.t.} \quad 1 - \min\{\epsilon_k\} \leq \alpha_0 \leq 1.$$

The solution of this problem is also obtained via a golden section search over α_0 if $\rho(\alpha_0 \mathbf{1})$ is quasiconcave in α_0. This equal prior optimization provides only slight performance losses compared to the sequential search if the outage limits, rate targets, and the channel's fading parameters are equal for the terminals. A comparison for these optimization strategies is also a part of the numerical results.

6.6 Results for Outage Constrained Rate Balancing

For the simulation results corresponding to the outage based RB optimization, the spotbeam SatCom system with either $N = K = 3$ or $N = K = 7$ antennas and terminals from Sect. 6.1 is employed, where the communication takes place in the S-band (see Table 6.1). The small number of antennas and mobile terminals is to keep the outage constrained optimization tractable for the sequential outer probability search. If not stated otherwise, the target rates are fixed to $\rho_k = 1$ and the outage probability is limited to $\varepsilon_k = 0.1$ for all terminals. Moreover, per-antenna constraints with $P_\ell = P/N$ are employed to model the per-feed power constraints at the satellite. For this setup, ten random positions of the terminals are created and, therewith, ten realizations of the beamgain matrices $\boldsymbol{B}_k$ and FSLs $g_{\mathrm{FSL},k}$ according to the model in Sect. 6.1.2. For the Rician fading, $\kappa = 10\,\mathrm{dB}$ and $\kappa = 15\,\mathrm{dB}$ are considered as typical values and $\sigma_{\mathrm{rain},k} = 1.63\,\mathrm{dB}$ and $\sigma_{\mathrm{rain},k} = 3.26\,\mathrm{dB}$ serve as weak and strong attenuation uncertainty, respectively, for rain fading (e.g., see [52]).

6.6.1 Equal Fading Conditions for the Terminals

Figure 6.7 shows the average of the jointly achievable rate ρ versus the transmit power P in dB. The averaging is over the optima of all ten channel realizations. There are four lines within each subfigure. The first line is for only rain fading. The

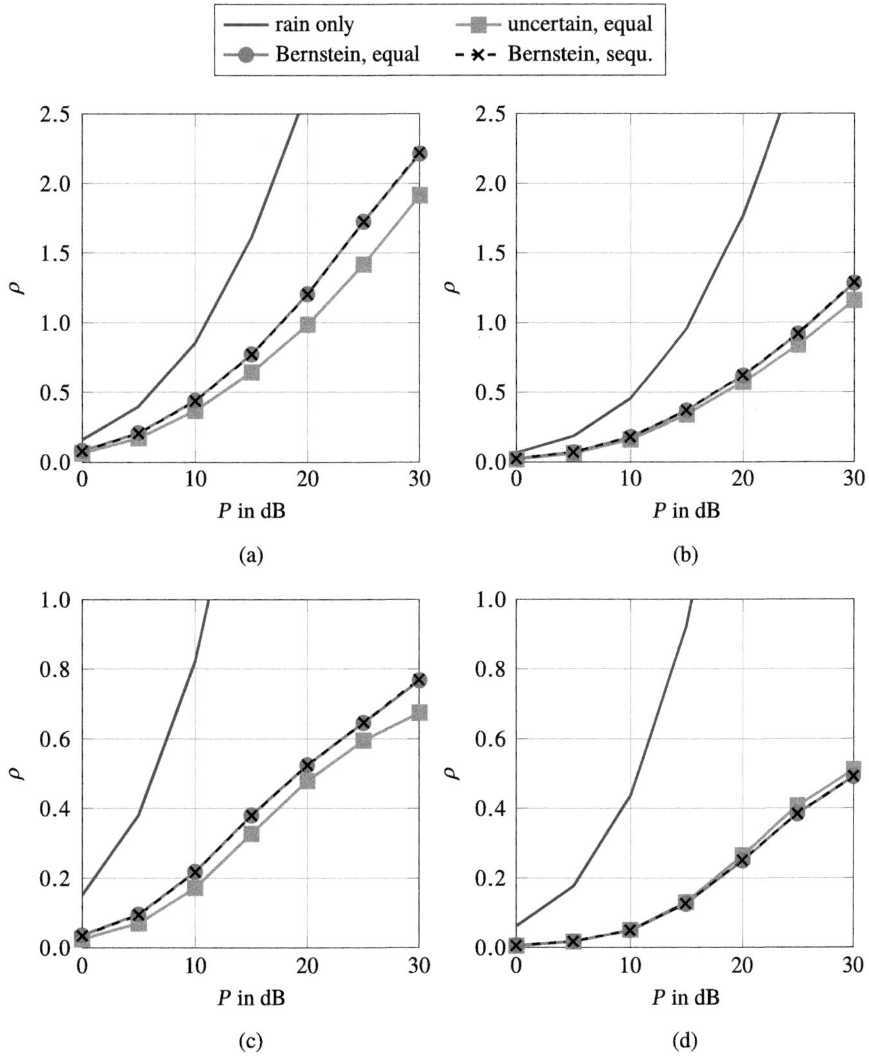

Fig. 6.7 Maximum ρ versus P for an S-band system with $N = K = 3$ spotbeams, targets $\rho_k = 1$, outage limits $\epsilon_k = 0.1$, and per-antenna power limits $P_\ell = P/N$. (**a**) $\kappa = 15\,\text{dB}$, $\sigma^2_{\text{rain},k} = 1.63\,\text{dB}$. (**b**) $\kappa = 15\,\text{dB}$, $\sigma^2_{\text{rain},k} = 3.26\,\text{dB}$. (**c**) $\kappa = 10\,\text{dB}$, $\sigma^2_{\text{rain},k} = 1.63\,\text{dB}$. (**d**) $\kappa = 10\,\text{dB}$, $\sigma^2_{\text{rain},k} = 3.26\,\text{dB}$

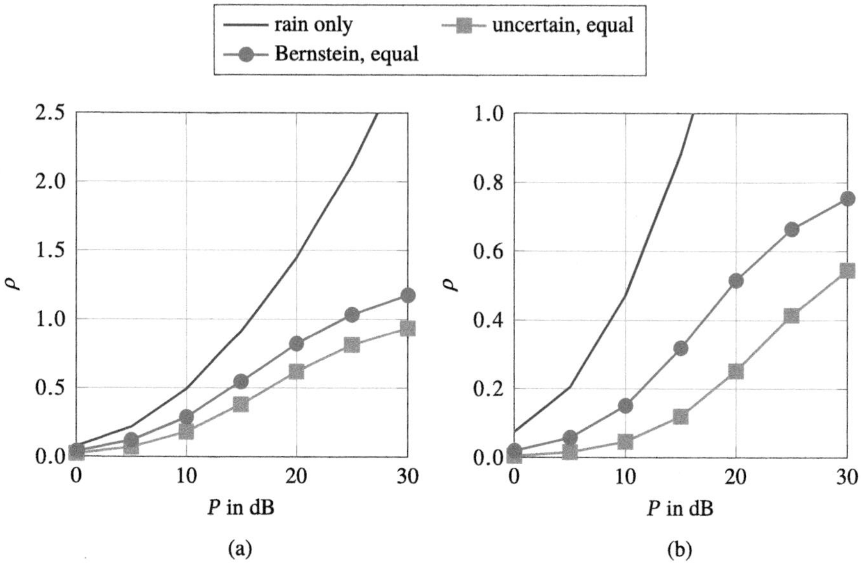

Fig. 6.8 Maximum ρ versus P for an S-band SatCom system with $N = K = 7$, equal targets $\rho_k = 1$, outage requirements $\epsilon_k = 0.1$, and per-antenna constraints $P_\ell = P/N$. (**a**) $\kappa = 15\,\text{dB}$, $\sigma^2_{\text{rain},k} = 1.63\,\text{dB}$. (**b**) $\kappa = 10\,\text{dB}$, $\sigma^2_{\text{rain},k} = 1.63\,\text{dB}$

other lines are for the model with rain attenuation and full-rank Rician fading. The lines represent the solutions based on the Bernstein-type inequality and the spherical uncertainty approximations for the inner problem and equal prior probabilities or the sequential prior probability optimization for the outer problem.

The smaller the additive channel uncertainty, i.e., the larger κ, and the smaller the attenuation variance $\sigma^2_{\text{rain},k}$, the larger is the reliably achievable data rate. The results for only rain fading lie strictly above the results with the additional Rician fading. Thus, the scattering that leads to a non-rank one covariance matrix of the SatCom channel reduces the reliably achievable data rates. The Bernstein-type inequality approximation again outperforms the spherical uncertainty approximation (cf. [36]).

The same simulations have been performed for ten channel mean realizations in the $K = N = 7$ antenna and terminal setup. Figure 6.8 sketches the average of the achievable rate ρ for $\kappa = 15\,\text{dB}$ and $\kappa = 10\,\text{dB}$. The sequential outer optimization for $\boldsymbol{\alpha}$ becomes computationally too demanding for this setup. Therefore, the figure only presents the results with the equal rain attenuation probability bounds $\boldsymbol{\alpha} = \alpha_0 \mathbf{1}$.

Changing from $K = 3$ to $K = 7$ terminals, the achieved data rate is tremendously reduced for all curves in case of weak Rician fading (see Fig. 6.8a). Additionally, the difference between the results with and without Rician fading reduces. Both these effects are due to the increased interference. This decrease in data rate almost vanishes for $\kappa = 10\,\text{dB}$ (see Fig. 6.8b), where the robustness with respect to the additive channel errors becomes dominant over the interference limitation. The gain of the inner Bernstein-type inequality approximation to the spherical uncertainty approximation increases with the system also for this SatCom scenario.

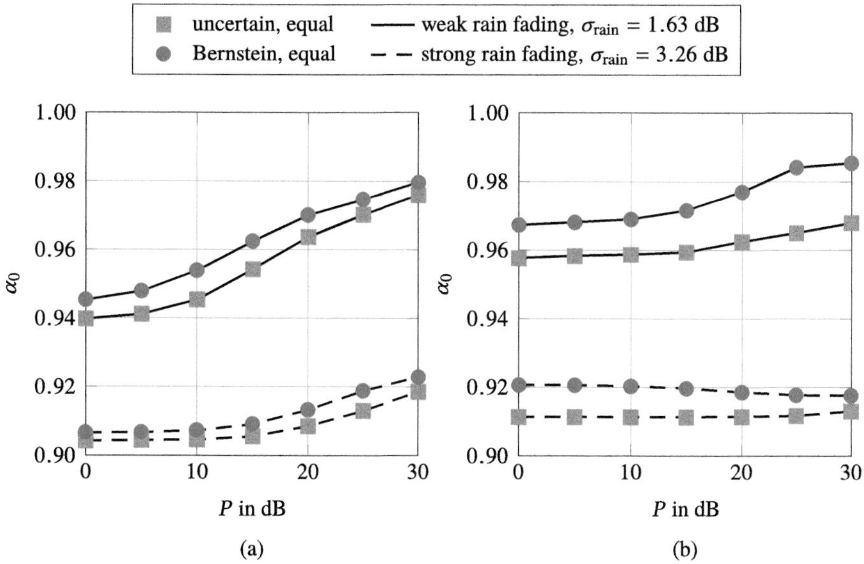

Fig. 6.9 Optimized α_0 versus P for SatCom with $N = K = 3$, targets $\rho_k = 1$, outage requirements $\epsilon_k = 0.1$, and $\boldsymbol{\alpha} = \alpha_0\mathbf{1}$ for limiting outages due to rain fading. (**a**) Rician fading, $\kappa = 15\,\text{dB}$. (**b**) Rician fading, $\kappa = 10\,\text{dB}$

Figure 6.9 sketches the common probability α_0 for limiting the outages due to rain fading at the solution for the $K = 3$ terminal SatCom system.[11] The probability naturally lies between 0.9 and 1.0 due to the fixed outage limitation $\epsilon_k = 0.1$ for all $k = 1, \ldots, K$. This solution probability α_0 increases with the transmit power. For weak rain fading, i.e., $\sigma^2_{\text{rain},k} = 1.63\,\text{dB}$, and Rician fading with $\kappa = 15\,\text{dB}$ (see Fig. 6.9a), the increase is from about 94 % to 98 % in the observed range. These values change only slightly with the type of the inner approximation, i.e., the optimal α_0 is only about 0.5 % smaller for the spherical uncertainty approximation than for the Bernstein-type inequality approximation. When the rain fading is strong instead, i.e., $\sigma^2_{\text{rain},k} = 3.26\,\text{dB}$, the optimal α_0 is smaller than for weak rain fading and only increases from 90.7 % to 92.2 %. For $\kappa = 10\,\text{dB}$ (see Fig. 6.9b), α_0 is larger than for $\kappa = 15\,\text{dB}$. In particular, α_0 increases now from 96.5 % to 98.5% with increasing P and weak rain fading. For strong rain fading, α_0 almost remains constant in the observed range for P.

Summarizing this, the probability value α_0 shifts from an almost equal division of the maximal allowed outage with respect to rain and Rician fading, i.e., $\alpha_0 \approx \sqrt{1 - \epsilon_k}$ with $\epsilon_k = 0.1$, towards a stricter outage limitation for the rain attenuation when the transmit power increases. When the variance for the rain fading increases to $\sigma_{\text{rain},k} = 3.26\,\text{dB}$, the outage limitations are stricter for the additive channel errors

[11] The optimal α_0 for $K = 7$ terminals and low rain fading shows the same behavior as for $K = 3$.

and less strict with respect to the attenuation uncertainty, i.e., α_0 is between 90.5 % and 92.5 %. In other words, only small multi-path channel errors are allowed to compensate for the increased variance of the rain attenuation.

6.6.2 Effects for Distinct Fading Conditions

A comparison of the performance and the solutions α_k's of the equal and the sequential outer optimization from Sect. 6.5.2 are shown in Figs. 6.10 and 6.11, respectively. In contrast to the case of equal fading conditions, the sequential search strategy outperforms the equal choice for $\boldsymbol{\alpha}$ when the rain attenuation variance and the outage limitation are increased for only one terminal (see Fig. 6.10). Here, only terminal 2 experiences rain fading with $\sigma_{\text{rain},2} = 3.26\,\text{dB}$ and has the outage requirement $\epsilon_2 = 0.20$, while terminals 1 and 3 are subject to weak rain fading $\sigma_{\text{rain},1} = \sigma_{\text{rain},3} = 1.63\,\text{dB}$ and have the outage limits $\epsilon_1 = \epsilon_3 = 0.10$. The rates for the sequentially optimized α_k are about 0.6 bit above that for $\alpha_k = \alpha_0$ if $P = 30\,\text{dB}$.

The sequential optimization takes advantage of the available range for α_2. Its resulting value is 0.85 for the inner uncertainty and Bernstein-type inequality approximations, while the resulting α_0 is restricted to lie above 0.90. The results for α_1 and α_3 are about 0.97 for both inner approximation methods. Their values increase only slightly with the transmit power P. Therefore, the sequential outer optimization benefits if the statistics of the multiplicative channel errors strongly differ among the served terminals while the choice $\boldsymbol{\alpha} = \alpha_0\mathbf{1}$ performs sufficiently well for terminals with equal outage requirements and similar fading parameters.

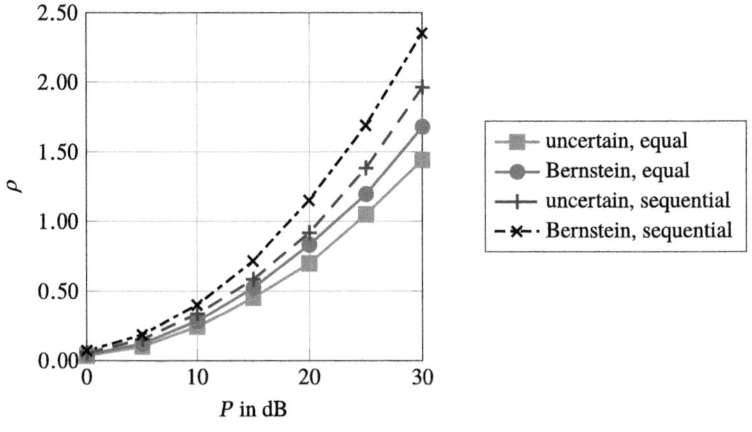

Fig. 6.10 Performance comparison for the sequential and equal optimization of α_k for $N = K = 3$, $\rho_k = 1$, and increased rain fading and outage margin for user 2, i.e., $\sigma_{\text{rain},2} = 3.26\,\text{dB}$ and $\epsilon_2 = 0.20$, and $\sigma_{\text{rain},1} = \sigma_{\text{rain},3} = 1.63\,\text{dB}$ and $\epsilon_1 = \epsilon_3 = 0.10$

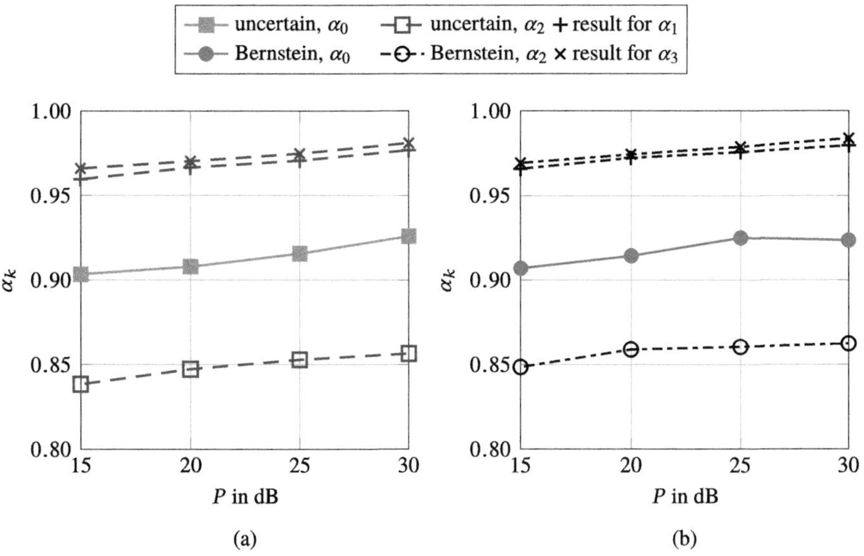

Fig. 6.11 Comparison of the α_k for the sequential and equal optimization. Basis is the system with $N = K = 3$, $\rho_k = 1$, increased rain fading and outage margin for terminal 2, i.e., $\sigma_{\text{rain},2} = 3.26\,\text{dB}$, $\epsilon_2 = 0.20$, $\sigma_{\text{rain},1} = \sigma_{\text{rain},3} = 1.63\,\text{dB}$, and $\epsilon_1 = \epsilon_3 = 0.10$. (**a**) Uncertainty bound. (**b**) Bernstein-type inequality bound

6.6.3 *Outage Probabilities at the Terminals*

After maximizing the lower bounds for the achievable rates, the terminals' actual outage probabilities $\text{p}_k = \Pr(r_k \leq \rho\rho_k)$ were computed. Figure 6.12 shows the empirical CDF of the resulting probabilities p_k for the SatCom scenario with $N = K = 3$ antennas and terminals, targets $\rho_k = 0.1$, and equal fading parameters for all terminals. The probability computation via a numerical evaluation of (6.12) is with the transmit covariance matrices from the inner RB optimization (6.20).[12]

The figures show that both inner approximation types, the Bernstein-type inequality and the uncertainty approximation, are loose also for the SatCom scenario. They fail to exploit the allowed outage probability. The Bernstein-type inequality approximation is a little less conservative than directly restricting the additive channel errors to lie in a sphere. In contrast, if only rain fading is taken into account and the Rician fading is neglected, the result is overoptimistic in the sense that the required outage limits are missed. The calculated outage probability can be far above the outage restriction.

For the figures, the fading parameter κ was varied from medium to weak Rician fading, i.e., from $\kappa = 10\,\text{dB}$ to $\kappa = 15\,\text{dB}$. The differences show that the inner

[12]The equal probability restriction is used for the outer optimization of $\boldsymbol{\alpha} = \alpha_0\mathbf{1}$.

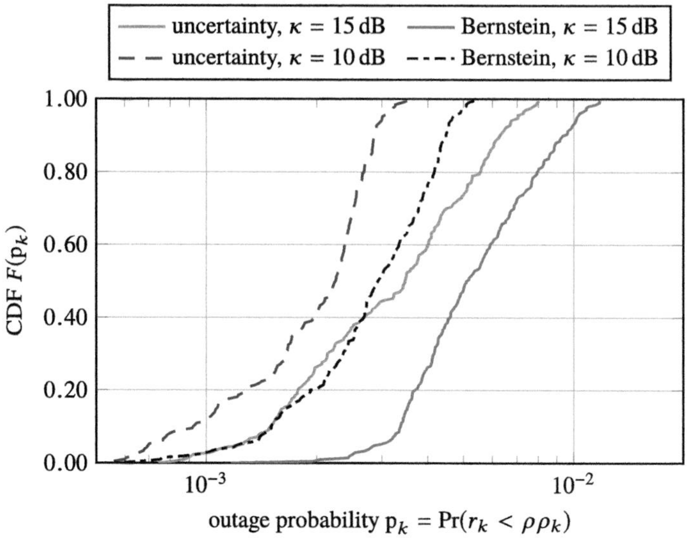

Fig. 6.12 Achieved outage probabilities for $N = K = 3$, $\rho_k = 1$, $\varepsilon_k = 0.1$, $\sigma_{\text{rain},k} = 1.63\,\text{dB}$, and equal requirements for the rain fading, i.e., $\alpha_k = \alpha_0$, $k = 1, \ldots, K$

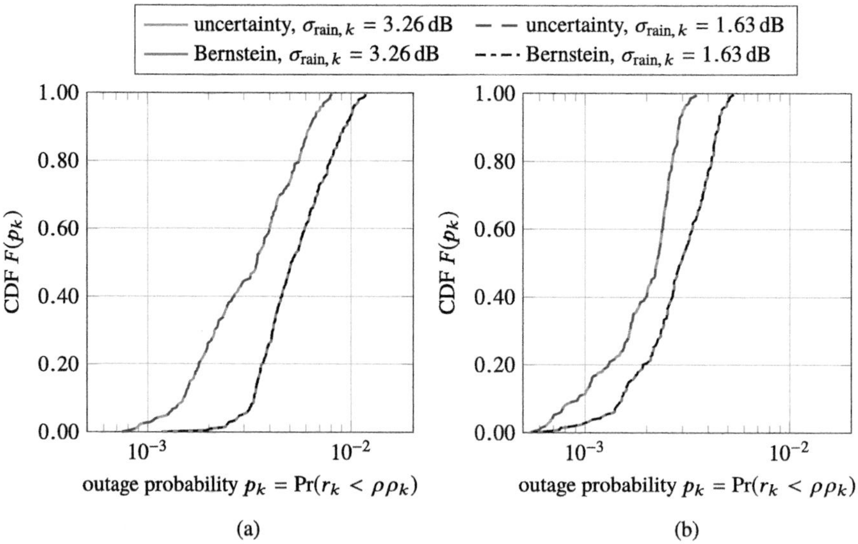

Fig. 6.13 Dependence of the outage probabilities with regard to the rain fading for the SatCom system with $N = K = 3$, $\rho_k = 1$, $\varepsilon_k = 0.1$, and $\alpha_k = \alpha_0$, $k = 1, \ldots, K$. (**a**) Rician factor $\kappa = 15\,\text{dB}$. (**b**) Rician factor $\kappa = 10\,\text{dB}$

convex approximation method becomes less tight, the larger the Rician fading is. In other words, the inner optimizations are more conservative when the likelihood of a large additive channel error increases. In contrast, changes in the rain fading parameter $\sigma^2_{\mathrm{rain},k}$ have almost no influence on the obtained outage probability. This is shown in Fig. 6.13, where $\sigma^2_{\mathrm{rain},k}$ is varied from the low to medium rain fading example. Hence, the restriction of the outages due to rain attenuation is sufficiently taken into account by the probability separation strategy.

Chapter 7
Summary, Conclusions, and Open Research

Ergodic and outage robust multi-user downlink beamforming designs with only partial channel knowledge have been investigated for future wireless communication. The modeling accuracy and computational tractability for these designs strongly depend on the shape of the channels' estimates and the statistics of the estimation errors. Two types of channel models have been analyzed:

1. the standard additive Gaussian error model from literature, and
2. a multiplicative error model, e.g., to approximate additive errors.

We have provided conservative deterministic formulations for the ergodic rate, the average MSE, and the epsilon-outage rate. Thereof, we have derived algorithms for solving the respective statistically robust versions of the QoS and RB optimizations.

7.1 Research Results for Robust Beamforming

7.1.1 Achievements and Open Problems for Multiplicative Channel Errors

For downlink communication with the previous multiplicative error model, we have extended the QoS feasibility results and the optimization algorithms from the perfect CSI case to ergodic and outage robust designs. The calculated epsilon-outage rate and also the derived tight upper and lower ergodic rate bounds have the classical Shannon form. The channel uncertainty has only enhanced the noise power for these metrics. This has enabled an approximate beamformer computation with the respective algorithms from perfect CSI. Moreover, we have proved that the resulting feasible ergodic rate region coincides with the perfect CSI rate region from [65]. A two-step beamformer design with the ergodic rate bounds and power allocation

A. Gründinger, *Statistical Robust Beamforming for Broadcast Channels and Applications in Satellite Communication*, Foundations in Signal Processing, Communications and Networking 22, https://doi.org/10.1007/978-3-030-29578-3_7

achieves any point in its interior. The novel SCS algorithms guarantee convergence to (local) solutions of the QoS and RB optimizations within few iterations. These algorithms have even achieved the global optimum in benchmarks.

With these properties, the multiplicative channel model provides a great trade-off between modeling accuracy, approximation capabilities, and computational tractability for spatially restricted scattering environments. This makes the model predestined for rural communication and SatCom with hundreds of receivers.

An open question for multiplicative channel errors is regarding existence and the form of a direct convex constraint reformulation with respect to the beamformers. Existence of such convex formulations will generally depend on the distribution of the channel error. Even if no convex constraint reformulation exists, the herein derived Sequential convex search (SCS) still ensures convergence to the global optimum for an appropriate starting point. This strategy is also interesting for implementations due to the computational efficient fixed point solution.

7.1.2 Summary of Results for Average MSE Optimization

For additive channel errors, the close connection between the rate and the MSE metric has let us analyze the ergodic rate constrained problems via average MSE constrained QoS and max–min optimizations. A new aspect for the min–max average MSE optimization has been explored for distinct MSE targets: When power limitations are very restrictive, receivers with too large MSE targets are switched off. This aspect is especially important for coordinating and scheduling transmission at higher layers. Otherwise, the high demands of video-streaming could cause zero data transmission for low demand telephone services if both serviced are provided simultaneously and in the same frequency bands.

Moreover, this property has required changes within the standard ACS algorithm for the above problems, i.e., that alternates between the equalizer and precoder update. In this context, we have derived a general duality relation for the MSE based precoder update via Lagrangian multiplier theory. The relation holds for active and inactive weighted sum MSE constraints of disjoint receiver groups, conic quadratic (power) constraints, and imperfect transmitter CSI. The obtained uplink problems include an inner power allocation and MMSE equalizer design and an outer maximization with respect to the Lagrangian multipliers for the downlink power constraints. This maximization has been proven to be a convex program for the QoS optimization and a quasiconvex program for the MSE optimization. This has enabled a PG solution with improved convergence in comparison to previous subgradient methods. This solution can serve as a strategy to cope with the dimensions of future wireless systems.

The duality framework is sufficiently general to cover also weighted MSE approximations for the data rate, e.g., the weighted sum rate maximization via the weighted sum MSE from [155]. For the sum MSE, we have also analyzed the influence of a sum power restriction, per-antenna constraints, and constraints for antenna groups. While a sum power limitation provides close performance benchmarks for systems with per-antenna limitations, it is insufficient to predict an appropriate precoder. Power limits for small antenna arrays provide a trade-off between the sum-power and per-antenna constraints for the precoder computation, when the latter per-antenna requirements are computationally intractable.

The duality framework has let us also include conic models of stochastic interference temperature limitations in [165]. It can even incorporate shaping constraints [319] via a subspace model for the beamformers. Alternatively, one can use an approximation with small limits for the corresponding conic quadratic constraints. This shows the versatile application of the proposed MSE duality.

Analyzing the effects of non-linear power limitations and extending the algorithms for these constraints remains an open topic. General non-linear power constraints have been considered by [57]. However, more important are constraints with a concave input-to-output mapping to model the saturation of RF amplifiers. A repeated linearization and MSE optimization may fit for these constraints, but also increases the computational complexity by the number of repetitions.

7.1.3 Contributions for Outage Rate Requirements

Conservative constraint approximations have also been required for chance-constrained QoS and RB beamformer optimizations with the additive error model. The focus has been on approximations with uncertainty constraints and concentration inequalities. A novel approach has been derived for each of these classes: an uncertainty approximation that bounds the channel error orthogonal to the channel's mean and a MSE based approximation via a Bernstein-type inequality. The orthogonal uncertainty restriction has outperformed the classical spherical uncertainty region for the channel error. The MSE approximation enables an ACS based beamformer design and improves existing MSE approaches based on Markov's inequality or ZF precoding. However, their average performance has remained below that of the second approximation from [36].

When accurate probability computations have been available, a subsequent power allocation has tremendously improved the performance of the conservative designs. It even tries to compensate for non-conservative beamformers, e.g., which have been based on a multiplicative error approximation of the additive channel model. This has shown to increase the performance and even extend the feasible range of the conservative beamformer designs for appropriate scenarios.

However, neither the approximations nor the power allocation have been tractable for increasing system dimensions. For low-complexity beamforming, another notion of robustness has been introduced by [45], where the noise variance is maximized subject to rate constraints. Future approximations may also exploit statistical limit distributions for outage analysis and approximations in massive MIMO.

7.1.4 Summary and Conclusions for Satellite Communication

For an improved interference management in multi-spotbeam SatCom, we have adapted the robust beamforming schemes to account for the joint multiplicative rain fading and additive multi-path fading. When the distortion by the antenna characteristics degrades the scattering to be rank-one, e.g., for fixed terminals, the ergodic rate bounds and outage rate formulations for multiplicative channel errors are valid. This has enabled adaptive beamforming for more than a hundred spotbeams, which has substantially outperformed the restrictive ZF and RZF beamforming schemes.

For mobile SatCom, we have handled the multiplicative channel fading with a tight lower bound for the ergodic rate and applied the MMSE approximation only with respect to the multi-path fading. With this conservative two-step bounding strategy, the balancing solution has remained tractable for a hundred spotbeams and per-antenna constraints when restricting to deterministic equalizers.

Also epsilon-outage rate approximations required a separation of the outages due to the multiplicative attenuation and additive fading for mobile SatCom. The separation can be chosen equally for all receivers if the rain fading statistics and the outage requirements are similar. However, this separation for the beamforming still encounters limitations in approximation accuracy. The achieved outage probabilities have been far below the actual requirements. Tighter approximations will be required for future research on outage constrained beamforming in SatCom. These approximations will have to jointly bound the two types of channel errors.

Future research in SatCom will also consider multi-satellite systems. Beamhopping between different illumination patterns has recently been suggested to keep beamforming tractable for coexisting satellites that serve the same area [300]. Then, the beamformer design must be robust against inter-satellite interference.

7.2 Other Research on Robust Beamforming

Ongoing research on ergodic robust precoding is undoubtedly for massive MIMO and the millimeter wave technology. The aim is to exploit the law of large numbers by excessively increasing the number of base station antennas in favor for simplified ergodic rate expressions and beamformer designs (e.g., see [20, 21]). The design shall then manage the interference from neighboring base stations. For TDD systems, this interference is due to a superposition of pilot sequences in the training phase [320, 321]. Therefore, the precoder optimization must be jointly with the choice for the training strategy [22]. The basis for this optimization has been an uplink–downlink duality that relates to the shown SINR and MSE dualities.

For uncertainty robust beamformer designs, the application of LPMs [105, 266], the advantages and limitations compared to the SDR based reformulations of quadratic uncertainty constraints, and other use cases than for BCs have not been sufficiently analyzed. Another topic for statistically robust designs are applications in utility maximization [17]. Research in this area will again have to fall back on deterministic approximations for the chance constraints to keep the computational complexity tractable even for small benchmark scenarios. The presented approximations can serve as a basis for such investigations.

A widely uncovered topic is outage robust cooperative communication. Investigations on relay aided systems (e.g., see [23, 24, 322]) provide a first comparison for the performance under outage limitations. When cooperation between receivers in a downlink system is allowed, joint chance constraints for the cooperating receivers will become important. Then, an outage is declared if either of the receivers' data rates falls below its presumed threshold. Literature on chance constrained programming with joint constraints is rich for linear stochastic inequalities [254, Chapter 4], but only few is known about joint inequalities with quadratic terms in random variables. Therefore, tractable approximations for the chance constraints will also be of interest for the beamformer designs in relay aided systems.

All these topics confirm the importance of statistically robust beamforming and are use cases for the presented ergodic and outage rate optimizations.

Appendix A
Additional Information

A.1 Basic Properties of the Rate Based Optimizations

This appendix shows that the optimum of the QoS problem (1.9) and the RB problem (1.12) is strictly positive and monotonic in a joint scaling of the rate targets and the power limitations, respectively. Furthermore, it presents the proof for the inversion property between (1.9) and (1.12).

A.1.1 Positivity and Monotonicity for the Optimum of the QoS Problem

The optimum of the QoS optimization (1.9) is strictly *positive* because any feasible precoder is $\boldsymbol{t} \neq \mathbf{0}$ and $\mathcal{P}$ is compact, i.e., closed and bounded. The rate point $\boldsymbol{r}(\mathbf{0}) = \mathbf{0}$ was excluded from the target region $\mathcal{R}$ such that $\boldsymbol{t} = \mathbf{0}$ is infeasible.

Furthermore, the optimum of (1.9) is *monotonic* in a joint scaling of the target rates, e.g., $(1-\epsilon)^{-1}\boldsymbol{r} \in \mathcal{R}$ with $\epsilon \in (0, 1)$.[1] To show this, let p and $\boldsymbol{t}$ be the solution of (1.9), that is, $p^{-1}\boldsymbol{t} \in \mathcal{P}$ and $\boldsymbol{r}(\boldsymbol{t}) \in \mathcal{R}$. Then, the loosened requirement $(1-\epsilon)^{-1}\boldsymbol{r}(\boldsymbol{t}') \in \mathcal{R}$ is also achieved by a $\boldsymbol{t}' = (1+\delta)^{-1}\boldsymbol{t}$ with a sufficiently small $\delta > 0$. This follows from the monotonicity of the data rates $r_k(\boldsymbol{t}')$ in $(1+\delta)$. Using $\boldsymbol{t}'$ instead of $\boldsymbol{t}$, the objective can be decreased to $p' = (1+\delta)^{-1}p$ such that $p'^{-1}\boldsymbol{t}' \in \mathcal{P}$. Hence, the optimum of (1.9) decreases when increasing the target region by downscaling the rates.

[1] The scaling of the rates results in the weakened rate requirements $r_k \geq (1-\epsilon)\rho_k$, $k = 1, \ldots, K$, for the exemplary per-user constrained target region (1.10).

A. Gründinger, *Statistical Robust Beamforming for Broadcast Channels and Applications in Satellite Communication*, Foundations in Signal Processing, Communications and Networking 22, https://doi.org/10.1007/978-3-030-29578-3

A.1.2 Positivity and Monotonicity for the Optimum of the Rate Balancing Problem

The optimum of the balancing problem (1.12) is also *positive* for regular channel conditions. The optimum would only be zero if at least one of the rates (1.3) was zero for all $\boldsymbol{t} \in \mathcal{P}$. Since $\mathcal{P} \in \mathbb{C}^{NK}$ is convex and has non-empty interior, $r_k(\boldsymbol{t}) = 0$ for all $\boldsymbol{t} \in \mathcal{P}$ only if $\boldsymbol{h}_k = \boldsymbol{0}$. For non-zero rates $\boldsymbol{r}(\boldsymbol{t})$, the positive but unbounded scaling $\rho^{-1} \in (0, \infty)$ allows us to reach any point on the line segment $z = \rho^{-1}\boldsymbol{r}(\boldsymbol{t})$. Hence, (1.12) is feasible and has a strictly positive optimum.

To show that the maximum of (1.12) is *monotonic* in a scaling of the precoder $\boldsymbol{t} \in \mathcal{P}$, let ρ and $\boldsymbol{t}$ be the solution of (1.12), i.e., $\rho^{-1}\boldsymbol{r}(\boldsymbol{t}) \in \mathcal{R}$ and $\boldsymbol{t} \in \mathcal{P}$. When we loosen the power constraint set by the joint scaling $(1+\delta)^{-1}\boldsymbol{t}' \in \mathcal{P}$ with $\delta > 0$ and replace $\boldsymbol{t}$ by $\boldsymbol{t}' = (1+\delta)\boldsymbol{t}$, the rates are increased. There is a $\epsilon \in (0, 1)$ that satisfies $\rho^{-1}(1-\epsilon)\boldsymbol{r}(\boldsymbol{t}') \in \mathcal{R}$, which increases the objective to $\rho' = (1-\epsilon)^{-1}\rho$.

A.1.3 Relation Between the QoS Optimization and the Rate Balancing Problem

The QoS problem (1.9) and the RB problem (1.12) are inverse to each other in that

$$p(\rho(p_0)) = p_0$$

$$\rho(p(\rho_0)) = \rho_0$$

holds if $p(\rho)$ denotes the QoS optimum (1.13) and $\rho(p)$ the RB optimum (1.14).

To prove this statement, we assume the opposite. Let $p_0 = p(\rho_0)$ with optimizer $\boldsymbol{t}$ be the solutions for the QoS problem. Furthermore, assume that $\rho(p_0) < \rho_0$ with $\boldsymbol{t}'$ were the solutions for the balancing problem. This contradicts the optimality of $\boldsymbol{t}'$ for the balancing problem since $\boldsymbol{t}$ is feasible and achieves ρ_0. Otherwise, assume that $\rho(p_0) > \rho_0$ with $\boldsymbol{t}'$ were the solutions for the balancing problem. This contradicts the optimality of $\boldsymbol{t}$ for the QoS problem since we can find a scaling $(1+\delta)^{-1}\boldsymbol{t}'$, with sufficiently small $\delta > 0$, that satisfies $\rho_0^{-1}\boldsymbol{r}(\boldsymbol{t}') \in \mathcal{R}$ and achieves the objective value $p' = (1+\delta)^{-1}p_0$.

A.2 Interference Functions and Property Preserving Transforms

According to [75], *positivity*, *monotonicity*, and *sublinearity* have to be verified to show that $I_k : \mathbb{R}_+^K \to \mathbb{R}_{++}$, $k = 1, \ldots, K$ form a standard interference function

$$\boldsymbol{I} : \mathbb{R}_+^K \to \mathbb{R}_+^K, \quad \boldsymbol{I}(\boldsymbol{p}) = [I_1(\boldsymbol{p}), \ldots, \mathrm{I}_K(\boldsymbol{p})]^{\mathrm{T}}. \tag{A.1}$$

Alternatively, one can exploit the structure of I_k and show that it is based on functions that satisfy the properties, e.g., affine maps $I_k(\boldsymbol{p}) = \boldsymbol{a}_k^{\mathrm{T}}\boldsymbol{p} + b_k$ with $\boldsymbol{a} \geq \boldsymbol{0}$ and $b_k > 0$ [75] or concave increasing functions, or combinations of these functions and property preserving transforms.

Lemma A.1 *Let* $I_k : \mathbb{R}_+^K \to \mathbb{R}_{++}$, $k = 1, \ldots, K$ *be concave increasing. Then,* (A.1) *defines a standard interference function* [323, Proposition 1].

Proof Since positivity and monotonicity are assumed, it remains to show the sublinearity property. Let I_k be concave and $\alpha = 1/\lambda$ for $\lambda \in (0, 1)$. Then,

$$\alpha I_k(\boldsymbol{p}) = \frac{1}{\lambda} I_k\Big(\lambda \frac{1}{\lambda}\boldsymbol{p} + (1-\lambda)\boldsymbol{0}\Big) \geq I_k\Big(\frac{1}{\lambda}\boldsymbol{p}\Big) + (1-\lambda) I_k(\boldsymbol{0}) > I_k(\alpha \boldsymbol{p}).$$

The inequalities are due to concavity of I_k and $I_k(\boldsymbol{0}) > 0$. □

There is a class of operations that preserve the interference function properties of the above basic blocks (cf. [75, 323]). Namely, the standard interference property is preserved for (1) non-negative affine mappings, (2) pointwise minimum (or maximum) with respect to a parameter of a closed set, (3) composition with a sublinear non-decreasing function, and (4) composition of standard interference functions. We also identify a concept like the perspective function in convex optimization [102, Section 3.2.6, p. 89].[2]

Lemma A.2 *Let* $f_k : \mathbb{R}_+ \to \mathbb{R}_+$, $x \mapsto f(x)$ *be sublinear increasing and* $I'_k : \mathbb{R}_+^K \to \mathbb{R}_+$ *with* $\boldsymbol{p} \mapsto I'_k(\boldsymbol{p})$ *be an interference function. The perspective* $I_k : \mathbb{R}_+^{K+1} \to \mathbb{R}_+$,

$$I_k([t, \boldsymbol{p}^{\mathrm{T}}]^{\mathrm{T}}) = t f_k(I'_k(\boldsymbol{p})/t), \quad t > 0$$

is an interference function. This even holds if t *is any entry of the vector* $\boldsymbol{p}$, *e.g.,* $I_k : \mathbb{R}_+^K \to \mathbb{R}_+$ *with*

$$I_k(\boldsymbol{p}) = p_k f_k(I'_k(\boldsymbol{p})/p_k), \quad p_k > 0.$$

Proof Positivity of I_k and monotonicity in $\boldsymbol{p}$ are by definition of f_k and I'_k. Monotonicity in t follows with sublinearity of f_k. Let $\alpha > 1$, then the inequality $t' = \alpha t > t$ implies

$$I_k([\alpha t, \boldsymbol{p}^{\mathrm{T}}]^{\mathrm{T}}) = \alpha t f_k(I'_k(\boldsymbol{p})/(\alpha t)) > t f_k(\alpha I'_k(\boldsymbol{p})/(\alpha t)) = I_k([t, \boldsymbol{p}^{\mathrm{T}}]^{\mathrm{T}}). \qquad \text{(A.2)}$$

For $t = p_k$, let $\boldsymbol{p}' = [p_1, \ldots, \alpha p_k, \ldots, p_K]^{\mathrm{T}}$ with $\alpha > 1$. Then, (A.2) becomes

$$I_k(\boldsymbol{p}') > p_k f_k(I'_k(\boldsymbol{p}')/p_k) \geq p_k f_k(I'_k(\boldsymbol{p})/p_k),$$

[2]The standard perspective $t I_k(\boldsymbol{p}_{\setminus\{k\}}/t)$ fails ($sublinearity$) as it is *scale invariant*.

where the first inequality is as in (A.2) and the second is due to monotonicity of I_k and f_k. Finally, I_k is sublinear as the interference function I'_k is sublinear and f_k is increasing, i.e., $I'_k(\boldsymbol{p}) > I'_k(\alpha \boldsymbol{p})/\alpha$ imposes

$$\alpha I_k([t, \boldsymbol{p}]) = \alpha t f_k(I'_k(\boldsymbol{p})/t) > \alpha t f_k(I'_k(\alpha \boldsymbol{p})/\alpha t) = \alpha I_k(\alpha[t, \boldsymbol{p}]). \qquad \square$$

Besides exploiting concavity, also this transforms can be used to prove that the following data rate like function results in a standard interference function for (A.1):

$$r_k = \log_2\left(1 + p_k(\boldsymbol{a}_k^{\mathrm{T}}\boldsymbol{p} + b_k)^{-1}\right), \quad \boldsymbol{a}_k \geq 0, \quad b_k > 0, \quad k = 1, \ldots, K. \tag{A.3}$$

Lemma A.3 *The function $I_k(\boldsymbol{p}) = p_k/r_k$ with r_k in* (A.3) *is concave increasing.*

Proof First, the function $f : \mathbb{R}_+ \to \mathbb{R}_+$ with $x \mapsto f(x)$ and

$$f(x) = \frac{1}{\ln\left(1 + \frac{1}{x}\right)} \tag{A.4}$$

is concave increasing. Its first and second order derivatives read as

$$f'(x) = f^2(x)\frac{1}{x^2 + x} \geq 0$$

$$f''(x) = 2f'(x)\Big(f(x) - x - \frac{1}{2}\Big) \leq 0,$$

respectively, where positivity of f' is due to $x > 0$. To show that $f''(x) \leq 0$ for $x \geq 0$, we substitute $x = 1/y$ and remark that $g : \mathbb{R}_+ \to \mathbb{R}_+$ with $y \mapsto g(y)$ and

$$g(y) = \frac{1}{\ln(1 + y)} - \frac{1}{y}$$

has the limited function values $g(y) \in (0, 1/2]$ for $y \geq 0$.

Therewith, I_k can be constructed with the perspective $h([t, x]^{\mathrm{T}}) = t/f(x/t)$ and a composition with an affine map in $\boldsymbol{p}$. This proves concavity also for $I_k(\boldsymbol{p})$. $\square$

Theorem A.1 *The function $\boldsymbol{I} : \mathbb{R}_+^K \to \mathbb{R}_{++}^K$ with $\boldsymbol{p} \mapsto \boldsymbol{I}(\boldsymbol{p}) = [I_1(\boldsymbol{p}), \ldots, I_K(\boldsymbol{p})]^{\mathrm{T}}$ and I_k from Lemma A.3 is a standard interference function.*

Proof It suffices to note that any positive concave increasing functions I_k, $k = 1, \ldots, K$ yield a standard interference function (see Lemma A.1).[3] An alternative proof can be based on the perspective of an interference function. As f in (A.4) is sublinear increasing, we obtain a standard interference function with Lemma A.2:

[3]Also Cavalcante et al. [323, Proposition 1] mentioned that strictly concave functions are standard interference functions, but a proof is missing in this work.

$$I_k(\boldsymbol{p}) = p_k f(I'_k(\boldsymbol{p})/p_k), \quad k = 1, \ldots, K,$$

where I'_k is affine in $\boldsymbol{p}$ for f from (A.4). □

A mapping that does not preserve the standard interference function properties is

$$\boldsymbol{I}(\boldsymbol{p}) = \boldsymbol{\Gamma}\boldsymbol{I}'(\boldsymbol{p}), \quad \boldsymbol{\Gamma} = \operatorname{diag}(\vartheta_1, \ldots, \vartheta_K) \in \mathbb{R}_+^{K\times K} \tag{A.5}$$

if $\vartheta_i > 0$ for only a subset of the entries $i \in \mathcal{I} \subset \{1, \ldots, K\}$, with $|\mathcal{I}| < K$. The resulting function $\boldsymbol{I}$ is non-standard because it violates strict positivity with $\vartheta_k I_k(\boldsymbol{p}) = 0$ for $k \notin \mathcal{I}$. Nevertheless, the iteration $\boldsymbol{p}^{(n+1)} = \boldsymbol{I}(\boldsymbol{p}^{(n)})$ shares the monotonicity properties of (1.18) and, moreover, either converges to the unique point $\boldsymbol{p} = \boldsymbol{I}(\boldsymbol{p})$, which minimizes $\mathbf{1}^{\mathrm{T}}\boldsymbol{p}$, or diverges (cf. Theorem 1.1) in some entries of $\boldsymbol{p}$. The reason is that those entries p_k with $k \notin \mathcal{I}$ are immediately zero and the reduced interference function $\boldsymbol{J}(\boldsymbol{s}) = \boldsymbol{\Gamma}_{\mathcal{I}}\boldsymbol{I}_{\mathcal{I}}(\boldsymbol{p}_{\mathcal{I}}(\boldsymbol{s}))$ is standard if it comprises the entries $i \in \mathcal{I} = \{j(1), \ldots, j(|\mathcal{I}|)\}$ and $\boldsymbol{p}_{\mathcal{I}} : \mathbb{R}_+^{|\mathcal{I}|} \to \mathbb{R}_+^K$ defines the mapping $p_{\mathcal{I},j(i)}(\boldsymbol{s}) = s_i$ and $p_{\mathcal{I},k}(\boldsymbol{s}) = 0$ for $k \notin \mathcal{I}$.

The non-negative linear map (A.5) is amongst others important for balancing optimizations, where ϑ_k, $k = 1, \ldots, K$ defines monotonically increasing or decreasing functions of a common balancing level that become zero beyond some lower or upper bound, respectively. Then, the following statement holds.[4]

Proposition A.1 *Let $\boldsymbol{I} : \mathbb{R}_+^K \to \mathbb{R}_+^K$ be a standard interference function and $\boldsymbol{\Gamma}(\rho) \in \mathbb{R}^{K\times K}$, $\boldsymbol{\Gamma}(\rho) = \operatorname{diag}(\vartheta_1(\rho), \ldots, \vartheta_K(\rho))$ comprises the values of the monotonically increasing functions $\vartheta_k : \mathbb{R}_+ \to \mathbb{R}_+$, $k = 1, \ldots, K$, with $\vartheta_i(\rho') > \vartheta_i(\rho)$ if $\rho' > \rho$ and $\vartheta_i(\rho) > 0$. Furthermore, consider existence of $\hat{\rho} > \check{\rho}$ and $\hat{\boldsymbol{\lambda}}, \check{\boldsymbol{\lambda}} \in \mathbb{R}_+^K$ such that*

(i) *$\vartheta_i(\check{\rho}) > 0$ for $i \in \check{\mathcal{I}} \neq \emptyset$, and $\check{\boldsymbol{\lambda}} = \boldsymbol{\Gamma}(\check{\rho})\boldsymbol{I}(\check{\boldsymbol{\lambda}})$ with $\mathbf{1}^{\mathrm{T}}\check{\boldsymbol{\lambda}} < c$*[5]*;*
(ii) *$\vartheta_i(\hat{\rho}) > 0$ for $i \in \hat{\mathcal{I}} \neq \emptyset$ and $\hat{\boldsymbol{\lambda}} = \boldsymbol{\Gamma}(\hat{\rho})\boldsymbol{I}(\hat{\boldsymbol{\lambda}})$ with $\mathbf{1}^{\mathrm{T}}\hat{\boldsymbol{\lambda}} > c$.*

Then, there is a unique $(\rho^\star, \boldsymbol{\lambda}^\star)$ that satisfies $\boldsymbol{\lambda}^\star = \boldsymbol{\Gamma}(\rho^\star)\boldsymbol{I}(\boldsymbol{\lambda}^\star)$ and $c = \mathbf{1}^{\mathrm{T}}\boldsymbol{\lambda}^\star$.

Proof The relation $\hat{\rho} > \check{\rho}$ and monotonically increasing functions ϑ_k imply $\check{\mathcal{I}} \subseteq \hat{\mathcal{I}}$. Moreover, it imposes $\hat{\boldsymbol{\lambda}} \geq \boldsymbol{\Gamma}(\hat{\rho})\boldsymbol{I}(\check{\boldsymbol{\lambda}}) \geq \boldsymbol{\Gamma}(\check{\rho})\boldsymbol{I}(\check{\boldsymbol{\lambda}}) = \check{\boldsymbol{\lambda}}$, where the second inequality only holds with equality for entries $k \notin \hat{\mathcal{I}}$ and is strict for $i \in \hat{\mathcal{I}}$.

Based on these relations, let $\rho > 0$ reside between the bounds, i.e., $\hat{\rho} > \rho > \check{\rho}$. Then, $\hat{\boldsymbol{\lambda}} \in \mathbb{R}_+^K$ is feasible, i.e., $\hat{\boldsymbol{\lambda}} \geq \boldsymbol{\Gamma}(\rho)\boldsymbol{I}(\hat{\boldsymbol{\lambda}})$, and the sequence $\boldsymbol{\lambda}^{(n)}$ monotonically decreases when starting with $\boldsymbol{\lambda}^{(0)} = \hat{\boldsymbol{\lambda}}$. Similarly, the sequence $\boldsymbol{\lambda}^{(n)}$ monotonically increases when $\boldsymbol{\lambda}^{(0)} = \check{\boldsymbol{\lambda}}$ because $\check{\boldsymbol{\lambda}} \leq \boldsymbol{\Gamma}(\rho)\boldsymbol{I}(\check{\boldsymbol{\lambda}})$. The support set $\mathcal{I}$, which is defined by ρ, is the same in both sequences after $n = 1$ iterations, that is, $\lambda_i^{(1)} > 0$ for $i \in \mathcal{I}$

[4] A similar statement can be employed for the reversed monotonicity, i.e., when the function $\vartheta_k(\rho)$ is decreasing and becomes zero if ρ exceeds some value $\rho_{\max,k}$ for each $k = 1, \ldots, K$.

[5] This condition can be relaxed to $\check{\boldsymbol{\lambda}} \geq \boldsymbol{\Gamma}(\check{\rho})\boldsymbol{I}(\check{\boldsymbol{\lambda}})$ without changing the statement of the lemma.

and $\lambda_i^{(1)} = 0$ for $i \notin \mathcal{I}$. Moreover, the convergence point of the reduced iterations, i.e., based on the non-zero entries of $\boldsymbol{\lambda}^{(n)}$, is equal and unique because $\mathcal{J}(\cdot; \rho)$ is standard. Hence, there is a unique $\boldsymbol{\lambda}^\star$ for any $\rho \in [\check{\rho}, \hat{\rho}]$, with strictly increasing entries λ_i, $i \in \mathcal{I}$ for increasing balancing level ρ.

This shows in turn that there is only one unique $\rho^\star$ and the corresponding solution $\boldsymbol{\lambda}^\star$ that satisfies the requirement $c = \mathbf{1}^{\mathrm{T}}\boldsymbol{\lambda}^\star$ for given $c \in \mathbb{R}_+$. □

A.3 Ergodic Rate Bounds for Multiplicative Fading

This appendix details the derivation of the ergodic rate bounds in Sect. 3.2.2, with the structure of (3.11) and the noise and the offset parameters from Table 3.1. The basis is Jensen's inequality (e.g., [102, p. 77]): If $f : \mathcal{X} \to \mathbb{R}$ is convex and the random variable z is in the domain of f, i.e., $\Pr(z \in \mathcal{X}) = 1$, the mean of the function values $\mathrm{E}[f(z)]$ is bounded as

$$\mathrm{E}[f(z)] \geq f(\mathrm{E}[z]).$$

The inequality is strict if the function f is strictly convex, and it is reversed for a (strictly) concave function f.

A.3.1 Derivation of Lower and Upper Bounds on the Ergodic Rate

The ergodic rate is lower and upper bounded by the bounds LB1 and UB1, that is,

$$R_k^{(\mathrm{LB1})}(\boldsymbol{t}) \leq R_k(\boldsymbol{t}) \leq R_k^{(\mathrm{UB1})}(\boldsymbol{t}), \tag{A.6}$$

where the two bounds have the structure in (3.11) and the parameters from Table 3.1. The lower bound LB1 follows directly with Jensen's inequality and the fact that $r_k(\boldsymbol{t}, \zeta_k^{-1/2}\bar{\boldsymbol{h}}_k)$ is convex decreasing in ζ_k [cf. (3.11)].[6] The derivation of $R_k^{(\mathrm{UB1})}$ additionally exploits the structure of $A_k : \mathbb{R}_+ \to \mathbb{R}_+$ (see Footnote 5 of Chap. 3):

$$A_k(x) = \mathrm{E}[a_k(\zeta_k, x)], \quad a_k(\zeta_k, x) = \log_2(\zeta_k + x).$$

If the expectation $\mathrm{E}[\zeta_k]$ of the random variable $\zeta_k \in \mathbb{R}_+$ exists and is finite, then $A_k(x) \leq a_k(\mathrm{E}[\zeta_k], x)$ follows from Jensen's inequality since $a_k(\zeta_k, x)$ is concave

[6] The function $\log_2(1 + \frac{a}{\zeta_k + b})$, with $a, b \in \mathbb{R}_+$ and $\zeta_k \in \mathbb{R}_+$, is the composition of the convex function $\log_2(1 + z^{-1})$ with the affine mapping $z = \frac{\zeta_k + b}{a}$, and thus convex in ζ_k (cf. [102, p.̃79]).

in ζ_k. The difference $a_k(\mathrm{E}[\zeta_k], x) - A_k(x)$ decreases with increasing $x \in \mathbb{R}_+$.[7] Its maximum is $\mathrm{E}[\log_2(\mathrm{E}[\zeta_k]/\zeta_k)]$ at $x = 0$ and becomes zero for $x \to \infty$. Therefore,

$$\begin{aligned} &a_k(\mathrm{E}[\zeta_k], x+y) - A_k(x+y) + \mathrm{E}[\log_2(\mathrm{E}[\zeta_k]/\zeta_k)] \\ &\geq a_k(\mathrm{E}[\zeta_k], x) - A_k(x) \\ &\geq a_k(\mathrm{E}[\zeta_k], x+y) - A_k(x+y) \end{aligned} \tag{A.7}$$

holds for all $x, y \in \mathbb{R}_+$. The first inequality provides the rate bound $R_k^{(\mathrm{UB1})}(\boldsymbol{t}) \geq R_k(\boldsymbol{t})$ when adding $A_k(x+y)$ and subtracting $a_k(\mathrm{E}[\zeta_k], x)$ at both sides of (A.7) and replacing y and x with the useful signal power $\mathtt{S}_k$ (3.14) and the interference $\mathtt{I}_k$ (3.15), respectively. The second inequality of (A.7) in turn proves validness of the lower bound LB1 in (A.6). This bound is tight when the second inequality in (A.7) becomes an equality, i.e., if either $y = 0$ or $x \to \infty$. In contrast, the first inequality of (A.7) and thus UB1 are tight only if both $x = 0$ and $y \to \infty$ hold.

Similar derivation steps show that UB2 and LB2 bound the ergodic rate as

$$R_k^{(\mathrm{UB2})}(\boldsymbol{t}) \geq R_k(\boldsymbol{t}) \geq R_k^{(\mathrm{LB2})}(\boldsymbol{t}). \tag{A.8}$$

Now, Jensen's inequality and concavity of $r_k(\boldsymbol{t}, \zeta_k^{-1/2}\bar{\boldsymbol{h}})$ in ζ_k^{-1} are sufficient to find UB2.[8] To obtain LB2, Jensen's inequality is applied to the function [cf. (3.5)]

$$B_k(x) = \mathrm{E}[b_k(\zeta_k^{-1}, x)], \quad b_k(\zeta_k^{-1}, x) = \log_2(1 + \zeta_k^{-1} x).$$

As a result $b_k(\mathrm{E}[\zeta_k^{-1}], x) - B_k(x) \geq 0$, because $b_k(\zeta_k^{-1}, x)$ is concave in ζ_k^{-1}. This difference is zero for $x = 0$ and increases with $x \in \mathbb{R}_+$ up to $\mathrm{E}[\log_2(\zeta_k^{-1}/\,\mathrm{E}[\zeta_k^{-1}])]$ for $x \to \infty$.[9] Therewith, the following series of inequalities are valid for $x, y \in \mathbb{R}_+$:

$$\begin{aligned} b_k(\mathrm{E}[\zeta_k^{-1}], x) - B_k(x) &\leq b_k(\mathrm{E}[\zeta_k^{-1}], x+y) - B_k(x+y) \\ &\leq b_k(\mathrm{E}[\zeta_k^{-1}], x) - B_k(x) + \mathrm{E}[\log_2(\zeta_k^{-1}/\,\mathrm{E}[\zeta_k^{-1}])]. \end{aligned} \tag{A.9}$$

Here, the second inequality results in the lower bound LB2 when adding $B_k(x+y)$, subtracting $b_k(\mathrm{E}[\zeta_k^{-1}], x)$ and $\mathrm{E}[\log_2(\zeta_k^{-1}/\,\mathrm{E}[\zeta_k^{-1}])]$ at both sides of the inequality, and replacing y and x with the signal power $\mathtt{S}_k$ and the interference $\mathtt{I}_k$, respectively.

---

[7]The derivatives for both sides of $\mathrm{E}[\ln(\zeta_k + x)] \leq \ln(\mathrm{E}[\zeta_k] + x)$ satisfy $\mathrm{E}\left[\frac{1}{\zeta_k + x}\right] \geq \frac{1}{\mathrm{E}[\zeta_k] + x}$.

[8]The expression $\log_2(1 + sa/(1+sb))$ is composed of the two concave increasing functions $\log_2(1+z)$ and $z = \frac{a}{1/s+b}$ in s. Thus, it is also concave increasing in $s = \zeta_k^{-1}$ (cf. [102, p. 79]).

[9]The derivatives of $B_k(x)$ and $b_k(\zeta_k^{-1}, x)$ with respect to x satisfy $\mathrm{E}\left[\frac{1}{1/\zeta_k^{-1} + x}\right] \leq \frac{1}{1/\,\mathrm{E}\left[\zeta_k^{-1}\right] + x}$.

For the alternative bounds ALB and AUB, we substitute $\zeta_k = e^{z_k}$ into A_k:

$$A_k(x) = \mathrm{E}[a_k(e^{z_k}, x)] \geq a_k(e^{\mathrm{E}[z_k]}, x) = \tilde{A}_k(x), \tag{A.10}$$

where the inequality is due to convexity of $\log_2(e^{z_k} + x)$ in z_k. Equality holds in (A.10) for $x = 0$ and $x \to \infty$, but the first order derivatives of A_k and $\tilde{A}_k$,

$$A_k'(x) = \log_2(e)\,\mathrm{E}\left[\frac{1}{e^{z_k} + x}\right] \qquad \tilde{A}_k'(x) = \log_2(e)\frac{1}{e^{\mathrm{E}[z_k]} + x},$$

respectively, are only equal to zero for $x \to \infty$, while

$$\tilde{A}_k'(0) = \log_2(e)e^{-\mathrm{E}[z_k]} \leq \log_2(e)\,\mathrm{E}[e^{-z_k}] = A_k'(0).$$

Since both functions furthermore smoothly increase with $x \in \mathbb{R}_+$, we expect the difference $A_k(x) - \tilde{A}_k(x)$ to be pseudo-concave. This allows a numerical computation of the distance

$$d_k = \max\{d \in \mathbb{R}_+ : A_k(x) - \tilde{A}_k(x),\ x \in \mathbb{R}_+\}$$

via a golden section search [324, Chapter 8] and bounding $A_k(x + y) - A_k(y)$, $x, y \geq 0$ with the inequalities

$$\tilde{A}_k(x+y) - \tilde{A}_k(x) - d_k \leq A_k(x+y) - A_k(x) \leq \tilde{A}_k(x+y) - \tilde{A}_k(x) + d_k. \tag{A.11}$$

The bounds $R_k^{(\mathrm{ALB})}$ and $R_k^{(\mathrm{AUB})}$ with parameters in Table 3.1 follow again from replacing y and x with $\mathtt{S}_k$ and $\mathtt{I}_k$, respectively, and $z_k = \ln(\zeta_k)$ within (A.11).

A.4 Feasible QoS Region with Ergodic Rate Bounds

The proof of Lemma 3.1 exploits the MMSE feasible region from Lemma 1.1. Therewith, it is sufficient to rewrite the rate constraints of (3.16) as uplink MMSE requirements. Since the ergodic rate bounds (3.11) have the same logarithmic structure as the perfect CSI rates (1.3), the standard (SINR) dual formulation by[10]

$$\begin{aligned} R_k^{(\mathrm{B}),(\mathrm{ul})} &= \log_2\left(1 + \mathrm{SINR}_k^{(\mathrm{B}),(\mathrm{ul})}\right) + \mu_k^{(\mathrm{B})}, \\ \mathrm{SINR}_k^{(\mathrm{B}),(\mathrm{ul})} &= \frac{\nu_k^{(\mathrm{B}),-1}|\bar{\boldsymbol{h}}_k^{\mathrm{H}}\boldsymbol{u}_k|^2\lambda_k}{1 + \sum_{i=1}^K \nu_i^{(\mathrm{B}),-1}|\bar{\boldsymbol{h}}_i^{\mathrm{H}}\boldsymbol{u}_k|^2\lambda_i}. \end{aligned} \tag{A.12}$$

[10]The duality and feasibility test are without loss of generality based on a sum power restriction.

Rewriting (A.12) as MSEs and inserting the equalizers, the MMSEs read as

$$\mathrm{MMSE}_k^{(\mathrm{B}),(\mathrm{ul})} = 1 - \nu_k^{(\mathrm{B})} \bar{\boldsymbol{h}}_k^{\mathrm{H}} \Big(\mathbf{I} + \sum_{i=1}^{K} \bar{\boldsymbol{h}}_i \lambda_i \nu_i^{(\mathrm{B})} \bar{\boldsymbol{h}}_i^{\mathrm{H}} \Big) \bar{\boldsymbol{h}}_k \tag{A.13}$$

and the uplink MMSE constraints for the rate constraints of (3.16) are[11]

$$\mathrm{MMSE}_k^{(\mathrm{B}),(\mathrm{ul})} \leq 2^{-\max\left\{\rho_k - \mu_k^{(\mathrm{B})}, 0\right\}} = \varepsilon_k. \tag{A.14}$$

With $\rho_k \in \mathbb{R}_+$, feasible targets ε_k still satisfy the box constraint $0 < \varepsilon_k \leq 1$. The proof for $\sum_{k=1}^{K} \varepsilon_k \geq K - N$ is equivalent to [65, Proof of Theorem 1].

A.5 On Uniqueness of the QoS Optimal Power Allocation

This appendix discusses uniqueness for the solution of the QoS optimization (3.48) with ergodic rate requirements and fixed beamformers. By definition, any solution of (3.48) resides in the intersection of the feasible set

$$\mathcal{S} = \left\{ \boldsymbol{p} \in \mathbb{R}_+^K : \boldsymbol{p} \geq \boldsymbol{I}(\boldsymbol{p}; \boldsymbol{\rho}) \right\}$$

and the objective's sub-level set, i.e., the compact convex polytope

$$\mathcal{T}(P^\star) = \left\{ \boldsymbol{p} \in \mathbb{R}_+^K : \tilde{\boldsymbol{A}} \boldsymbol{p} \leq P^\star \mathbf{1} \right\}, \tag{A.15}$$

where $P^\star$ is the minimal scaling of the polytope such that $\mathcal{T}(P^\star) \cap \mathcal{S}$ is non-empty. In other words, the solutions reside on the tangential plane, where the upper right boundary of $\mathcal{T}(P^\star)$ touches the lower left boundary of $\mathcal{S}$.

Due to the standard interference property of $\boldsymbol{I}(\cdot; \boldsymbol{\rho})$, this lower boundary of $\mathcal{S}$ is characterized by the unique fixed point

$$\boldsymbol{p}^\star = \boldsymbol{I}(\boldsymbol{p}^\star; \boldsymbol{\rho}), \tag{A.16}$$

which is the *minimum point* of $\mathcal{S}$ [230, Theorem 5.57],[12] i.e., $\boldsymbol{p}^\star \leq \boldsymbol{p}'$ for every $\boldsymbol{p}' \in \mathcal{S}$. Therefore, $\boldsymbol{p}^\star \in \mathcal{T}(P^\star) \cap \mathcal{S}$ is a solution of (3.48) and provides[13]

$$P^\star = \max_{\ell} \mathbf{e}_\ell^{\mathrm{T}} \tilde{\boldsymbol{A}} \boldsymbol{p}^\star. \tag{A.17}$$

[11] The maximum operation is due to $\mathrm{MMSE}_k^{(\mathrm{B}),(\mathrm{ul})} \leq 1$, which holds for $\lambda_k = 0$.

[12] A $\boldsymbol{x} \in \mathcal{A}$ is a *(strict) minimum point* of $\mathcal{A}$ if $\boldsymbol{x} \leq \boldsymbol{y}$ ($\boldsymbol{x} < \boldsymbol{y}$) for every $\boldsymbol{y} \neq \boldsymbol{x}$, $\boldsymbol{y} \in \mathcal{A}$.

[13] It is impossible to further reduce the objective of (3.48) as $\tilde{\boldsymbol{A}} \geq \mathbf{0}$ and $\boldsymbol{p}' \notin \mathcal{S}$ if $\exists i : p_i' < p_i^\star$.

While (3.48) can have multiple solutions in general, $\boldsymbol{p}^\star$ is the unique solution for cases where $\mathcal{S} \cap \mathcal{T}(P^\star)$ becomes a singleton. This occurs if $\tilde{\boldsymbol{A}} > \boldsymbol{0}$ [230, Corollary 5.58]. An example is the total sum power minimization [230, p. 182], where the objective degenerates to $\tilde{\boldsymbol{a}}^{\mathrm{T}}\boldsymbol{p}$.

The solution of (3.48) is also unique if $\boldsymbol{p}^\star$ is a *strict minimum point* of $\mathcal{S}$, i.e., $\boldsymbol{p}^\star < \boldsymbol{p}'$ for $\boldsymbol{p}' \neq \boldsymbol{p}^\star$, $\boldsymbol{p}' \in \mathcal{S}$. Then, there is no feasible $\boldsymbol{p}' \geq \boldsymbol{p}^\star$, $\boldsymbol{p}' \neq \boldsymbol{p}^\star$ that still resides in $\mathcal{T}(P^\star)$ because $\tilde{\boldsymbol{A}}\boldsymbol{p}' > \tilde{\boldsymbol{A}}\boldsymbol{p}^\star$. A strict minimum point is caused by strict (local) monotonicity of $\boldsymbol{I}(\cdot;\boldsymbol{\rho})$ at $\boldsymbol{p}^\star$. Then, forcing either element of $\boldsymbol{p}'$ to lie strictly above $\boldsymbol{p}^\star$, e.g., $p_i' = p_i^\star + \delta_i$ with $\delta_i > 0$, induces $p_k' > p_k^\star$ for $k \neq i$ if $\boldsymbol{p}' \in \mathcal{S}$ [11, Section 2.2.2].

In contrast, $\mathcal{S} \cap \mathcal{T}(P^\star)$ contains also other solutions than $\boldsymbol{p}^\star$ if there is a $\boldsymbol{p}' \in \mathcal{S}$, $\boldsymbol{p}' \geq \boldsymbol{p}^\star$, that satisfies the inequality condition

$$\mathbf{e}_n^{\mathrm{T}}\tilde{\boldsymbol{A}}\boldsymbol{p}^\star < \mathbf{e}_n^{\mathrm{T}}\tilde{\boldsymbol{A}}\boldsymbol{p}' < \max_\ell \mathbf{e}_\ell^{\mathrm{T}}\tilde{\boldsymbol{A}}\boldsymbol{p}^\star.$$

This property depends on the structure of $\boldsymbol{I}(\cdot;\boldsymbol{\rho})$ and $\tilde{\boldsymbol{A}}$. For example, let $\boldsymbol{p} = [\boldsymbol{p}_1^{\mathrm{T}}, \boldsymbol{p}_2^{\mathrm{T}}]^{\mathrm{T}}$ and $\boldsymbol{I}(\boldsymbol{p};\boldsymbol{\rho}) = [\boldsymbol{I}_1^{\mathrm{T}}(\boldsymbol{p}_1), \boldsymbol{I}_2^{\mathrm{T}}(\boldsymbol{p}_2)]^{\mathrm{T}}$ such that $\boldsymbol{p} \in \mathcal{S}$ is equivalent to $\boldsymbol{p}_i \in \mathcal{S}_i = \{\boldsymbol{p}_i \in \mathbb{R}_+^{K_i} : \boldsymbol{p}_i \geq \boldsymbol{I}_i(\boldsymbol{p}_i)\}$, $i = 1, 2$. Furthermore, let $\tilde{\boldsymbol{A}}$ be block diagonal with blocks $\tilde{\boldsymbol{A}}_1, \tilde{\boldsymbol{A}}_2$, such that $\tilde{\boldsymbol{A}}\boldsymbol{p} = [(\tilde{\boldsymbol{A}}_1\boldsymbol{p}_1)^{\mathrm{T}}, (\tilde{\boldsymbol{A}}_2\boldsymbol{p}_2)^{\mathrm{T}}]^{\mathrm{T}}$ and assume that the following inequality holds:

$$\max_n \mathbf{e}_n^{\mathrm{T}}\tilde{\boldsymbol{A}}_2\boldsymbol{p}_2^\star < \max_\ell \mathbf{e}_\ell^{\mathrm{T}}\tilde{\boldsymbol{A}}_1\boldsymbol{p}_1^\star.$$

Then, any $\boldsymbol{p}' \in \mathcal{T}(P^\star)$, $\boldsymbol{p}' \neq \boldsymbol{p}^\star$, where $\boldsymbol{p}_1' = \boldsymbol{p}_1^\star$ and $\boldsymbol{p}_2' = \alpha\boldsymbol{p}_2^\star$ with

$$\alpha \in \big(1, \max_\ell \mathbf{e}_\ell^{\mathrm{T}}\tilde{\boldsymbol{A}}_1\boldsymbol{p}_1^\star / \max_n \mathbf{e}_n^{\mathrm{T}}\tilde{\boldsymbol{A}}_2\boldsymbol{p}_2^\star\big],$$

is also an element of $\mathcal{S}$ due to the independence of $\boldsymbol{I}_1$ and $\boldsymbol{I}_2$ and sublinearity of $\boldsymbol{I}_2$. Hence, any $\boldsymbol{p}' \in \mathcal{S} \cap \mathcal{T}(P^\star)$ is a solution of the optimization in (3.48).

A.6 Duality for Second Order Cone Programs

A standard (dual) form of the general SOCP reads as[14]

$$\max_{\boldsymbol{y},\boldsymbol{z}} f(\boldsymbol{y}) \quad \text{s.t.} \quad \boldsymbol{c}_i - \boldsymbol{A}_i^{\mathrm{T}}\boldsymbol{y} = \boldsymbol{z}_i, \quad \boldsymbol{z}_i \in \mathcal{L}^{q_i}, \quad i = 1, \ldots, m, \tag{A.18}$$

where $f(\boldsymbol{y})$ is a concave function in $\boldsymbol{y} \in \mathbb{R}^n$ and $\boldsymbol{z} = [\boldsymbol{z}_1^{\mathrm{T}}, \ldots, \boldsymbol{z}_{q_i}^{\mathrm{T}}]^{\mathrm{T}} \in \mathbb{R}^q$, $\boldsymbol{z}_i = [z_{i1}, \boldsymbol{z}_{i2}^{\mathrm{T}}]^{\mathrm{T}}$. An example for $f(\boldsymbol{y})$ is $-\boldsymbol{b}^{\mathrm{T}}\boldsymbol{y}$, but we focus on $-|\boldsymbol{b}^{\mathrm{T}}\boldsymbol{y}|^2$ instead. Using multipliers $\boldsymbol{x}_i \in \mathbb{R}^{q_i}$ and $\mu_i \geq 0$, $i = 1, \ldots, m$, the Lagrangian function reads as

[14] Alternatively, the equality and conic constraints of (A.18) may be combined to $\boldsymbol{c}_i - \boldsymbol{A}_i^{\mathrm{T}}\boldsymbol{y} \in \mathcal{L}^{q_i}$.

$$L(\boldsymbol{y}, \boldsymbol{z}, \boldsymbol{x}, \boldsymbol{\mu}) = f(\boldsymbol{y}) + \sum_{i=1}^{m} \boldsymbol{x}_i^{\mathrm{T}}(\boldsymbol{c}_i - \boldsymbol{A}_i^{\mathrm{T}}\boldsymbol{y}) - \boldsymbol{x}_i^{\mathrm{T}}\boldsymbol{z}_i + \mu_i(z_{i1} - \|\boldsymbol{z}_{i2}\|_2). \tag{A.19}$$

This function allows to recast (A.18) as the max–min optimization

$$\max_{\boldsymbol{y},\boldsymbol{z}} \inf_{\boldsymbol{x},\boldsymbol{\mu}} L(\boldsymbol{y}, \boldsymbol{z}, \boldsymbol{x}, \boldsymbol{\mu}) \quad \text{s.t.} \quad \boldsymbol{z}_i \in \mathcal{L}^{q_i}, \quad \mu_i \geq 0, \quad i = 1, \ldots, m \tag{A.20}$$

and the Lagrangian dual problem with exchanged min- and maximization as[15]

$$\min_{\boldsymbol{x},\boldsymbol{\mu}} \sup_{\boldsymbol{y},\boldsymbol{z}} L(\boldsymbol{y}, \boldsymbol{z}, \boldsymbol{x}, \boldsymbol{\mu}) \quad \text{s.t.} \quad \boldsymbol{z}_i \in \mathcal{L}^{q_i}, \quad \mu_i \geq 0, \quad i = 1, \ldots, m. \tag{A.22}$$

Due to convexity of (A.18), the duality gap between (A.20) and (A.22) is zero, that is, strong duality holds at the solutions of these problem formulations [103, Chapter 2]. Otherwise, the objective of (A.22) upper bounds the optimum of (A.18).

The Karush–Kuhn–Tucker (KKT) optimality conditions provide a necessary and sufficient characterization for the solutions (e.g., [102, 227]):

1. *Feasibility:* The vectors $\boldsymbol{y}$, $\boldsymbol{z}$, and $\boldsymbol{\mu}$ are primal and dual feasible, respectively, if

$$\boldsymbol{c}_i - \boldsymbol{A}_i^{\mathrm{T}}\boldsymbol{y} = \boldsymbol{z}_i, \quad \boldsymbol{z}_i \in \mathcal{L}^{q_i}, \quad \mu_i \geq 0, \quad i = 1, \ldots, m. \tag{A.23}$$

2. *Complementary Slackness:* The following products including $\boldsymbol{x}$ and $\boldsymbol{\mu}$ are zero:

$$\boldsymbol{x}_i^{\mathrm{T}}\left(\boldsymbol{c}_i - \boldsymbol{A}_i^{\mathrm{T}}\boldsymbol{y} - \boldsymbol{z}_i\right) = \boldsymbol{0}, \quad \mu_i(z_{i1} - \|\boldsymbol{z}_{i2}\|_2) = 0, \quad i = 1, \ldots, m. \tag{A.24}$$

3. *First Order Conditions:* The derivatives of the Lagrangian function are zero, i.e.,

$$\frac{\partial}{\partial \boldsymbol{y}} f(\boldsymbol{y}) = \sum_{i=1}^{m} \boldsymbol{A}_i\boldsymbol{x}_i, \quad \boldsymbol{x}_i = \mu_i[1, -\boldsymbol{z}_{i2}^{\mathrm{T}}\|\boldsymbol{z}_{i2}\|_2^{-1}]^{\mathrm{T}}, \quad i = 1, \ldots, m. \tag{A.25}$$

Lemma A.4 *In particular, solutions* $\boldsymbol{x}_i \in \mathcal{L}^{q_i}$ *for* (A.21) *have the structure* [73, 74]

$$\boldsymbol{x}_i = \lambda_i \operatorname{bdiag}(1, -\mathbf{I})\left(\boldsymbol{c}_i - \boldsymbol{A}_i^{\mathrm{T}}\boldsymbol{y}\right), \quad \lambda_i \in \mathbb{R}_+, \quad i = 1, \ldots, m, \tag{A.26}$$

where $\lambda_i = 0$ *if* $c_{i1} - \boldsymbol{a}_{i1}^{\mathrm{T}}\boldsymbol{y} < \|\boldsymbol{c}_{i2} - \boldsymbol{A}_{i2}^{\mathrm{T}}\boldsymbol{y}\|_2$, $\boldsymbol{c} = \left[c_{i1}, \boldsymbol{c}_{i2}^{\mathrm{T}}\right]^{\mathrm{T}}$ *and* $\boldsymbol{A}_i = [\boldsymbol{a}_{i1}, \boldsymbol{A}_{i2}]$.

---

[15] For the linear objective $f(\boldsymbol{y}) = -\boldsymbol{b}^{\mathrm{T}}\boldsymbol{y}$, a conic form representation of (A.22) reads as

$$\min_{\boldsymbol{x},\boldsymbol{\mu}} \sum_{i=1}^{m} \boldsymbol{c}_i^{\mathrm{T}}\boldsymbol{x}_i \quad \text{s.t.} \quad \sum_{i=1}^{m} \boldsymbol{A}_i\boldsymbol{x}_i = \boldsymbol{b}, \quad \boldsymbol{x}_i \in \mathcal{L}^{q_i}, \quad i = 1, \ldots, m. \tag{A.21}$$

Proof The latter condition in (A.24) shows that $\mu_i > 0$ only if $z_{i1} = \|z_{i2}\|_2$ and $\mu_i = 0$ exactly if $z_{i1} > \|z_{i2}\|_2$. This allows to write the structure for $\boldsymbol{x}_i$ in (A.25) as

$$\boldsymbol{x}_i = \mu_i \|z_{i2}\|_2^{-1}[z_{i1}, -z_{i2}^{\mathrm{T}}]^{\mathrm{T}}, \tag{A.27}$$

where z_{i1} replaces $\|z_{i2}\|_2$. Now, defining the non-negative variable $\lambda_i = \mu_i \|z_{i2}\|_2^{-1} \in \mathbb{R}_+$ and replacing z_i according to (A.23), we obtain (A.26). □

Therewith, we aim at a dual representation when $f(\boldsymbol{y}) = -|\boldsymbol{b}^{\mathrm{T}}\boldsymbol{y}|^2$ and the parameters $\boldsymbol{A}_i$, $\boldsymbol{b}$, and $\boldsymbol{c}_i$ have similar properties as for (4.30):[16]

$$c_{i1} = 0, \qquad \boldsymbol{A}_{i2}\boldsymbol{c}_{i2} = \boldsymbol{0}, \quad i = 1, \ldots, m, \tag{A.28}$$

$$\boldsymbol{A}_{i2}^{\mathrm{T}}\boldsymbol{b} = \boldsymbol{0}, \qquad \boldsymbol{A}_{i2}^{\mathrm{T}}\boldsymbol{a}_{\ell 1} = \boldsymbol{0}, \quad i = 1, \ldots, m, \quad \ell = r+1, \ldots, m, \tag{A.29}$$

$$\boldsymbol{a}_{j1}^{\mathrm{T}}\boldsymbol{a}_{i1} = 0, \qquad \boldsymbol{a}_{i1} \in \mathrm{range}\{\boldsymbol{A}_{i2}\}, \quad i \neq j, \quad i = 1, \ldots, r, \quad j = 1, \ldots, m. \tag{A.30}$$

Theorem A.2 *Given* (A.28)–(A.30) *and Lemma A.4, the minimization problem*

$$\min_{\boldsymbol{\lambda} \geq \boldsymbol{0}} -\sum_{i=1}^{\kappa} \lambda_i \|\boldsymbol{c}_{i2}\|_2^2 \quad \text{s.t.} \quad 1 - \lambda_k \boldsymbol{a}_{k1}^{\mathrm{T}} \Big(\sum_{\ell=1}^{\kappa} \lambda_\ell \boldsymbol{A}_{\ell 2} \boldsymbol{A}_{\ell 2}^{\mathrm{T}} \Big)^{\dagger} \boldsymbol{a}_{k1} \geq 0, \quad k = 1, \ldots, r,$$
$$\|\boldsymbol{b}\|_2^2 - \sum_{\ell=r+1}^{\kappa} \lambda_\ell \|\boldsymbol{a}_{\ell 1}\|_2^2 \geq 0 \tag{A.31}$$

is a Lagrangian dual of (A.22) *if* $\boldsymbol{a}_{\ell 1}$, $\ell = r+1, \ldots, \kappa$, $\kappa \leq m$ *are collinear to* $\boldsymbol{b}$.

Proof Inserting (A.26) and in (A.25) for $\boldsymbol{x}_i$, at the first and second position in (A.19), respectively, the Lagrangian function becomes[17]

[16]These properties also apply for the SINR constrained power minimization of [67, 73, 74].

[17]Note that (A.33) corresponds to the Quadratically constrained program (QCP)

$$\max_{\boldsymbol{y}} f(\boldsymbol{y}) \quad \text{s.t.} \quad (c_{i1} - \boldsymbol{a}_{i1}^{\mathrm{T}}\boldsymbol{y})^2 \geq \|\boldsymbol{c}_{i2} - \boldsymbol{A}_{i2}^{\mathrm{T}}\boldsymbol{y}\|_2^2, \quad i = 1, \ldots, m, \tag{A.32}$$

whose necessary KKT optimality conditions are sufficient for satisfying (A.23)–(A.25) by construction of (A.33) with (A.27) if the KKT points additionally fulfill $c_{i1} - \boldsymbol{a}_{i1}^{\mathrm{T}}\boldsymbol{y} \geq 0$, $i = 1, \ldots, m$ [73, Appendix A]. While KKT points can exist for (A.32), that violate this necessary requirement, these points are obviously infeasible for (A.18) [see (A.23)]. Hence, solving (A.32) generally provides a valid solution for (A.18) only if there is a mapping for KKT points $\boldsymbol{y}$ with $c_{i1} - \boldsymbol{a}_{i1}^{\mathrm{T}}\boldsymbol{y} < 0$ into KKT points $\boldsymbol{y}'$ with $c_{i1} - \boldsymbol{a}_{i1}^{\mathrm{T}}\boldsymbol{y}' > 0$.

$$L(\boldsymbol{y}, \boldsymbol{\lambda}) = f(\boldsymbol{y}) + \sum_{i=1}^{m} \lambda_i \big((c_{i1} - \boldsymbol{a}_{i1}^{\mathrm{T}} \boldsymbol{y})^2 - \|\boldsymbol{c}_{i2} - \boldsymbol{A}_{i2}^{\mathrm{T}} \boldsymbol{y}\|_2^2 \big). \tag{A.33}$$

With property (A.28) and $f(\boldsymbol{y}) = -|\boldsymbol{b}^{\mathrm{T}} \boldsymbol{y}|^2$, this Lagrangian function becomes

$$L(\boldsymbol{y}, \boldsymbol{\lambda}) = -\sum_{i=1}^{m} \lambda_i \|\boldsymbol{c}_{i2}\|_2^2 + \boldsymbol{y}^{\mathrm{T}} \Big(-\boldsymbol{b}\boldsymbol{b}^{\mathrm{T}} + \sum_{i=1}^{m} \lambda_i \big(\boldsymbol{a}_{i1} \boldsymbol{a}_{i1}^{\mathrm{T}} - \boldsymbol{A}_{i2} \boldsymbol{A}_{i2}^{\mathrm{T}} \big) \Big) \boldsymbol{y}. \tag{A.34}$$

Here, the dual objective $g(\boldsymbol{\lambda}) = \sup_{\boldsymbol{y}} L(\boldsymbol{y}, \boldsymbol{\lambda})$ in (A.22) is unbounded above unless

$$\boldsymbol{b}\boldsymbol{b}^{\mathrm{T}} - \sum_{i=1}^{m} \lambda_i \big(\boldsymbol{a}_{i1} \boldsymbol{a}_{i1}^{\mathrm{T}} - \boldsymbol{A}_{i2} \boldsymbol{A}_{i2}^{\mathrm{T}} \big) \succeq \boldsymbol{0}, \tag{A.35}$$

where $g(\boldsymbol{\lambda}) = -\sum_{i=1}^{m} \lambda_i \|\boldsymbol{c}_{i2}\|_2^2$ and $\boldsymbol{y} = \boldsymbol{0}$ if the linear matrix inequality is strict and $\boldsymbol{y}$ lies in the nullspace of the singular matrix. Due to orthogonality (A.29) and positive semidefiniteness, this condition holds if and only if

$$\boldsymbol{b}\boldsymbol{b}^{\mathrm{T}} - \sum_{\ell=r+1}^{m} \lambda_\ell \boldsymbol{a}_{\ell 1} \boldsymbol{a}_{\ell 1}^{\mathrm{T}} \succeq \boldsymbol{0}, \tag{A.36}$$

$$\sum_{i=1}^{r} \lambda_i \boldsymbol{a}_{i1} \boldsymbol{a}_{i1}^{\mathrm{T}} - \sum_{i=1}^{m} \lambda_i \boldsymbol{A}_{i2} \boldsymbol{A}_{i2}^{\mathrm{T}} \succeq \boldsymbol{0}. \tag{A.37}$$

Due to (A.36), $\lambda_\ell > 0$ only holds for collinear $\boldsymbol{a}_{\ell 1}$ and $\boldsymbol{b}$ and $\lambda_\ell = 0$ otherwise. Let $\ell = r+1, \ldots, \kappa$ be the indices of these vectors. Then, (A.36) becomes the second constraint in (A.31) and $g(\boldsymbol{y})$ reduces to the first $\kappa - r + 1$ summands.

Furthermore, because of mutual independence (orthogonality) between the vectors $\boldsymbol{a}_{ij}$, $j = 1, \ldots, r$, (A.37) holds if and only if

$$\sum_{j=1}^{m} \lambda_j \boldsymbol{A}_{j2} \boldsymbol{A}_{j2}^{\mathrm{T}} - \lambda_k \boldsymbol{a}_{k1} \boldsymbol{a}_{k1}^{\mathrm{T}} \succeq 0, \quad k = 1, \ldots, r.$$

Applying Schur's complement relation [293], these positive semidefiniteness conditions transform to the second constraints in (A.31). The range-space property in (A.30), which can be relaxed to $\boldsymbol{a}_{i1} \in \operatorname{range}\left\{\sum_{j=1}^{m} \lambda_j \boldsymbol{A}_{j2}\right\}$, ensures sufficiency of the reformulated constraints in (A.31) for (A.37). Otherwise, the multiplier $\lambda_i = 0$ if $\boldsymbol{a}_{i1} \notin \operatorname{range}\left\{\sum_{j=1}^{m} \lambda_j \boldsymbol{A}_{j2}\right\}$. □

A.6.1 Application to Uplink–Downlink Duality for MSE Based QoS Optimization

With the extended precoder $\tilde{t} = [p, \mathrm{Re}(t^{\mathrm{T}}), \mathrm{Im}(t^{\mathrm{T}})]^{\mathrm{T}}$, the real valued equivalent standard SOC constraint formulation of (4.30) reads as [cf. (A.18)][18]

$$\begin{aligned} \max_{\tilde{t},z}\ -(\mathbf{e}_1^{\mathrm{T}}\tilde{t})^2 \quad \text{s.t.} \quad -\tilde{A}_\ell^{\mathrm{T}}\tilde{t} &= z_{\ell+K}, \quad z_{\ell+K} \in \mathcal{L}^{NK}, \quad \ell = 1,\dots,L, \\ c_k - \tilde{R}_k^{\mathrm{T}}\tilde{t} &= z_k, \quad\quad z_k \in \mathcal{L}^{NK+1}, \quad k = 1,\dots,K, \end{aligned} \tag{A.38}$$

where we introduced $c_k = [c_{1k}, c_{2k}^{\mathrm{T}}]^{\mathrm{T}}$ and $\tilde{R}_k = [\tilde{r}_{1k}, \tilde{R}_{2,k}]$ as

$$c_k = \begin{bmatrix} \mathbf{0} \\ \bar{\sigma}_k \end{bmatrix}, \quad \tilde{R}_k^{\mathrm{T}} = \begin{bmatrix} 0 & \mathbf{e}_k^{\mathrm{T}} \otimes (1-\varepsilon_k)^{-1/2}\,\mathrm{Re}(\bar{h}_k^{\mathrm{H}}) & \mathbf{e}_k^{\mathrm{T}} \otimes (1-\varepsilon_k)^{-1/2}\,\mathrm{Im}(\bar{h}_k^{\mathrm{T}}) \\ \mathbf{0} & \mathrm{Re}(\hat{R}_k^{1/2}) & -\mathrm{Im}(\hat{R}_k^{1/2}) \\ \mathbf{0} & \mathrm{Im}(\hat{R}_k^{1/2}) & \mathrm{Im}(\hat{R}_k^{1/2}) \\ 0 & \mathbf{0}^{\mathrm{T}} & \mathbf{0}^{\mathrm{T}} \end{bmatrix}$$

for the MSE constraints and, for the power constraints, $\tilde{A}_\ell = [\tilde{a}_{1\ell}, \tilde{A}_{2\ell}]$ reads as

$$\tilde{A}_\ell^{\mathrm{T}} = \begin{bmatrix} \sqrt{P_\ell} & \mathbf{0} & \mathbf{0} \\ \mathbf{0} & \mathrm{Re}(A_\ell^{1/2}) & -\mathrm{Im}(A_\ell^{1/2}) \\ \mathbf{0} & \mathrm{Im}(A_\ell^{1/2}) & \mathrm{Re}(A_\ell^{1/2}) \end{bmatrix}, \quad \ell = 1,\dots,L.$$

Since c_{2k}, $\tilde{r}_{1k}$, and $\tilde{a}_{1\ell}$ meet the orthogonality requirements [cf. (A.28)–(A.30)]

$$\begin{aligned} \tilde{R}_{2k}c_{2k} &= \mathbf{0}, \quad k = 1,\dots,K, \\ \tilde{R}_{2k}\mathbf{e}_1 = \mathbf{0}, \quad \tilde{A}_{2m}\mathbf{e}_1 &= \mathbf{0}, \quad k = 1,\dots,K, \quad m = 1,\dots,L, \\ \tilde{R}_{2k}a_{1\ell} = \mathbf{0}, \quad \tilde{A}_{2m}a_{1\ell} &= \mathbf{0}, \quad k = 1,\dots,K, \quad m,\ell = 1,\dots,L, \\ \tilde{r}_{2k}^{\mathrm{T}}\tilde{a}_{1\ell} = \mathbf{0}, \quad \tilde{r}_{1k}^{\mathrm{T}}\tilde{r}_{1j} &= \mathbf{0}, \quad k \neq j, \quad k,j = 1,\dots,K, \quad \ell = 1,\dots,L, \end{aligned}$$

we can apply Theorem A.2 to write the dual optimization of (A.38) as

$$\begin{aligned} \min_{\mu \geq \mathbf{0}, \lambda \geq \mathbf{0}} -\sum_{i=1}^{K} \lambda_i \|c_{i2}\|_2^2 \quad \text{s.t.} \quad 1 - \sum_{\ell=1}^{L} \mu_\ell \|\tilde{a}_{\ell 1}\|_2^2 \geq 0 \\ 1 - \lambda_k \tilde{r}_{k1}^{\mathrm{T}} \Big(\sum_{i=1}^{K} \lambda_i \tilde{R}_{i2}\tilde{R}_{i2}^{\mathrm{T}} + \sum_{\ell=1}^{L} \mu_\ell \tilde{A}_{\ell 2}\tilde{A}_{\ell 2}^{\mathrm{T}} \Big)^{\dagger} \tilde{r}_{k1} \geq 0, \quad k = 1,\dots,K. \end{aligned} \tag{A.39}$$

[18] This is actually the dual form of a standard SOCP [102, Section 5.2.1] as shown in (A.18).

Inserting the substitutes for $\boldsymbol{c}_k$, $\tilde{\boldsymbol{R}}_k$, and $\tilde{\boldsymbol{A}}_\ell$, the equivalent complex form of (A.39) reads as[19]

$$\begin{aligned}
&\min_{\boldsymbol{\mu}\geq\mathbf{0},\boldsymbol{\lambda}\geq\mathbf{0}} -\sum_{i=1}^{K}\lambda_i\tilde{\sigma}_i^2 \quad \text{s.t.} \quad 1-\sum_{\ell=1}^{L}\mu_\ell P_\ell \geq 0 \\
&1-\frac{\lambda_k}{1-\varepsilon_k}\bar{\boldsymbol{h}}_k^{\mathrm{H}}\Big(\sum_{i=1}^{K}\lambda_i\boldsymbol{R}_k+\sum_{\ell=1}^{L}\mu_\ell\boldsymbol{A}_{k,\ell}\Big)^{\dagger}\bar{\boldsymbol{h}}_k \geq 0, \quad k=1,\ldots,K.
\end{aligned} \tag{A.40}$$

This formulation is equivalent to the dual formulation of the per-user MSE constrained QoS optimization in Corollary 4.1.

A.6.2 Reconstruction of the Primal Variables

The duality gap between the optima of the dual and the primal problem is zero. Both problems have conic convex formulations and the solution $\boldsymbol{\lambda}^\star$ for (A.31) satisfies the KKT conditions (A.23)–(A.25). Given $\boldsymbol{\lambda}^\star$, reconstructing the primal solution $\boldsymbol{y}$ is by the zero-derivative condition (A.25). We separate $\boldsymbol{y}$ into the sum of two vectors

$$\boldsymbol{y}=\boldsymbol{\Pi}_b\boldsymbol{y}+(\mathbf{I}-\boldsymbol{\Pi}_b)\boldsymbol{y}=\boldsymbol{y}_1+\boldsymbol{y}_2, \tag{A.41}$$

where $\boldsymbol{y}_1=\boldsymbol{\Pi}_b\boldsymbol{y}$, $\boldsymbol{\Pi}_b=\|\boldsymbol{b}\|_2^{-1}\boldsymbol{b}\boldsymbol{b}^{\mathrm{T}}$ is the projection into the span of $\boldsymbol{b}$, and $\boldsymbol{y}_2=(\mathbf{I}-\boldsymbol{\Pi}_b)\boldsymbol{y}$ is the projection into the orthogonal complement space.

Inserting (A.28)–(A.30), (A.25), and the zero duality gap provides the solution for $\boldsymbol{y}_1$ and the structure of $\boldsymbol{y}_2$, which read as

$$\boldsymbol{y}_1=\frac{\sqrt{\sum_{i=1}^{m}\lambda_i\|\boldsymbol{c}_{i2}\|_2^2}}{\|\boldsymbol{b}\|_2}\boldsymbol{b}, \tag{A.42}$$

$$\boldsymbol{y}_2=\sum_{j=1}^{r}\beta_j\boldsymbol{y}_{j2}=\sum_{j=1}^{r}\boldsymbol{a}_{j1}^{\mathrm{T}}\boldsymbol{y}_2\lambda_j\Big(\sum_{i=1}^{m}\lambda_i\boldsymbol{A}_{i2}\boldsymbol{A}_{i2}^{\mathrm{T}}\Big)^{\dagger}\boldsymbol{a}_{j1}. \tag{A.43}$$

That is, $\boldsymbol{y}_2$ is a linear combination of the vectors $\boldsymbol{y}_{j2}=\lambda_j(\sum_{i=1}^{m}\lambda_i\boldsymbol{A}_{i2}\boldsymbol{A}_{i2}^{\mathrm{T}})^{\dagger}\boldsymbol{a}_{j1}$, with weights β_j, $j=1,\ldots,r$. Finding the weights β_j, $j=1,\ldots,r$ is in turn the solution of the conic inequality conditions from primal feasibility (A.23), that is,

$$-\boldsymbol{a}_{i1}^{\mathrm{T}}\bar{\boldsymbol{Y}}_2\boldsymbol{\beta}\geq\sqrt{\|\boldsymbol{c}_{i2}\|_2^2+\|\boldsymbol{A}_{i2}^{\mathrm{T}}\bar{\boldsymbol{Y}}_2\boldsymbol{\beta}\|_2^2}, \quad i=1,\ldots,r, \tag{A.44}$$

[19] For the complex reconstruction, we used the equivalent complex expressions for real valued representations and matrix vector multiplications in [325, Lemma 1], for example.

where $\bar{Y}_2 = [y_{12}, \ldots, y_{r2}]$, $\beta = [\beta_1, \ldots, \beta_r]^{\mathrm{T}}$, and we inserted the orthogonality of (A.28). Since existence of such a β is ensured, its computation is a feasibility problem over a set of conic convex constraints.

Computing β simplifies for the considered applications, where we can exploit the block structures of $a_{i1}, i = 1, \ldots, r$ and $A_{i2}, i = 1, \ldots, m$, e.g.,

$$a_{i1} = \mathbf{e}_i \otimes \tilde{a}_{i1}, \quad i = 1, \ldots, r, \qquad A_{i2} = \mathbf{I}_m \otimes \tilde{A}_{i2}, \quad i = 1, \ldots, m, \tag{A.45}$$

with auxiliary vectors $\tilde{a}_{i1} \in \mathbb{R}^{n/m}$ and $\tilde{A}_{i2} \in \mathbb{R}^{n/m \times n/m}$ and the canonical unit vector $\mathbf{e}_i \in \mathbb{R}^m$. The structure particularly leads to

$$a_{i1}^{\mathrm{T}} y_{i2} > 0, \quad a_{i1}^{\mathrm{T}} y_{j2} = 0, \quad j \neq i, \quad i, j = 1, \ldots, r \tag{A.46}$$

$$\bar{Y}_2^{\mathrm{T}} A_{i2} A_{i2}^{\mathrm{T}} \bar{Y}_2 = \mathrm{diag}\left(\|A_{i2}^{\mathrm{T}} y_{12}\|_2^2, \ldots, \|A_{i2}^{\mathrm{T}} y_{r2}\|_2^2\right), \quad i = 1, \ldots, r. \tag{A.47}$$

The conic form equality condition in (A.44), therefore, becomes equivalent to

$$(a_{i1}^{\mathrm{T}} y_{i2})^2 \beta_i^2 = \|c_{i2}\|_2^2 + \sum_{j=1}^{r} \|A_{i2}^{\mathrm{T}} y_{j2}\|_2^2 \beta_j^2, \quad -a_{i1}^{\mathrm{T}} y_{i2} \beta_i > 0, \quad i = 1, \ldots, r. \tag{A.48}$$

Hence, the search for β is the solution to a linear equation system[20]

$$\Psi \beta^2 = \zeta, \tag{A.49}$$

where we substituted $\beta^2 = [\beta_1^2, \ldots, \beta_r^2]^{\mathrm{T}}$, $\zeta = [\|c_{12}\|_2^2, \ldots, \|c_{r2}\|_2^2]^{\mathrm{T}}$, and

$$[\Psi]_{i,j} = -\|A_{i2}^{\mathrm{T}} y_{j2}\|_2^2, \quad i \neq j,$$
$$[\Psi]_{i,i} = (a_{i1} y_{i2})^2 - \|A_{i2}^{\mathrm{T}} y_{i2}\|_2^2, \quad i, j = 1, \ldots, r.$$

A.7 Properties of the Dual Uplink MSE Optimizations

Lemma A.5 *The outer maximization of the dual max–min QoS optimization with weighted sum MSE constraints* (4.34) *is a convex optimization problem.*

Proof The constraint $\sum_{\ell=1}^{L} P_\ell \mu_\ell \leq 1$ is affine in μ and the objective $\sum_{j=1}^{M} \lambda_j^\star \hat{\sigma}_j^2$ is linear in the limit point $\lambda^\star$ of the sequence $\{\lambda^{(n)}\}_n$ with update rule (4.56). Hence, the outer maximization of (4.34) is a convex problem if the limit point

[20] This is similar to the available MSE dualities with a single sum power constraint [161].

$\lambda_j^\star = \lim_{n\to\infty} \lambda_j^{(n)}$ is concave in $\boldsymbol{\mu}$. The proof of this statement is by induction. Given a starting point $\boldsymbol{\lambda}'$, $\lambda_j^{(1)} = I_j(\boldsymbol{\lambda}', \boldsymbol{\mu})$ is the concatenation of the affine function

$$\boldsymbol{X}(\boldsymbol{\lambda}', \boldsymbol{\mu}) = \sum_{i=1}^{M} \lambda_i' \hat{\boldsymbol{R}}_i + \sum_{\ell=1}^{L} \mu_\ell \boldsymbol{A}_\ell \tag{A.50}$$

with the concave function[21]

$$f_j(\boldsymbol{X}) = \frac{m_j - \varepsilon_j}{\hat{\boldsymbol{h}}^{\mathrm{H}} \boldsymbol{G}_j \boldsymbol{X}^{\dagger} \boldsymbol{G}_j \hat{\boldsymbol{h}}} \tag{A.51}$$

and, hence, concave increasing in $\boldsymbol{\mu}$. By the same argument, $\lambda_j^{(n+1)} = I_j(\boldsymbol{\lambda}^{(n)}, \boldsymbol{\mu})$ is jointly concave increasing in $\boldsymbol{\lambda}^{(n)}$ and $\boldsymbol{\mu}$. Furthermore, since the concatenation of two concave increasing functions is again concave and increasing, also $\lambda_j^{(n+1)} = I_j(I_j(\boldsymbol{\lambda}^{(n-1)}, \boldsymbol{\mu}), \boldsymbol{\mu})$ is concave in $\boldsymbol{\mu}$ and $\boldsymbol{\lambda}^{(n-1)}$. Applying this, each sequence element is concave in $\boldsymbol{\mu}$, such that also the convergence point $\boldsymbol{\lambda}^\star$ shares this property. □

Lemma A.6 *Let $\boldsymbol{\lambda}^\star$ be a solution to either of the inner minimizations in* (4.91) *and* (4.89) *with given $\boldsymbol{\mu}$. Then,* $\mathrm{WAMSE}_j(\boldsymbol{\lambda}^\star, \boldsymbol{\mu}) = \varepsilon\varepsilon_j$ *for all indices $j \in \mathcal{J}$ within the active set*

$$\mathcal{J} = \{j \in \{1, \ldots, K\} : \varepsilon\varepsilon_j \leq m_j\}.$$

Proof Assume the contrary for the proof, i.e., a solution $\boldsymbol{\lambda}'$ exists with $\sum_{j=1}^{M} \lambda_j' \hat{\sigma}_j^2 = p$ and $\varepsilon_i^{-1}\,\mathrm{WAMSE}_i(\boldsymbol{\lambda}', \boldsymbol{\mu}) < \varepsilon$ for at least on $i \in \mathcal{J}$, where $\varepsilon = \max_j \varepsilon_j^{-1}\,\mathrm{WAMSE}_j(\boldsymbol{\lambda}', \boldsymbol{\mu})$. Constructing $\boldsymbol{\lambda} \leq \boldsymbol{\lambda}'$ as $\lambda_i = \lambda_i' - \epsilon$ with sufficiently small $\epsilon > 0$ and $\lambda_k = \lambda_k'$ for $k \neq i$, $\boldsymbol{\lambda}$ is in the ϵ-neighborhood of $\boldsymbol{\lambda}'$. However, $\mathrm{WAMSE}_j(\boldsymbol{\lambda}, \boldsymbol{\mu}) < \mathrm{WAMSE}_j(\boldsymbol{\lambda}', \boldsymbol{\mu}) = \varepsilon\varepsilon_j$ for $j \in \mathcal{J}$, $j \neq i$ and $\varepsilon\varepsilon_i > \mathrm{WAMSE}_i(\boldsymbol{\lambda}, \boldsymbol{\mu}) > \mathrm{WAMSE}_i(\boldsymbol{\lambda}', \boldsymbol{\mu})$, while the power is reduced to $\sum_{j=1}^{M} \lambda_j \hat{\sigma}_j^2 = p - \hat{\sigma}_j^2 \epsilon < p$ by $\boldsymbol{\lambda}$. This contradicts the initial optimality assumption of the power allocation $\boldsymbol{\lambda}'$. □

Lemma A.7 *The downlink optima $\rho^{(\mathrm{dl})}(p)$ and $p^{(\mathrm{dl})}(\boldsymbol{\varepsilon}(\rho))$ for* (4.109) *and* (4.19), *respectively, are inverse functions if the QoS targets are $\varepsilon_k(\rho)$, $k = 1, \ldots, K$. Then, also the uplink optima $\rho^{(\mathrm{ul})}(p)$ and $p^{(\mathrm{ul})}(\boldsymbol{\varepsilon}(\rho))$ of* (4.110) *and* (4.34), *respectively, are inverse to each other.*

[21] The function $g : \mathcal{H}_+^N \to \mathbb{R}_+$, with $g(\boldsymbol{X}) = (\boldsymbol{a}^{\mathrm{H}} \boldsymbol{X}^{-1} \boldsymbol{a})^{-1}$ is concave if $\frac{\mathrm{d}^2}{\mathrm{d}\alpha^2} g(\boldsymbol{A} + \alpha \boldsymbol{B}) \leq 0$. Writing

$$\frac{\mathrm{d}^2}{\mathrm{d}\alpha^2} g(\boldsymbol{A} + \alpha \boldsymbol{B}) = \frac{2}{(\boldsymbol{a}^{\mathrm{H}} \boldsymbol{X}^{-1} \boldsymbol{a})^3} \left(|(\boldsymbol{a}^{\mathrm{H}} \boldsymbol{X}^{-1/2})(\boldsymbol{X}^{-1/2} \boldsymbol{B} \boldsymbol{X}^{-1} \boldsymbol{a})|^2 - \|\boldsymbol{X}^{-1/2} \boldsymbol{B} \boldsymbol{X}^{-1} \boldsymbol{a}\|_2^2 \|\boldsymbol{X}^{-1/2} \boldsymbol{a}\|_2^2 \right) \Big|_{\boldsymbol{X} = \boldsymbol{A} + \alpha \boldsymbol{B}},$$

the inequality is by the Cauchy–Schwarz inequality. Therefore, also (A.51) is concave in $\boldsymbol{\mu}$.

Proof Since $\mathrm{AMSE}_k(\boldsymbol{t}) = \varepsilon_k(\rho)$ holds for all $k = 1, \ldots, K$ at the optimum of both optimizations and the entries of $\boldsymbol{\varepsilon}(\rho)$ are strict monotonically decreasing in ρ, $p^{(\mathrm{dl})}(\boldsymbol{\varepsilon}(\rho))$ strictly increases with ρ and vice versa $\rho^{(\mathrm{dl})}(p)$ strictly increases with p. This shows the inversion in the downlink. Since $\mathrm{AMSE}_k^{(\mathrm{ul})} = \varepsilon_k(\rho)$, $k = 1, \ldots, K$ also holds at the optimum of the uplink balancing and QoS optimization (cf. Proof of Lemma 4.2), the inversion property is also valid for these optimizations. □

With this lemma, the same argumentation as for the proof of Theorem 4.2 also provides strong duality between (4.109) and (4.110).

A.8 Some Distribution and Quantile Functions

The distribution function (CDF) $F_z : \mathbb{R} \rightarrow [0, 1]$ of a random $z \in \mathbb{R}$ reads as

$$F_z(x) = \Pr(z \leq x) \tag{A.52}$$

and its quantile $q \in \mathbb{R}$ at $\epsilon \in [0, 1]$ is $q = \mathrm{q}_z(\epsilon)$, where $\mathrm{q}_z : [0, 1] \rightarrow \mathbb{R}$ with

$$\mathrm{q}_z(\epsilon) = \inf\left\{x \in \mathbb{R} : F_z(x) \geq \epsilon\right\}. \tag{A.53}$$

If the CDF F_z is continuously increasing in its domain and $\epsilon \in (0, 1)$, the Quantile function (QF) simply becomes the inverse of the CDF, i.e.,

$$\mathrm{q}_z(\epsilon) = F_z^{-1}(\epsilon). \tag{A.54}$$

The CDF of a standard normal distributed random variable $z \sim \mathcal{N}(0, 1)$ shall be denoted by $\Phi : \mathbb{R} \rightarrow [0, 1]$, which is given by [203, Equation 26.2.2]

$$\Phi(x) = (2\pi)^{-1/2} \int_{-\infty}^{x} \mathrm{e}^{-t^2/2}\, \mathrm{d}t. \tag{A.55}$$

The related distribution of a log-normal random $z \in \mathbb{R}_+$, $\ln(z) \sim \mathcal{N}(\mu, \sigma^2)$ is

$$F_z(x) = \Phi\left(\frac{\ln(x) - \mu}{\sigma}\right). \tag{A.56}$$

If in turn $z_{\mathrm{dB}} = 10 \log_{10}(z)$ dB is log-normal distributed, the CDF reads as

$$F_z(x) = \Phi\left(\frac{\ln(10 \log_{10}(x)) - \mu}{\sigma}\right). \tag{A.57}$$

Since these CDFs are continuous, the corresponding quantile functions are completely defined by $\Phi^{-1} : (0, 1) \rightarrow \mathbb{R}$. The ϵ-quantile for the latter example is

$$F_z^{-1}(\epsilon) = 10^{10^{-1} F_{z_{\text{dB}}}^{-1}(\epsilon)}, \tag{A.58}$$

$$F_{z_{\text{dB}}}^{-1}(\epsilon) = \exp\left(\mu + \sigma \Phi^{-1}(\epsilon)\right). \tag{A.59}$$

The CDF of a (central) chi-square random variable $z \sim \mathcal{X}_k^2$, i.e., $z = \sum_{i=1}^k w_i^2 \in \mathbb{R}_+$ with i.i.d. normal distributed $w_i \sim \mathcal{N}(0, 1)$ and degree k, is denoted by $F_{\chi^2}(\cdot; k) : \mathbb{R}_+ \to [0, 1]$ and reads as (cf. [203, Equation 26.4.1])

$$F_{\chi^2}(x; k) = \frac{1}{\Gamma(k/2)} \int_0^{x/2} t^{k/2-1} \mathrm{e}^{-t} \, \mathrm{d}t. \tag{A.60}$$

For $k = 2$, i.e., $z \sim \mathcal{X}_2^2$ this distribution function simplifies to

$$F_{\chi^2}(x; 2) = 1 - \mathrm{e}^{-x/2}. \tag{A.61}$$

If z is exponentially distributed, i.e., $2z \sim \mathcal{X}_2^2$ for $z = |y|^2$ and $y \sim \mathcal{N}_{\mathbb{C}}(0, 1)$, then

$$F_z(x) = F_{\chi^2}(2x; 2) = 1 - \mathrm{e}^{-x}. \tag{A.62}$$

The corresponding ϵ-quantile is $F_z^{-1}(\epsilon) = -\ln(1 - \epsilon)$.

The series expansion for the CDF $F_{\chi^2(\lambda)}(\cdot; k) : \mathbb{R}_+ \to [0, 1]$ of a non-central chi-square-distributed random variable $z \sim \mathcal{X}_k^2(\lambda)$, i.e., $z = \sum_{i=1}^k (v_i - w_i)^2$ with $v_i \in \mathbb{R}_+$, non-centrality parameter $\lambda = \sum_{i=1}^k v_i^2$, and degree k, reads as [203, Equation 26.4.25]

$$F_{\chi^2(\lambda)}(x; k) = \mathrm{e}^{-(\lambda/2)} \sum_{j=0}^{\infty} \frac{(\lambda/2)^j}{j!} F_{\chi^2}(x; k + 2j). \tag{A.63}$$

This expression leads to an alternative series expansion for (5.28), which reads as

$$F_{\zeta_k^{-1}}(x) = \mathrm{e}^{-\sigma_{\xi_k}^{-2}} \sum_{j=0}^{\infty} \frac{\sigma_{\xi_k}^{-2j}}{j!} F_{\chi^2}(2\sigma_{\xi_k}^{-2} x; 2 + 2j). \tag{A.64}$$

The integral representation (5.28) in turn follows with an alternative representation of the non-central chi-square-distribution $F_{\chi^2(\lambda)}(x; k)$ from Nuttall [278]:

$$F_{\chi^2(\lambda)}(x; k) = 1 - Q_{k/2}(\sqrt{\lambda}, \sqrt{x}), \tag{A.65}$$

where $Q_M(a, b)$ denotes the Marcum-Q function

$$Q_M(a, b) = \frac{1}{a^{M-1}} \int_b^{\infty} s^M \mathrm{e}^{-\frac{x^2+a^2}{2}} I_{M-1}(as) \, \mathrm{d}s \tag{A.66}$$

and $I_N(x)$ is the modified Bessel function of the first kind, i.e., [203, Equation 9.6.10]

$$I_N(x) = \left(\frac{x}{2}\right)^N \sum_{n=0}^{\infty} \frac{(x/2)^{2n}}{n!\Gamma(N+n+1)}. \tag{A.67}$$

Inserting the latter two expressions with $M = 1$, $N = 0$, and $k = 2$ into (A.65), the function $F_{\chi^2(\lambda)}(x; 2)$ becomes

$$F_{\chi^2(\lambda)}(x; 2) = 1 - \int_{\sqrt{x}}^{\infty} s\, \mathrm{e}^{-\frac{s^2+\lambda}{2}} \sum_{n=0}^{\infty} \left(\frac{(\sqrt{\lambda} s/2)^n}{n!}\right)^2 \mathrm{d}s. \tag{A.68}$$

The CDF expression (5.28) in Sect. 5.2 is then obtained with $\lambda = 2\sigma_{\xi_k}^{-2}$. As for the computation of explicit values for the CDF, computing values for the quantile function of non-central chi-squared distributed random variables requires numerical evaluations, e.g., those from [203, Section 26.4] and [279, Section 4].

Moreover, the following lemma is useful to evaluate the quantile function for z^{-1}.

Lemma A.8 *Let the CDFs* $F_z, F_{z^{-1}} : \mathbb{R}_+ \to [0, 1]$ *of the random variables* $z, z^{-1} \in \mathbb{R}_{++}$ *be continuously increasing in their domain and* $\epsilon \in (0, 1)$. *Then, the relation* $F_z(x) = 1 - F_{z^{-1}}(x^{-1})$ *holds and* $q = F_z^{-1}(1 - \epsilon) = 1/F_{z^{-1}}^{-1}(\epsilon)$ *defines the* $1 - \epsilon$*-quantile of* F_z.

Proof Let the random variable $z \in \mathbb{R}_{++}$ induce a strict monotonically increasing CDF $F_z(x)$ and $\Pr(z > x) > 0$ for $z > 0$. Then, the CDF induced by z reads as

$$F_z(x) = \Pr(x^{-1} \leq z^{-1}) = 1 - \Pr(x^{-1} > z^{-1}) = 1 - F_{z^{-1}}(x^{-1}),$$

where the last equality is with continuity of the CDF $F_{z^{-1}} : \mathbb{R}_+ \to [0, 1]$.[22] Therewith, $F_z(x) = 1 - \epsilon$ is equivalent to $F_{z^{-1}}(x^{-1}) = \epsilon$ for $x > 0$ and the inverse CDFs satisfy

$$x = F_z^{-1}(1 - \epsilon) = (F_{z^{-1}}^{-1}(\epsilon))^{-1},$$

which proves the statement. □

Now, let $z = g(x, y)$ be a monotonic function $g : \mathbb{R}_+^2 \to \mathbb{R}$ of two independent continuous random variables $x, y \in \mathbb{R}_+$ with PDFs $f_x(x)$ and $f_y(y)$. The function g shall be monotonically increasing and invertible in y, that is, there is a continuous function $h : \mathbb{R}^2 \to \mathbb{R}_+$ such that $y \leq h(x, t)$ and $g(x, y) \leq t$ are equivalent. Examples are a quotient $g(x, y) = x^{-1}y$ and a linear function $g(x, y) = y - ax$,

[22] The last equality has to be replaced by a smaller or equal sign if the CDF is non-continuous.

with $h(x,t) = xt$ and $h(x,t) = t + ax$, respectively. The CDF of such a random variable z is generally unknown, but can be expressed via the integral representation

$$F_z(t) = \int_{D(X)} f_x(s) \Pr(y \le h(x,t) | x = s) \, \mathrm{d}s, \tag{A.69}$$

where $D(X)$ denotes the domain of f_x. Example $h(x,z) = xz$ is for a log-dB-normal random $y \in [1, \infty)$, where $\ln(y_{\mathrm{dB}}) \sim \mathcal{N}(\mu_{y_{\mathrm{dB}}}, \sigma^2_{y_{\mathrm{dB}}})$ and the PDF is

$$f_y(y) = \left(y \ln(y) \sqrt{(2\pi)} \sigma_{y_{\mathrm{dB}}}\right)^{-1} \exp\left(- \frac{(\ln(10 \log_{10}(y)) - \mu_{y_{\mathrm{dB}}})^2}{2\sigma^2_{y_{\mathrm{dB}}}} \right), \tag{A.70}$$

and a non-central chi-square random variable $x \sim \mathcal{X}_2^2(\lambda)$, which PDF reads as

$$f_x(x) = \frac{1}{2} \mathrm{e}^{-\frac{x+\lambda}{2}} \sum_{j=0}^{\infty} \frac{(\lambda x/4)^j}{(j!)^2}. \tag{A.71}$$

The CDF of the random variable $z = x^{-1} y \in \mathbb{R}_+$ is thus

$$F_z(t) = \int_0^{\infty} \frac{1}{2} \mathrm{e}^{-\frac{s+\lambda}{2}} \left(\sum_{j=0}^{\infty} \frac{(\lambda s/4)^j}{(j!)^2} \right) \Phi\left(\frac{\ln(10 \log_{10}(st)) - \mu_{y_{\mathrm{dB}}}}{\sigma_{y_{\mathrm{dB}}}} \right) \mathrm{d}s. \tag{A.72}$$

For $z = y - ax \in \mathbb{R}$, x is non-central chi-square distributed of degree 2, i.e., $x \sim \mathcal{X}_2^2(\lambda)$, while y is central chi-square distributed $y \sim \mathcal{X}^2_{2r}$. In this case, the CDF of z reformulates to

$$F_z(t) = \int_0^{\infty} \frac{1}{2} \mathrm{e}^{-\frac{s+\lambda}{2}} \left(\sum_{j=0}^{\infty} \frac{(\lambda s/4)^j}{(j!)^2} \right) F_{\chi^2}(t + as; 2r) \, \mathrm{d}s. \tag{A.73}$$

Remark A.1 The PDF of a non-central chi-squared distributed variable $x \sim \mathcal{X}_k^2(\lambda)$, with non-centrality parameter λ, reads as

$$f_x(x) = \frac{1}{2} \mathrm{e}^{-\frac{x+\lambda}{2}} \left(\frac{x}{\lambda} \right)^{\frac{k}{4} - \frac{1}{2}} I_{k/2-1}\left(\sqrt{\lambda x}\right) \tag{A.74}$$

with the modified Bessel function I_N from (A.67). For $k = 2$, this PDF simplifies to (A.71). Furthermore, if $\lambda = 0$ and thus $x \sim \mathcal{X}_k^2$, its PDF is

$$f_x(x) = \frac{1}{2^{k/2} \Gamma(k/2)} x^{k/2-1} \mathrm{e}^{-x/2}, \tag{A.75}$$

which becomes $f_x(x) = \frac{1}{2} \mathrm{e}^{-x/2}$ for $k = 2$.

References

1. D. Tse, P. Viswanath, *Fundamentals of Wireless Communications*, 4th edn. (Cambridge University Press, New York, NY, 2008)
2. P.D. Arapoglou, K. Liolis, M. Bertinelli, A. Panagopoulos, P. Cottis, R. De Gaudenzi, MIMO over satellite: a review. IEEE Commun. Surv. Tutorials **13**(1), 27 (2011)
3. T. Cover, J. Thomas, *Elements of Information Theory* (Wiley Interscience, Hoboken, NJ, 2006)
4. M.H.M. Costa, Writing on dirty paper (Corresp). IEEE Trans. Inf. Theory **29**(3), 439 (1983)
5. P. Viswanath, D.N.C. Tse, Sum capacity of the vector Gaussian broadcast channel and uplink-downlink duality. IEEE Trans. Inf. Theory **49**(8), 1912 (2003)
6. H. Weingarten, Y. Steinberg, S. Shamai, The capacity region of the Gaussian multiple-input multiple-output broadcast channel. IEEE Trans. Inf. Theory **52**(9), 3936 (2006)
7. A. Goldsmith, S. Jafar, N. Jindal, S. Vishwanath, Capacity limits of MIMO channels. IEEE J. Sel. Areas Commun. **21**(5), 684 (2003)
8. W. Zhang, S. Kotagiri, J. Laneman, On downlink transmission without transmit channel state information and with outage constraints. IEEE Trans. Inf. Theory **55**(9), 4240 (2009)
9. F.A. Dietrich, P. Breun, W. Utschick, Robust Tomlinson Harashima precoding for the wireless broadcast channel. IEEE Trans. Signal Process. **55**(2), 631 (2007)
10. F. Rashid-Farrokhi, K. Liu, L. Tassiulas, Transmit beamforming and power control for cellular wireless systems. IEEE J. Sel. Areas Commun. **16**(8), 1437 (1998)
11. M. Schubert, H. Boche, QoS-based resource allocation and transceiver optimization. Found. Trends Commun. Inf. Theory **2**(6), 383 (2005)
12. E. Karipidis, N. Sidiropoulos, Z.Q. Luo, Quality of service and max-min fair transmit beamforming to multiple cochannel multicast groups. IEEE Trans. Signal Process. **56**(3), 1268 (2008)
13. D. Christopoulos, S. Chatzinotas, B. Ottersten, Weighted fair multicast multigroup beamforming under per-antenna power constraints. IEEE Trans. Signal Process. **62**(19), 5132 (2014)
14. M. Hanif, L.N. Tran, A. Tölli, M. Juntti, S. Glisic, Efficient solutions for weighted sum rate maximization in multicellular networks with channel uncertainties. IEEE Trans. Signal Process. **61**(22), 5659 (2013)
15. W. Utschick, J. Brehmer, Monotonic optimization framework for coordinated beamforming in multi-cell networks. IEEE Trans. Signal Process. **59**(99) (2011)

A. Gründinger, *Statistical Robust Beamforming for Broadcast Channels and Applications in Satellite Communication*, Foundations in Signal Processing, Communications and Networking 22, https://doi.org/10.1007/978-3-030-29578-3

16. M. Rossi, A. Tulino, O. Simeone, A. Haimovich, Non-convex utility maximization in Gaussian MISO broadcast and interference channels, in *Proceedings of the 36th International Conference on Acoustics, Speech, and Signal Processing ICASSP* (2011), pp. 2960–2963
17. E. Björnson, G. Zheng, M. Bengtsson, B. Ottersten, Robust monotonic optimization framework for multicell MISO systems. IEEE Trans. Signal Process. **60**(5), 2508 (2012)
18. S. Srinivasa, S. Jafar, The optimality of transmit beamforming: a unified view. IEEE Trans. Inf. Theory **53**(4), 1558 (2007)
19. E. Larsson, E. Jorswieck, Competition versus cooperation on the MISO interference channel. IEEE J. Sel. Areas Commun. **26**(7), 1059 (2008)
20. E. Larsson, F. Tufvesson, O. Edfors, T. Marzetta, Massive MIMO for next generation wireless systems. IEEE Commun. Mag. **52**(2), 186 (2014)
21. F. Rusek, D. Persson, B.K. Lau, E. Larsson, T. Marzetta, O. Edfors, F. Tufvesson, Scaling up MIMO: opportunities and challenges with very large arrays. IEEE Signal Process. Mag. **30**(1), 40 (2013)
22. D. Neumann, A. Gründinger, M. Joham, W. Utschick, Rate-balancing in massive MIMO using statistical precoding, in *Proceedings of the 14th IEEE International Workshop on Signal Processing Advances in Wireless Communications SPAWC* (2015), pp. 226–230
23. E. Chiu, V. Lau, S. Zhang, B. Mok, Precoder design for multi-antenna partial decode-and-forward (PDF) cooperative systems with statistical CSIT and MMSE-SIC receivers. IEEE Trans. Wirel. Commun. **11**(4), 1343 (2012)
24. P. Wu, R. Schober, Cooperative beamforming for single-carrier frequency-domain equalization systems with multiple relays. IEEE Trans. Wirel. Commun. **11**(6), 2276 (2012)
25. M. Kang, M.S. Alouini, Capacity of correlated MIMO Rayleigh channels. IEEE Trans. Wirel. Commun. **5**(1), 143 (2006)
26. M. Kang, M.S. Alouini, Capacity of MIMO Rician channels. IEEE Trans. Wirel. Commun. **5**(1), 112 (2006)
27. M. Kießling, Unifying analysis of ergodic MIMO capacity in correlated Rayleigh fading environments. Eur. Trans. Telecommun. **16**(1), 17 (2005)
28. M. Matthaiou, G. Alexandropoulos, H.Q. Ngo, E. Larsson, Analytic framework for the effective rate of MISO fading channels. IEEE Trans. Commun. **60**(6), 1741 (2012)
29. N. Beaulieu, J. Hu, A closed-form expression for the outage probability of decode-and-forward Relaying in dissimilar Rayleigh fading channels. IEEE Commun. Lett. **10**(12), 813 (2006)
30. A. Avestimehr, D. Tse, Outage capacity of the fading relay channel in the low-SNR regime. IEEE Trans. Inf. Theory **53**(4), 1401 (2007)
31. A. Housfater, T.J. Lim, Outage probability of MISO broadcast systems with noisy channel side information, in *Proceedings of the 44th Asilomar Conference on Signals, Systems, and Computers* (2010), pp. 1232–1236
32. A. Moustakas, S. Simon, On the outage capacity of correlated multiple-path MIMO channels. IEEE Trans. Inf. Theory **53**(11), 3887 (2007)
33. D. Kontaxis, G. Tsoulos, S. Karaboyas, Ergodic capacity optimization for single-stream beamforming transmission in MISO Rician fading channels. IEEE Trans. Veh. Technol. **62**(2), 628 (2013)
34. M. Razaviyayn, M.S. Boroujeni, Z.Q. Luo, A stochastic weighted MMSE approach to sum rate maximization for a MIMO interference channel, in *Proceedings of the 12th IEEE International Workshop on Signal Processing Advances in Wireless Communications SPAWC* (2013), pp. 325–329
35. S. Vorobyov, H. Chen, A. Gershman, On the relationship between robust minimum variance beamformers with probabilistic and worst-case distortionless response constraints. IEEE Trans. Signal Process. **56**(11), 5719 (2008)
36. K.Y. Wang, A.C. So, T.H. Chang, W.K. Ma, C.Y. Chi, Outage constrained robust transmit optimization for multiuser MISO downlinks: tractable approximations by conic optimization. IEEE Trans. Signal Process. **62**(21), 5690 (2014)

37. S.A. Vorobyov, A.B. Gershman, Z.Q. Luo, Robust adaptive beamforming using worst-case performance optimization: a solution to the signal mismatch problem. IEEE Trans. Signal Process. **51**(2), 313 (2003)
38. S. Shahbazpanahi, A.B. Gershman, Z.Q. Luo, K.M. Wong, Robust adaptive beamforming for general-rank signal models. IEEE Trans. Signal Process. **51**(9), 2257 (2003)
39. J. Li, P. Stoica, Z. Wang, On robust capon beamforming and diagonal loading. IEEE Trans. Signal Process. **51**(7), 1702 (2003)
40. J. Li, P. Stoica (eds.), *Robust Adaptive Beamforming* (Wiley, New York, 2005)
41. F. Dietrich, *Robust Signal Processing for Wireless Communications* (Springer, New York, 2007)
42. H. Du, P.J. Chung, A probabilistic approach for robust leakage-based MU-MIMO downlink beamforming with imperfect channel state information. IEEE Trans. Wirel. Commun. **11**(3), 1239 (2012)
43. M. Joham, P. Castro, W. Utschick, L. Castedo, Robust precoding with limited feedback design based on precoding MSE for MU-MISO systems. IEEE Trans. Signal Process. **60**(6), 3101 (2012)
44. N. Vučić, H. Boche, A tractable method for chance-constrained power control in downlink multiuser MISO systems with channel uncertainty. IEEE Signal Process. Lett. **16**(5), 346 (2009)
45. M. Medra, Y. Huang, W.K. Ma, T.N. Davidson, Low-complexity robust MISO downlink precoder design under imperfect CSI. IEEE Trans. Signal Process. **64**(12), 3237 (2016)
46. N. Zorba, A. Pérez-Neira, Robust multibeam opportunistic schemes under quality of service constraints, in *Proceedings of the IEEE International Conference on Communications ICC* (2007), pp. 5371–5376
47. M. Botros Shenouda, L. Lampe, Quasi-convex designs of max-min linear BC precoding with outage QoS constraints, in *Proceedings of the 6th International Symposium on Wireless Communication Systems ISWCS* (2009), pp. 156–160
48. E. Björnson, R. Zakhour, D. Gesbert, B. Ottersten, Cooperative multicell precoding: rate region characterization and distributed strategies with instantaneous and statistical CSI. IEEE Trans. Signal Process. **58**(8), 4298 (2010)
49. C. Shen, T.H. Chang, K.Y. Wang, Z. Qiu, C.Y. Chi, Distributed robust multicell coordinated beamforming with imperfect CSI: an ADMM approach. IEEE Trans. Signal Process. **60**(6), 2988 (2012)
50. J. Wang, M. Bengtsson, B. Ottersten, D. Palomar, Robust MIMO precoding for several classes of channel uncertainty. IEEE Trans. Signal Process. **61**(12), 3056 (2013)
51. X. He, Y.C. Wu, Probabilistic QoS Constrained robust downlink multiuser MIMO transceiver design with arbitrarily distributed channel uncertainty. IEEE Trans. Wirel. Commun. **12**(12), 6292 (2013)
52. D. Christopoulos, S. Chatzinotas, G. Zheng, J. Grotz, B. Ottersten, Linear and nonlinear techniques for multibeam joint processing in satellite communications. EURASIP J. Wirel. Commun. Netw. **1**(162), 1 (2012)
53. L. Cottatellucci, M. Debbah, E. Casini, R. Rinaldo, R. Mueller, M. Neri, G. Gallinaro, Interference mitigation techniques for broadband satellite system, in *Proceedings of the 24th AIAA International Communications Satellite Systems Conference ICSSC 2006* (2006)
54. N. Zorba, A. Pérez-Neira, Robust power allocation schemes for multibeam opportunistic transmission strategies under quality of service constraints. IEEE J. Sel. Areas Commun. **26**(6), 1025 (2008)
55. J. Montesinos, O. Besson, C. Laure de Tournemine, Adaptive beamforming for large arrays in satellite communications systems with dispersed coverage. IET Commun. **5**(3), 350 (2011)
56. J. Arnau, B. Devillers, C. Mosquera, A. Pérez-Neira, Performance study of multiuser interference mitigation schemes for hybrid broadband multibeam satellite architectures. EURASIP J. Wirel. Commun. Netw. **1**(132), 1 (2012)
57. G. Zheng, S. Chatzinotas, B. Ottersten, Generic optimization of linear precoding in multibeam satellite systems. IEEE Trans. Wirel. Commun. **11**(6), 2308 (2012)

58. X. Lei, L. Cottatellucci, S. Ghanem, Adaptive beamforming in mobile, massively multiuser satellite communications: a system perspective, in *Proceedings of the 13th International Workshop on Signal Processing for Space Communications SPSC* (2014), pp. 51–58
59. P.D. Arapoglou, E. Michailidis, A. Panagopoulos, A. Kanatas, R. Prieto-Cerdeira, The land mobile earth-space channel. IEEE Veh. Technol. Mag. **6**(2), 44 (2011)
60. B. Devillers, A. Pérez-Neira, C. Mosquera, Joint linear precoding and beamforming for the forward link of multi-beam broadband satellite systems, in *Proceedings of the IEEE Global Communications Conference GLOBECOM* (2011)
61. S. Chatzinotas, G. Zheng, B. Ottersten, Energy-efficient MMSE beamforming and power allocation in multibeam satellite systems, in *Proceedings of the 45th Asilomar Conference on Signals, Systems, and Computers* (2011), pp. 1081–1085
62. D. Christopoulos, J. Arnau, S. Chatzinotas, C. Mosquera, B. Ottersten, MMSE performance analysis of generalized multibeam satellite channels. IEEE Commun. Lett. **17**(7), 1332 (2013)
63. A. Gründinger, A. Barthelme, M. Joham, W. Utschick, Mean square error beamforming in SatCom: uplink-downlink duality with per-feed constraints, in *Proceedings of the 11th International Symposium on Wireless Communication Systems ISWCS* (2014), pp. 600–605
64. A. Gharanjik, B. Shankar, P.D. Arapoglou, M. Bengtsson, B. Ottersten, Robust precoding with phase uncertainty, in *Proceedings of the 39th International Conference on Acoustics, Speech, and Signal Processing ICASSP* (2015), pp. 3083–3087
65. R. Hunger, M. Joham, A complete description of the QoS feasibility region in the vector broadcast channel. IEEE Trans. Signal Process. **57**(7), 3870 (2010)
66. R. Hunger, *Analysis and Transceiver Design for the MIMO Broadcast Channel*. Foundations in Signal Processing, Communications and Networking, vol. 8 (Springer, Berlin/Heidelberg, 2013)
67. M. Schubert, H. Boche, Solution of the multiuser downlink beamforming problem with individual SINR constraints. IEEE Trans. Veh. Technol. **53**(1), 18 (2004)
68. S. Shi, M. Schubert, H. Boche, Downlink MMSE transceiver optimization for multiuser MIMO systems: duality and sum-MSE minimization. IEEE Trans. Signal Process. **55**(11), 5436 (2007)
69. F. Rashid-Farrokhi, L. Tassiulas, K. Liu, Joint optimal power control and beamforming in wireless networks using antenna arrays. IEEE Trans. Commun. **46**(10), 1313 (1998)
70. E. Visotsky, U. Madhow, Optimum beamforming using transmit antenna arrays, in *Proceedings of the 49th IEEE Vehicular Technology Conference*, vol. 1 (1999), pp. 851–856
71. M. Bengtsson, B. Ottersten, Optimal downlink beamforming using semidefinite optimization, in *Proceedings of the 37th Annual Allerton Conference on Communication, Control, and Computing* (1999), pp. 987–996
72. M. Bengtsson, B. Ottersten, Optimal and suboptimal transmit beamorming, in *Handbook of Antennas in Wireless Communications*, Chap. 13, ed. by L. Godara (CRC Press, Boca Raton, FL, 2001)
73. A. Wiesel, Y. Eldar, S. Shamai, Linear precoding via conic optimization for fixed MIMO receivers. IEEE Trans. Signal Process. **54**(1), 161 (2006)
74. W. Yu, T. Lan, Transmitter optimization for the multi-antenna downlink with per-antenna power constraints. IEEE Trans. Signal Process. **55**(6), 2646 (2007)
75. R.D. Yates, A framework for uplink power control in cellular radio systems. IEEE J. Sel. Areas Commun. **13**(7), 1341 (1995)
76. E. Jorswieck, H. Boche, Rate balancing for the multi-antenna Gaussian broadcast channel, in *IEEE Sixth International Symposium on Spread Spectrum Techniques and Applications*, vol. 2 (2002), pp. 545–549
77. S. Shi, M. Schubert, H. Boche, Capacity balancing for multiuser MIMO systems, in *Proceedings of the 32th International Conference on Acoustics, Speech, and Signal Processing ICASSP*, vol. 3 (2007), pp. 397–400
78. K. Cumanan, J. Tang, S. Lambotharan, Rate balancing based linear transceiver design for multiuser MIMO system with multiple linear transmit covariance constraints, in *Proceedings of the IEEE International Conference on Communications ICC* (2011)

79. A. Tölli, M. Codreanu, M. Juntti, Linear multiuser MIMO transceiver design with quality of service and per-antenna power constraints. IEEE Trans. Signal Process. **56**(7), 3049 (2008)
80. W. Yang, G. Xu, Optimal downlink power assignment for smart antenna systems, in *Proceedings of the 23th International Conference on Acoustics, Speech, and Signal Processing ICASSP* (1998), pp. 3337–3340
81. M. Schubert, H. Boche, A unifying theory for uplink and downlink multiuser beamforming in *International Zurich Seminar on Broadband Communications, Access, Transmission, and Networking* (2002), pp. 27-1–27-6
82. D. Tse, P. Viswanath, Downlink-uplink duality and effective bandwidths, in *Proceedings of the IEEE International Symposium on Information Theory* (2002), p. 52
83. M. Schubert, H. Boche, Iterative multiuser uplink and downlink beamforming under SINR constraints. IEEE Trans. Veh. Technol. **53**(7), 2324 (2005)
84. J. Zander, Performance of optimum transmitter power control in cellular radio systems. IEEE Trans. Veh. Technol. **41**(1), 57 (1992)
85. M. Hirsch, H. Smith, Chapter 4: monotone dynamical systems, in *Handbook of Differential Equations: Ordinary Differential Equations*, vol. 2, ed. by P.D.A. Cañada, A. Fonda (North-Holland, Amsterdam, 2006), pp. 239–357
86. V. Berinde, *Iterative Approximation of Fixed Points* (Springer, New York, 2007)
87. U. Krause, P. Ranft, A limit set trichotomy for monotone nonlinear dynamical systems. Nonlinear Anal. Theory Methods Appl. **19**(4), 375 (1992)
88. J. Zander, Distributed cochannel interference control in cellular radio systems. IEEE Trans. Veh. Technol. **41**(3), 305 (1992)
89. M. Schubert, H. Boche, A generic approach to QoS-based transceiver optimization. IEEE Trans. Commun. **55**(8), 1557 (2007)
90. H. Boche, M. Schubert, A superlinearly and globally convergent algorithm for power control and resource allocation with general interference functions. IEEE/ACM Trans. Netw. **16**(2), 383 (2008)
91. M. Schubert, H. Boche, *Interference Calculus*. Foundations in Signal Processing, Communications and Networking, vol. 7, 1st edn. (Springer, Berlin/Heidelberg, 2012)
92. N. Vučić, M. Schubert, Fixed point iteration for max-min SIR balancing with general interference functions, in *Proceedings of the 36th International Conference on Acoustics, Speech, and Signal Processing ICASSP* (2011), pp. 3456–3459
93. S. Kandukuri, S. Boyd, Optimal power control in interference-limited fading wireless channels with outage-probability specifications. IEEE Trans. Wirel. Commun. **1**(1), 46 (2002)
94. S. Boyd, S.J. Kim, L. Vandenberghe, A. Hassibi, A tutorial on geometric programming. Optim. Eng. **8**(1), 67 (2007)
95. M. Chiang, C. wei Tan, D. Palomar, D. O'Neill, D. Julian, Power control by geometric programming. IEEE Trans. Wirel. Commun. **6**(7), 2640 (2007)
96. D. Cai, T. Quek, C.W. Tan, A unified analysis of max-min weighted SINR for MIMO downlink system. IEEE Trans. Signal Process. **59**(8), 3850 (2011)
97. J. Sturm, Using SeDuMi 1.02, a MATLAB toolbox for optimization over symmetric cones. Optim. Methods Softw. **11**(1), 625 (1999)
98. K. Toh, R. Tütüncü, M. Todd, SDPT3—a matlab software package for semidefinite programming. Optim. Methods Softw. **11**(1), 545 (1999)
99. M. Grant, S. Boyd, CVX: Matlab software for disciplined convex programming (2009), http://stanford.edu/~boyd/cvx
100. Y. Nesterov, A. Nemirovski, *Interior Point Polynomial Algorithms in Convex Programming* (Studies in Applied Mathematics (SIAM), Philadelphia, PA, 1994)
101. Z.Q. Luo, W. Yu, Z.Q. Luo, W. Yu, An introduction to convex optimization for communications and signal processing. IEEE J. Sel. Areas Commun. **24**(8), 1426 (2006)
102. S. Boyd, L. Vandenberghe, *Convex Optimization*, 1st edn. (Cambridge University Press, New York, NY, 2004)
103. A. Ben-Tal, A. Nemirovski, *Lectures on Modern Convex Optimization. Analysis, Algorithms, and Engineering Applications* (Society for Industrial and Applied Mathematics, Philadelphia,

PA, 2001)
104. A. Gershman, N. Sidiropoulos, S. Shahbazpanahi, M. Bengtsson, B. Ottersten, Convex optimization-based beamforming. IEEE Signal Process. Mag. **27**(3), 62 (2010)
105. Y. Huang, D. Palomar, S. Zhang, Lorentz-positive maps and quadratic matrix inequalities with applications to robust MISO transmit beamforming. IEEE Trans. Signal Process. **61**(5), 1121 (2013)
106. L. Vandenberghe, S. Boyd, Semidefinite programming. SIAM Rev. **3**(1), 49 (1996)
107. H. Wolkowicz, R. Saigal, L. Vandenberghe, *Handbook of Semidefinite Programming: Theory, Algorithms, and Applications*, 1st edn. (Kluwer, Norwell, MS, 2000)
108. Y. Huang, D. Palomar, Rank-constrained separable semidefinite programming with applications to optimal beamforming. IEEE Trans. Signal Process. **58**(2), 664 (2010)
109. Z.Q. Luo, W.K. Ma, A.C. So, Y. Ye, S. Zhang, Semidefinite relaxation of quadratic optimization problems. IEEE Signal Process. Mag. **27**(3), 20 (2010)
110. A. Wiesel, Y. Eldar, S. Shamai, Zero-forcing precoding and generalized inverses. IEEE Trans. Signal Process. **56**(9), 4409 (2008)
111. M. Joham, C. Hellings, R. Hunger, QoS feasibility for the MIMO broadcast channel: robust formulation and multi-carrier systems, in *Proceedings of the 8th International Symposium on Modeling and Optimization in Mobile, Ad Hoc, and Wireless Networks* (2010), pp. 610–614
112. H. Feyzmahdavian, M. Johansson, T. Charalambous, Contractive interference functions and rates of convergence of distributed power control laws. IEEE Trans. Wirel. Commun. **11**(12), 4494 (2012)
113. H. Boche, M. Schubert, The structure of general interference functions and applications. IEEE Trans. Inf. Theory **54**(11), 4980 (2008)
114. E. Jorswieck, B. Ottersten, A. Sezgin, A. Paulraj, Guaranteed performance region in fading orthogonal space-time coded broadcast channels. EURASIP J. Wirel. Commun. Netw. **2008**(39), 1 (2008)
115. M. Joham, R. Hunger, Feasible rate region of the MIMO broadcast channel with linear transceivers, in *Proceedings of the 14th International ITG Workshop on Smart Antennas WSA* (2010), pp. 342–349
116. T. Bogale, L. Vandendorpe, Robust sum MSE optimization for downlink multiuser MIMO systems with arbitrary power constraint: generalized duality approach. IEEE Trans. Signal Process. **60**(4), 1862 (2012)
117. L. Zhang, R. Zhang, Y.C. Liang, Y. Xin, H.V. Poor, On Gaussian MIMO BC-MAC duality with multiple transmit covariance constraints. IEEE Trans. Inf. Theory **58**(4), 2064 (2012)
118. G. Dartmann, X. Gong, W. Afzal, G. Ascheid, On the duality of the max-min beamforming problem with per-antenna and per-antenna-array power constraints. IEEE Trans. Veh. Technol. **62**(2), 606 (2013)
119. G.H. Golub, C.F. van Loan, *Matrix Computations*, 3rd edn. (The Johns Hopkins University Press, Baltimore, NY, 1996)
120. D. Cai, T. Quek, C.W. Tan, S. Low, Max-min weighted SINR in coordinated multicell MIMO downlink, in *Proceedings of the 9th International Symposium on Modeling and Optimization in Mobile, Ad Hoc, and Wireless Networks* (2011), pp. 286–293
121. S. He, Y. Huang, L. Yang, A. Nallanathan, P. Liu, A multi-cell beamforming design by uplink-downlink max-min SINR duality. IEEE Trans. Wirel. Commun. **11**(8), 2858 (2012)
122. D. Luenberger, Quasi-convex programming. SIAM J. Appl. Math. **16**(5), 1090 (1968)
123. J. Penot, M. Volle, Surrogate programming and multipliers in quasi-convex programming. SIAM J. Control Optim. **42**(6), 1994 (2004)
124. F. Glover, Surrogate constraint duality in mathematical programming. Oper. Res. **23**(3), 434 (1975)
125. A.B. Gershman, Z.Q. Luo, S. Shahbazpanahi, *Robust Adaptive Beamforming Based on Worst-Case Performance Optimization*, Chap. 2 (Wiley, New York, 2005), pp. 49–89
126. J. Lindblom, E. Larsson, E. Jorswieck, Parameterization of the MISO IFC rate region: the case of partial channel state information. IEEE Trans. Wirel. Commun. **9**(2), 500 (2010)

127. E. Karipidis, A. Gründinger, J. Lindblom, E. Larsson, Pareto-optimal beamforming for the MISO interference channel with partial CSI, in *Proceedings of the 3rd IEEE International Workshop on Computational Advances in Multi-Sensor Adaptive Processing* (2009), pp. 5–8
128. B. Hassibi, B.M. Hochwald, How much training is needed in multiple-antenna wireless links?. IEEE Trans. Inf. Theory **49**(4), 951 (2003)
129. N. Jindal, MIMO broadcast channels with finite-rate feedback. IEEE Trans. Inf. Theory **52**(11), 5045 (2006)
130. S. Ji, S. Wu, A.C. So, W.K. Ma, Multi-group multicast beamforming in cognitive radio networks via rank-two transmit beamformed alamouti space-time coding, in *Proceedings of the 38th International Conference on Acoustics, Speech, and Signal Processing ICASSP* (2013), pp. 4409–4413
131. L. Li, A. Goldsmith, Capacity and optimal resource allocation for fading broadcast channels: I. Ergodic capacity. IEEE Trans. Inf. Theory **47**(3), 1083 (2001)
132. G. Taricco, E. Riegler, On the ergodic capacity of correlated Rician fading MIMO channels with interference. IEEE Trans. Inf. Theory **57**(7), 4123 (2011)
133. T. Yoo, A. Goldsmith, Capacity and power allocation for fading MIMO channels with channel estimation error. IEEE Trans. Inf. Theory **52**(5), 2203 (2006)
134. F.A. Dietrich, W. Utschick, P. Breun, Linear precoding based on a stochastic MSE criterion, in *13th European Signal Processing Conference* (2005)
135. R. El Assir, F.A. Dietrich, M. Joham, W. Utschick, Min-max MSE precoding for broadcast channels based on statistical channel state information, in *Proceedings of the 7th IEEE International Workshop on Signal Processing Advances Wireless Communications SPAWC*, Cannes (2006)
136. M. Ding, S. Blostein, MIMO minimum total MSE transceiver design with imperfect CSI at both ends. IEEE Trans. Signal Process. **57**(3), 1141 (2009)
137. R. Henrion. Introduction to chance constrained programming. Introduction Tutorial in E-Collection https://stoprog.org (2003), https://stoprog.org
138. M. Shenouda, T. Davidson, Probabilistically-constrained approaches to the design of the multiple antenna downlink, in *Proceedings of the 42nd Asilomar Conference on Signals, Systems, and Computers* (2008), pp. 1120–1124
139. A. Gründinger, M. Joham, W. Utschick, Design of beamforming in the satellite downlink with static and mobile users, in *Proceedings of the 45th Asilomar Conference on Signals, Systems, and Computers*, Pacific Grove (2011)
140. A. Gründinger, M. Joham, W. Utschick, Feasibility test and globally optimal beamformer design in the satellite downlink based on instantaneous and ergodic rates, in *Proceedings of the 16th International ITG Workshop on Smart Antennas WSA*, Dresden (2012), pp. 217–224
141. A. Gründinger, M. Joham, W. Utschick, Bounds on optimal power minimization and rate balancing in the satellite downlink, in *Proceedings of the IEEE International Conference on Communications ICC* (2012), pp. 3600–3605
142. A. Gründinger, A. Butabayeva, M. Joham, W. Utschick, Chance constrained and ergodic robust QoS power minimization in the satellite downlink, in *Proceedings of the 46th Asilomar Conference on Signals, Systems, and Computers* (2012), pp. 1147–1151
143. A. Gründinger, D. Leiner, M. Joham, C. Hellings, W. Utschick, Ergodic robust rate balancing for rank-one vector broadcast channels via sequential approximations, in *Proceedings of the 38th International Conference on Acoustics, Speech, and Signal Processing ICASSP* (2013), pp. 4744–4748
144. A. Gründinger, R. Bethenod, M. Joham, M. Riemensberger, W. Utschick, Optimal power allocation for the chance-constrained vector broadcast channel and rank-one channel approximation, in *Proceedings of the 12th IEEE International Workshop on Signal Processing Advances in Wireless Communications SPAWC* (2013), pp. 31–35
145. A. Gründinger, M. Joham, J.P. González-Coma, L. Castedo, W. Utschick, Average sum MSE minimization in the multi-user downlink with multiple power constraints, in *Proceedings of the 48th Asilomar Conference on Signals, Systems, and Computers* (2014), pp. 1279–1285

146. A. Gründinger, J. Pickart, M. Joham, W. Utschick, A probabilistic downlink beamforming approach with multiplicative and additive channel errors, in *Proceedings of the 49th Annual Conference on Information Sciences and Systems*, Baltimore, MD (2015)
147. A. Gründinger, M. Joham, W. Utschick, *Balancing for Interference-Limited Multi-User Satellite Communications*, Chap. 8 (Springer, New York, 2015), pp. 165–198
148. G. Alfano, A.M. Tulino, A. Lozano, S. Verdú, Capacity of MIMO channels with one-sided correlation, in *IEEE Eighth International Symposium on Spread Spectrum Techniques and Applications* (2005), pp. 515–519
149. H. Shin, M.Z. Win, J.H. Lee, M. Chiani, On the capacity of doubly correlated MIMO channels. IEEE Trans. Wirel. Commun. **5**(8), 2253 (2006)
150. B. Marks, G. Wright, Technical note–a general inner approximation algorithm for nonconvex mathematical programs. Oper. Res. **26**(4), 681 (1978)
151. H. Tuy, F. Al-Khayyal, P. Thach, Monotonic optimization: branch and cut methods, in *Essays and Surveys in Global Optimization*, Chap. 2 (Springer, New York, NY, 2005), pp. 39–78
152. D. Guo, S. Shamai, S. Verdú, Mutual information and minimum mean-square error in Gaussian channels. IEEE Trans. Inf. Theory **51**(4), 1261 (2005)
153. P. Stoica, Y. Jiang, J. Li, On MIMO channel capacity: an intuitive discussion. IEEE Signal Process. Mag. **22**(3), 83 (2005)
154. S. Shi, M. Schubert, H. Boche, Rate optimization for multiuser MIMO systems with linear processing. IEEE Trans. Signal Process. **56**(8), 4020 (2008)
155. S. Christensen, R. Agarwal, E. Carvalho, J. Cioffi, Weighted sum-rate maximization using weighted MMSE for MIMO-BC beamforming design. IEEE Trans. Wirel. Commun. **7**(12), 4792 (2008)
156. Q. Shi, M. Razaviyayn, Z.Q. Luo, C. He, An Iteratively Weighted MMSE Approach to distributed sum-utility maximization for a MIMO interfering broadcast channel. IEEE Trans. Signal Process. **59**(9), 4331 (2011)
157. A. Gründinger, M. Joham, W. Utschick, Stochastic transceiver design in point-to-point MIMO channels with imperfect CSI, in *Proceedings of the 15th International ITG Workshop on Smart Antennas WSA*, Aachen (2011)
158. A. Gründinger, M. Joham, W. Utschick, Stochastic transceiver design in multi-antenna channels with statistical channel state information, in *Proceedings of the 36th International Conference on Acoustics, Speech, and Signal Processing ICASSP*, Prague (2011)
159. F. Wang, X. Yuan, S.C. Liew, D. Guo, Wireless MIMO switching: weighted sum mean square error and sum rate optimization. IEEE Trans. Inf. Theory **59**(9), 5297 (2013)
160. M. Kießling, *Statistical Analysis and Transmit Prefiltering for MIMO Wireless Systems in Correlated Fading Environments*, Dissertation, University of Stuttgart (2004)
161. R. Hunger, M. Joham, W. Utschick, On the MSE-duality of the broadcast channel and the multiple access channel. IEEE Trans. Signal Process. **57**(2), 698 (2009)
162. M. Joham, M. Vonbun, W. Utschick, MIMO BC/MAC MSE duality with imperfect transmitter and perfect receiver CSI, in *Proceedings of the 11th IEEE International Workshop on Signal Processing Advances in Wireless Communications SPAWC* (2010)
163. J.P. González-Coma, A. Gründinger, M. Joham, L. Castedo Ribas, W. Utschick, Multi-stream MIMO MSE balancing with generalized power constraints, in *Proceedings of the 15th IEEE International Workshop on Signal Processing Advances in Wireless Communications SPAWC* (2016)
164. J.P. González-Coma, A. Gründinger, M. Joham, L. Castedo Ribas, W. Utschick, On MSE balancing in the MIMO broadcast channel with unequal targets, in *Proceedings of the 19th International ITG Workshop on Smart Antennas WSA*, Munich (2016)
165. J.P. González-Coma, A. Gründinger, M. Joham, L. Castedo Ribas, MSE balancing in the MIMO BC: unequal targets and probabilistic interference constraints. IEEE Trans. Signal Process. **65**(12), 3293 (2017)
166. F. Boccardi, H. Huang, M. Trivellato, Multiuser eigenmode transmission for MIMO broadcast channels with limited feedback, in *Proceedings of the 7th IEEE International Workshop on Signal Processing Advances in Wireless Communications SPAWC* (2007)

167. M. Kobayashi, G. Caire, Joint beamforming and scheduling for a multi-antenna downlink with imperfect transmitter channel knowledge. IEEE J. Sel. Areas Commun. **25**(7), 1468 (2007)
168. S. Vorobyov, Y. Rong, A. Gershman, Robust adaptive beamforming using probability-constrained optimization, in *IEEE/SP 13th Workshop on Statistical Signal Processing* (2005), pp. 934–939
169. A. Mutapcic, S.J. Kim, S. Boyd, A tractable method for robust downlink beamforming in wireless communications, in *Proceedings of the 41st Asilomar Conference on Signals, Systems, and Computers* (2007), pp. 1224–1228
170. N. Vučić, H. Boche, Robust QoS-constrained optimization of downlink multiuser MISO systems. IEEE Trans. Signal Process. **57**(2), 714 (2009)
171. K.Y. Wang, T.H. Chang, W.K. Ma, A.C. So, C.Y. Chi, Probabilistic SINR constrained robust transmit beamforming: a Bernstein-type inequality based conservative approach, in *Proceedings of the 36th International Conference on Acoustics, Speech, and Signal Processing ICASSP* (2011), pp. 3080–3083
172. I. Bechar, A Bernstein-type inequality for stochastic processes of quadratic forms Gaussian variables (2009). https://arxiv.org/abs/0909.3595
173. K.Y. Wang, T.H. Chang, W.K. Ma, C.Y. Chi, A semidefinite relaxation based conservative approach to robust transmit beamforming with probabilistic SINR constraints, in *Proceedings of the 18th European Signal Processing Conference EUSIPCO* (2010), pp. 407–411
174. J.P. Imhof, Computing the distribution of quadratic forms in normal variables. Biometrika **48**(3/4), 419 (1961)
175. A. Goldsmith, *Wireless Communications* (Cambridge University Press, Cambridge, 2005)
176. L. Tong, B. Sadler, M. Dong, Pilot-assisted wireless transmissions: general model, design criteria, and signal processing. IEEE Signal Process. Mag. **21**(6), 12 (2004)
177. D.S. Shiu, G.J. Foschini, M.J. Gans, J.M. Kahn, Fading correlation and its effect on the capacity of multielement antenna systems. IEEE Trans. Commun. **48**(3), 502 (2000)
178. M. Ding, S.D. Blostein, Maximum mutual information design for MIMO systems with imperfect channel knowledge. IEEE Trans. Inf. Theory **56**(10), 4793 (2010)
179. M. Castaneda Garcia, A. Mezghani, J. Nossek, Design of single user limited feedback systems. IEEE Trans. Wirel. Commun. **13**(10), 5812 (2014)
180. A. Coulson, A. Williamson, R. Vaughan, A statistical basis for lognormal shadowing effects in multipath fading channels. IEEE Trans. Commun. **46**(4), 494 (1998)
181. A. Goldsmith, L. Greenstein, G. Foschini, Error statistics of real-time power measurements in cellular channels with multipath and shadowing. IEEE Trans. Veh. Technol. **43**(3), 439 (1994)
182. J. Roh, B. Rao, Efficient feedback methods for MIMO channels based on parameterization. IEEE Trans. Wirel. Commun. **6**(1), 282 (2007)
183. D. Hammarwall, M. Bengtsson, B. Ottersten, Acquiring partial CSI for spatially selective transmission by instantaneous channel norm feedback. IEEE Trans. Signal Process. **56**(3), 1188 (2008)
184. E. Björnson, D. Hammarwall, B. Ottersten, Exploiting quantized channel norm feedback through conditional statistics in arbitrarily correlated MIMO systems. IEEE Trans. Signal Process. **57**(10), 4027 (2009)
185. F. Dietrich, W. Utschick, Pilot-assisted channel estimation based on second-order statistics. IEEE Trans. Signal Process. **53**(3), 1178 (2005)
186. M. Vanderveen, A.J. van der Veen, A. Paulraj, Estimation of multipath parameters in wireless communications. IEEE Trans. Signal Process. **46**(3), 682 (1998)
187. T. Kürner, D. Cichon, W. Wiesbeck, Concepts and results for 3D digital terrain-based wave propagation models: an overview. IEEE J. Sel. Areas Commun. **11**(7), 1002 (1993)
188. K. Remley, H. Anderson, A. Weisshar, Improving the accuracy of ray-tracing techniques for indoor propagation modeling. IEEE Trans. Veh. Technol. **49**(6), 2350 (2000)
189. J. Tarng, W.S. Liu, Y.F. Huang, J.M. Huang, A novel and efficient hybrid model of radio multipath-fading channels in indoor environments. IEEE Trans. Antennas Propag. **51**(3), 585 (2003)

190. K. Rizk, J. Wagen, F. Gardiol, Two-dimensional ray-tracing modeling for propagation prediction in microcellular environments. IEEE Trans. Veh. Technol. **46**(2), 508 (1997)
191. C. Tzaras, B. Evans, S. Saunders, Physical-statistical analysis of land mobile-satellite channel. Electron. Lett. **34**(13), 1355 (1998)
192. M. Dottling, A. Jahn, D. Didascalou, W. Wiesbeck, Two- and three-dimensional ray tracing applied to the land mobile satellite (LMS) propagation channel. IEEE Antennas Propag. Mag. **43**(6), 27 (2001)
193. A. Lehner, A. Steingass, Time series multipath modeling of suburban environments in landmobile satellite navigation, in *Proceedings of the 2nd European Conference on Antennas and Propagation EuCAP* (2007)
194. E. Larsson, P. Stoica, *Space-Time Block Coding for Wireless Communications* (Cambridge University Press, Cambridge, 2003)
195. S.M. Kay, *Fundamentals of Statistical Signal Processing: Estimation Theory* (Prentice Hall, New Jersey, NJ, 1993)
196. C. Farsakh, J. Nossek, Spatial covariance based downlink beamforming in an SDMA mobile radio system. IEEE Trans. Commun. **46**(11), 1497 (1998)
197. Y.C. Liang, F. Chin, Downlink channel covariance matrix (DCCM) estimation and its applications in wireless DS-CDMA systems. IEEE J. Sel. Areas Commun. **19**(2), 222 (2001)
198. D. Love, R. Heath, W. Santipach, M. Honig, What is the value of limited feedback for MIMO channels?. IEEE Commun. Mag. **42**(10), 54 (2004)
199. C.K. Au-Yeung, D. Love, On the performance of random vector quantization limited feedback beamforming in a MISO system. IEEE Trans. Wirel. Commun. **6**(2), 458 (2007)
200. D. Love, R. Heath, V. Lau, D. Gesbert, B. Rao, M. Andrews, An overview of limited feedback in wireless communication systems. IEEE J. Sel. Areas Commun. **26**(8), 1341 (2008)
201. K. Mukkavilli, A. Sabharwal, E. Erkip, B. Aazhang, On beamforming with finite rate feedback in multiple-antenna systems. IEEE Trans. Inf. Theory **49**(10), 2562 (2003)
202. T. Zemen, C. Mecklenbräuker, Time-variant channel estimation using discrete prolate spheroidal sequences. IEEE Trans. Signal Process. **53**(9), 3597 (2005)
203. M. Abramowitz, I.A. Stegun, *Handbook of Mathematical Functions*, 1st edn. (Dover, New York, 1964)
204. A. Panagopoulos, P.D. Arapoglou, P. Cottis, Satellite communications at Ku, Ka, and V bands: propagation impairments and mitigation techniques. IEEE Commun. Surv. Tutorials **6**(3), 2 (2004)
205. ITU-R Recommendation P.618-11. Propagation data and prediction methods required for the design of earth-space telecommunication systems, Geneva (2013), http://www.itu.int/rec/R-REC-P.618/en
206. D. Christopoulos, S. Chatzinotas, M. Matthaiou, B. Ottersten, Capacity analysis of multibeam joint decoding over composite satellite channels, in *Proceedings of the 45th Asilomar Conference on Signals, Systems, and Computers* (2011), pp. 1795–1799
207. M. Nakagami, The m-distribution—a general formula of intensity distribution of rapid fading, in *Statistical Methods in Radio Wave Propagation*, ed. by W. Hoffman (Pergamon, Oxford, 1960), pp. 3–36
208. N. Beaulieu, C. Cheng, Efficient nakagami-m fading channel simulation. IEEE Trans. Veh. Technol. **54**(2), 413 (2005)
209. A. Abdi, W. Lau, M.S. Alouini, M. Kaveh, A new simple model for land mobile satellite channels: first- and second-order statistics. IEEE Trans. Wirel. Commun. **2**(3), 519 (2003)
210. J. Griffin, R. Guy, Zoom antennas for multiple spotbeam coverage of Europe, in *Proceedings of the IEE Colloquium on Satellite Antenna Technology in the 21st Century* (1991)
211. H.O. Georgii, *Stochastik: Einführung in die Wahrscheinlichkeitstheorie und Statistik*, 4th edn. (De Gruyter, Berlin, 2009)
212. M. Nicoli, O. Simeone, U. Spagnolini, Multislot estimation of fast-varying space-time communication channels. IEEE Trans. Signal Process. **51**(5), 1184 (2003)

213. D. Neumann, M. Joham, L. Weiland, W. Utschick, Low-complexity computation of LMMSE channel estimates in massive MIMO, in *Proceedings of the 19th International ITG Workshop on Smart Antennas WSA* (2015)
214. B. Yang, Projection approximation subspace tracking. IEEE Trans. Signal Process. **43**(1), 95 (1995)
215. E. Jorswieck, Lack of duality between SISO Gaussian MAC and BC with statistical CSIT. Electron. Lett. **42**(25), 1466 (2006)
216. H. Boche, M. Schubert, A general duality theory for uplink and downlink beamforming, in *IEEE 56th Vehicular Technology Conference, 2002. Proceedings. VTC 2002-Fall*, vol. 1 (2002), pp. 87–91
217. J.P. González-Coma, M. Joham, P.M. Castro, L. Castedo, QoS constrained power minimization in the MISO broadcast channel with imperfect CSI. Signal Process. **131**, 447 (2017)
218. A. Pastore, M. Joham, Mutual information bounds for MIMO channels under imperfect receiver CSI, in *Proceedings of the 43rd Asilomar Conference on Signals, Systems, and Computers*, Pacific Grove, CA (2009), pp. 1456–1460
219. M. Joham, A. Gründinger, A. Pastore, J. Fonollosa, W. Utschick, Rate balancing in the vector BC with erroneous CSI at the receivers, in *Proceedings of the 47th Annual Conference on Information Sciences and Systems*, Baltimore, MD (2013)
220. A. Cuyt, V. Petersen, B. Verdonk, H. Waadeland, W. Jones, Confluent hypergeometric functions, in *Handbook of Continued Fractions for Special Functions* (Springer, Amsterdam, 2008), pp. 319–341
221. J. Lindblom, E. Larsson, E. Jorswieck, Parameterization of the MISO interference channel with transmit beamforming and partial channel state information, in *Signals, Systems and Computers, 2008 42nd Asilomar Conference on* (2008), pp. 1103–1107
222. A. Lapidoth, S. Moser, Capacity bounds via duality with applications to multiple-antenna systems on flat-fading channels. IEEE Trans. Inf. Theory **49**(10), 2426 (2003)
223. S. Moser, Some expectations of a non-central Chi-square distribution with an even number of degrees of freedom, in *2007 IEEE Region 10 Conference (TENCON)* (2007)
224. S. Moser, The fading number of memoryless multiple-input multiple-output fading channels. IEEE Trans. Inf. Theory **53**(7), 2652 (2007)
225. C. Hellings, M. Joham, W. Utschick, Gradient-based rate balancing for MIMO broadcast channels with linear precoding, in *Proceedings of the 15th International ITG Workshop on Smart Antennas WSA* (2011)
226. C. Nuzman, Contraction approach to power control, with non-monotonic applications, in *Proceedings of the IEEE Global Communications Conference GLOBECOM* (2007), pp. 5283–5287
227. D.P. Bertsekas, *Nonlinear Programming* (Athena Scientific, Belmont, MA, 1999)
228. R. Hunger, D.A. Schmidt, M. Joham, W. Utschick, A general covariance-based optimization framework using orthogonal projections, in *Proceedings of the 9th IEEE International Workshop on Signal Processing Advances in Wireless Communications SPAWC* (2008), pp. 76–80
229. A. Bermon, R. Plemmons, *Nonnegative Matrices in the Mathematical Sciences* (Society for Industrial and Applied Mathematics, Philadelphia, 1994)
230. S. Stańczak, M. Wiczanowski, H. Boche, *Fundamentals of Resource Allocation in Wireless Networks*. Foundations in Signal Processing, Communications and Networking, vol. 3, 2nd edn. (Springer, Berlin/Heidelberg, 2008)
231. S. Stańczak, M. Kaliszan, N. Bambos, M. Wiczanowski, A characterization of max-min SIR-balanced power allocation with applications, in *Proceedings of the IEEE International Symposium on Information Theory ISIT* (2009), pp. 2747–2751
232. M. Grant, S. Boyd, Y. Ye, Disciplined convex programming, in *Global Optimization: From Theory to Implementation, Nonconvex Optimization and Its Applications*, ed. by L. Liberti, N. Maculan (Springer, New York, 2006), pp. 155–210
233. S.A. Vavasis, Complexity issues in global optimization: a survey, in *Handbook of Global Optimization* (Kluwer, Dordrecht, 1995), pp. 27–41

234. X. Zhang, D.P. Palomar, B. Ottersten, Statistically robust design of linear MIMO transceivers. IEEE Trans. Signal Process. **56**(8), 3678 (2008)
235. J. Wang, D.P. Palomar, Robust MMSE precoding in MIMO channels with pre-fixed receivers. IEEE Trans. Signal Process. **58**(11), 5802 (2010)
236. T. Endeshaw, B. Chalise, L. Vandendorpe, MSE uplink-downlink duality of MIMO systems under imperfect CSI, in *International Workshop on Computational Advances in Multi-Sensor Adaptive Processing (CAMSAP)*, Aruba, Dutch Antilles (2009), pp. 384–387
237. M. Razaviyayn, M. Hong, Z.Q. Luo, Linear transceiver design for a MIMO interfering broadcast channel achieving max-min fairness, in *Proceedings of the 45th Asilomar Conference on Signals, Systems, and Computers*, Pacific Grove, CA (2011), pp. 1309–1313
238. T.M. Kim, F. Sun, A. Paulraj, Low-complexity MMSE precoding for coordinated multipoint with per-antenna power constraint. IEEE Signal Process. Lett. **20**(4), 395 (2013)
239. F. Negro, I. Ghauri, D.T.M. Slock, Sum rate maximization in the noisy MIMO interfering broadcast channel with partial CSIT via the expected weighted MSE, in *Proceedings of the 9th International Symposium on Wireless Communication Systems ISWCS* (2012), pp. 576–580
240. D.P. Palomar, V. S., Representation of mutual information via input estimates. IEEE Trans. Inf. Theory **53**(2), 453 (2007)
241. M. Ding, S.D. Blostein, Uplink-downlink duality in normalized MSE or SINR under imperfect channel knowledge, in *Proceedings of the IEEE Global Communications Conference GLOBECOM*, Washington, DC (2007), pp. 3786–3790
242. J.R. Magnus, The exact moments of a ratio of quadratic forms in normal variables. Annales D'Économie et de Statistique **1**(4), 95 (1986)
243. M.S. Paolella, Computing moments of ratios of quadratic forms in normal variables. Comput. Stat. Data Anal. **42**(3), 313 (2003). Computational Econometrics
244. Y. Bao, R. Kan, On the moments of ratios of quadratic forms in normal random variables. J. Multivar. Anal. **117**, 229 (2013)
245. O. Lieberman, A Laplace approximation to the moments of a ratio of quadratic forms. Biometrica **81**(4), 681 (1994)
246. J. Gorski, F. Pfeuffer, K. Klamroth, Biconvex sets and optimization with biconvex functions: a survey and extensions. Math. Methods Oper. Res. **66**(3), 373 (2007)
247. M.S. Lobo, L. Vandenberghe, S. Boyd, H. Lebret, Applications of second-order cone programming. Linear Algebra Appl. **284**(1), 193 (1998)
248. S. Shi, M. Schubert, N. Vučić, H. Boche, MMSE optimization with per-base-station power constraints for network MIMO systems, in *Proceedings of the IEEE International Conference on Communications ICC* (2008), pp. 4106–4110
249. M. Ulbrich, S. Ulbrich, *Nichtlineare Optimierung*. Mathematik Kompakt (Springer, Basel, 2012)
250. J.P. Gonzalez-Coma, M. Joham, P.M. Castro, L. Castedo, Power minimization in the multiuser MIMO-OFDM broadcast channel with imperfect CSI, in *Proceedings of the 22nd European Signal Processing Conference EUSIPCO* (2014), pp. 815–819
251. J.P. González-Coma, M. Joham, P.M. Castro, L. Castedo, Power minimization and QoS feasibility region in the multiuser MIMO broadcast channel with imperfect CSI, in *Proceedings of the 12th IEEE International Workshop on Signal Processing Advances in Wireless Communications SPAWC*, Darmstadt (2013), pp. 619–623
252. H. Boche, M. Schubert, A general theory for SIR balancing. EURASIP J. Wirel. Commun. Netw. **2006**(2), 10 (2006)
253. S. Shi, M. Schubert, H. Boche, Downlink MMSE transceiver optimization for multiuser MIMO systems: MMSE balancing. IEEE Trans. Signal Process. **56**(8), 3702 (2008)
254. A. Sharpio, D. Dentcheva, A. Ruszczyński, *Lectures on Stochastic Programming* (Society for Industrial and Applied Mathematics, Philadelphia, 2009)
255. N. Letzepis, A. Grant, Capacity of the multiple spot beam satellite channel with Rician fading. IEEE Trans. Inf. Theory **54**(11), 5210 (2008)

256. Y. Ye, Interior-Point Algorithm: Theory and Application. Tech. rep., Department of Management Sciences, University of Iowa City (1995)
257. L. Ozarow, S. Shamai, A. Wyner, Information theoretic considerations for cellular mobile radio. IEEE Trans. Veh. Technol. **43**(2), 359 (1994)
258. A. Ben-Tal, L. El Ghaoui, A. Nemirovski, *Robust Optimization*. Princeton Series in Applied Mathematics (Princeton University Press, Princeton, 2009)
259. C.M. Lagoa, X. Li, M. Sznaier, Probabilistically constrained linear programs and risk-adjusted controller design. SIAM J. Optim. **15**(3), 938 (2005)
260. R. Henrion, Structural properties of linear probabilistic constraints. Optimization **56**(4), 425 (2007)
261. D. Raphaeli, Distribution of noncentral indefinite quadratic forms in complex normal variables. IEEE Trans. Inf. Theory **42**(3), 1002 (1996)
262. T. Al-Naffouri, B. Hassibi, On the distribution of indefinite quadratic forms in Gaussian random variables, in *Proceedings of the IEEE International Symposium on Information Theory ISIT* (2009), pp. 1744–1748
263. B. Chalise, S. Shahbazpanahi, A. Czylwik, A. Gershman, Robust downlink beamforming based on outage probability specifications. IEEE Trans. Wirel. Commun. **6**(10), 3498 (2007)
264. X. He, Y.C. Wu, Tight probabilistic MSE constrained multiuser MISO transceiver design under channel uncertainty, in *Proceedings of the IEEE International Conference on Communications ICC* (2015)
265. D. Bertsimas, M. Sim, Tractable approximations to robust conic optimization problems. Math. Program. **107**, 5 (2006)
266. R. Hildebrand, An LMI description for the cone of Lorentz-positive maps. Linear Multilinear Algebra **55**(6), 551 (2007)
267. R. Hildebrand, Linear Multilinear Algebra **59**(7), 719 (2011)
268. D. Hsu, S. Kakade, T. Zhang, An LMI description for the cone of Lorentz-positive maps II. Electron. Commun. Probab. **17**(52), 1 (2012)
269. P. Massart, *Concentration Inequalities and Model Selection*. Lecture Notes in Mathematics, vol. 1896 (Springer, Berlin/Heidelberg, 2007)
270. A. Ben-Tal, A. Nemirovski, On safe tractable approximations of chance-constrained linear matrix inequalities. Math. Oper. Res. **34**(1), 1 (2009)
271. N.L. Johnson, S. Kotz, N. Balakrishnan, *Continuous Univariate Distributions*. Wiley Series in Probability and Mathematical Statistics, vol. 2, 2nd edn. (Wiley, New York, NY, 1995)
272. L. Xu, K. Wong, J.K. Zhang, D. Ciochina, M. Pesavento, A Riemannian distance for robust downlink beamforming, in *Proceedings of the 12th IEEE International Workshop on Signal Processing Advances in Wireless Communications SPAWC* (2013), pp. 465–469
273. J. Lindblom, E. Karipidis, E. Larsson, Outage rate regions for the MISO IFC, in *Proceedings of the 43rd Asilomar Conference on Signals, Systems, and Computers* (2009), pp. 1120–1124
274. C. Lin, C.J. Lu, W.H. Chen, Outage-constrained coordinated beamforming with opportunistic interference cancellation. IEEE Trans. Signal Process. **62**(16), 4311 (2014)
275. N. Zlatanov, Z. Hadzi-Velkov, G. Karagiannidis, R. Schober, Cooperative diversity with mobile nodes: capacity outage rate and duration. IEEE Trans. Inf. Theory **57**(10), 6555 (2011)
276. A. Gründinger, L. Gerdes, M. Joham, W. Utschick, Bounds on the outage constrained capacity of the Gaussian relay channel, in *Communications in Interference Limited Networks* (Springer, New York, 2016)
277. R. Di Taranto, P. Popovski, Outage performance in cognitive radio systems with opportunistic interference cancelation. IEEE Trans. Wirel. Commun. **10**(4), 1280 (2011)
278. A. Nuttall, Some integrals involving the Marcum-Q function. IEEE Trans. Inf. Theory **21**(1), 95 (1975)
279. D. Benton, K. Krishnamoorthy, Computing discrete mixtures of continuous distributions: noncentral chisquare, noncentral t, and the distribution of the square of the sample multiple correlation coefficient. Comput. Stat. Data Anal. **43**(2), 249 (2003)

280. M. Kang, M.S. Alouini, Quadratic forms in complex Gaussian matrices and performance analysis of MIMO systems with cochannel interference. IEEE Trans. Wirel. Commun. **3**(2), 418 (2004)
281. A. Mathai, S. Provost (eds.), *Quadratic Forms in Random Variables: Theory and Applications*. Statistics: Textbooks and Monographs, vol. 126 (Marcel Dekker, New York, 1992). Various authors
282. B.K. Shah, Distribution of definite and of indefinite quadratic forms from a non-central normal distribution. Ann. Math. Stat. **34**(1), 186 (1963)
283. S.J. Press, Linear combinations of non-central Chi-square variates. Ann. Math. Stat. **37**(2), 480 (1966)
284. J. Gil-PeLaez, Note on the inversion theorem. Biometrika **38**(3–4), 481 (1951)
285. P. Duchesne, P.L. De Micheaux, Computing the distribution of quadratic forms: further comparisons between the Liu–Tang–Zhang approximation and exact methods. Comput. Stat. Data Anal. **54**(4), 858 (2010)
286. R. Henrion, C. Strugarek, Convexity of chance constraints with independent random variables. Comput. Optim. Appl. **41**(2), 263 (2008)
287. R. Lucchetti, G. Salinetti, Uniform convergence of probability measures: topological criteria. J. Multivar. Anal. **51**(2), 254 (1994)
288. M. Payaró, A. Pascual-Iserte, M. Lagunas, Robust power allocation designs for multiuser and multiantenna downlink communication systems through convex optimization. IEEE J. Sel. Areas Commun. **25**(7), 1390 (2007)
289. A. Shaverdian, M. Nakhai, Robust distributed beamforming with interference coordination in downlink cellular networks. IEEE Trans. Commun. **62**(7), 2411 (2014)
290. D. Bertsimas, D. Brown, C. Caramanis, Theory and applications of robust optimization. SIAM Rev. **53**(3), 464 (2011)
291. E. Song, Q. Shi, M. Sanjabi, R. Sun, Z.Q. Luo, Robust SINR-constrained MISO downlink beamforming: when is semidefinite programming relaxation tight?, in *Proceedings of the 36th International Conference on Acoustics, Speech, and Signal Processing ICASSP* (2011), pp. 3096–3099
292. G. Zheng, K.K. Wong, B. Ottersten, Robust cognitive beamforming with bounded channel uncertainties. IEEE Trans. Signal Process. **57**(12), 4871 (2009)
293. R.W. Cottle, Manifestations of the Schur complement. Linear Algebra Appl. **8**(3), 189 (1974)
294. J.D. Gayrard, Terabit satellite: myth or reality?, in *Proceedings of the 1st International Conference on Advances in Satellite and Space Communications SPACOMM* (2009)
295. P. Thompson, B. Evans, L. Castenet, M. Bousquet, T. Mathiopoulos, Concepts and technologies for a terabit/s satellite, in *Proceedings of the 3rd International Conference on Advances in Satellite and Space Communications* (IARIA XPS Press, Budapest, 2011), pp. 12–19
296. O. Vidal, G. Verelst, J. Lacan, E. Alberty, J. Radzik, M. Bousquet, Next generation high throughput satellite system, in *Proceedings of the 1st AESS European Conference on Satellite Telecommunications ESTEL* (2012)
297. A. Duflos, B. Evans, P. Thompson, A. Tomatis, S. Amos, A. Laurent, C. Noeldeke, G. Verlest, Approaching the Terabit/s Satellite: A System Study. Executive Summary 1, Revision 1, ESA Contract No:. 4000103563 (2012)
298. P.D. Arapoglou, K. Liolis, A. Panagopoulos, Railway satellite channel at Ku band and above: composite dynamic modeling for the design of fade mitigation techniques. Int. J. Satell. Commun. Netw. **30**(1), 1 (2012)
299. K. Liolis, A. Panagopoulos, S. Scalise, On the combination of tropospheric and local environment propagation effects for mobile satellite systems above 10 GHz. IEEE Trans. Veh. Technol. **59**(3), 1109 (2010)
300. S.K. Sharma, S. Chatzinotas, B. Ottersten, Cognitive beamhopping for spectral coexistence of multibeam satellites. Int. J. Satell. Commun. Netw. **33**(1), 69–91 (2014)

301. M. Diaz, N. Courville, C. Mosquera, G. Liva, G. Corazza, Non-linear interference mitigation for broadband multimedia satellite systems, in *Proceedings of the International Workshop on Satellite and Space Communications* (2007), pp. 61–65
302. N. Zorba, M. Realp, A. Pérez-Neira, An improved partial CSIT random beamforming for multibeam satellite systems, in *Proceedings of the 10th International Workshop on Signal Processing for Space Communications SPSC* (2008)
303. C. Hellings, S. Herrmann, W. Utschick, Carrier cooperation can reduce the transmit power in parallel MIMO broadcast channels with zero-forcing. IEEE Trans. Signal Process. **61**(12), 3021 (2013)
304. C. Hellings, W. Utschick, Linear precoding in parallel MIMO broadcast channels: separable and inseparable scenarios, in *Proceedings of the 17th International ITG Workshop on Smart Antennas WSA* (2013)
305. P. Kelley, Overview of the DVB-SH specifications. Int. J. Satell. Commun. Netw. **27**(4–5), 198 (2009)
306. A. Morello, V. Mignone, DVB-S2: the second generation standard for satellite broad-band services. Proc. IEEE **94**(1), 210 (2006)
307. S. Rao, Parametric design and analysis of multiple-beam reflector antennas for satellite communications. IEEE Antennas Propag. Mag. **45**(4), 26 (2003)
308. T. Braun, Antenna, Chap. 3, in *Satellite Communications Payload and System* (Wiley/IEEE Press, New York, 2012), pp. 33–77
309. G. Maral, M. Bousquet, *Satellite Communications Systems*, 5th edn. (Wiley, England, 2009)
310. P.D. Arapoglou, P. Burzigotti, M. Bertinelli, A. Bolea Alamanac, R. De Gaudenzi, To MIMO or not to MIMO in mobile satellite broadcasting systems. IEEE Trans. Wirel. Commun. **10**(9), 2807 (2011)
311. ITU-R Recommendation P.840-3. Attenuation due to clouds and fog. Geneva (1999), http://www.itu.int/rec/R-REC-P.840/en
312. ITU-R Recommendation P.676-10. Attenuation by atmospheric gases. Geneva (2013), http://www.itu.int/rec/R-REC-P.676-10-201309-I/en
313. X. Boulanger, L. Feral, L. Castanet, N. Jeannin, G. Carrie, F. Lacoste, A rain attenuation time-series synthesizer based on a Dirac and lognormal distribution. IEEE Trans. Antennas Propag. **61**(3), 1396 (2013)
314. E. Lutz, A Markov model for correlated land mobile satellite channels. Int. J. Satell. Commun. **14**(4), 333 (1996)
315. ITU-R Recommendation P.681-7. Propagation data required for the design of earth-space land mobile telecommunication systems. Geneva (2009), http://www.itu.int/rec/R-REC-P.1815/en
316. L. Xiao, L. Cottatellucci, Parametric least squares estimation for nonlinear satellite channels, in *Proceedings of the 76nd IEEE Vehicular Technology Conference, Fall* (2012)
317. S. Enserink, M. Fitz, Constrained capacities of DVB-S2 constellations in log-normal channels at Ka band, in *Proceedings of the 2nd International Conference on Advances in Satellite and Space Communications* (2010), pp. 93–99
318. G. Zheng, S. Chatzinotas, B. Ottersten, Multi-gateway cooperation in multibeam satellite systems, in *Proceedings of the IEEE 23rd International Symposium on Personal Indoor and Mobile Radio Communications PIMRC* (2012), pp. 1360–1364
319. D. Hammarwall, M. Bengtsson, B. Ottersten, On downlink beamforming with indefinite shaping constraints. IEEE Trans. Signal Process. **54**(9), 3566 (2006)
320. D. Neumann, A. Gründinger, M. Joham, W. Utschick, Pilot coordination for large-scale multi-cell TDD systems, in *Proceedings of the 18th International ITG Workshop on Smart Antennas WSA* (2014)
321. J. Jose, A. Ashikhmin, T.L. Marzetta, S. Vishwanath, Pilot contamination and precoding in multi-cell TDD systems. IEEE Trans. Wirel. Commun. **10**(8), 2640 (2011)
322. A. Gründinger, L. Gerdes, M. Joham, W. Utschick, *Bounds on the Outage Constrained Capacity of the Gaussian Relay Channel*, Chap. 7 (Springer, New York, 2015), pp. 145–164

323. R. Cavalcante, S. Stanczak, M. Schubert, A. Eisenblaetter, U. Tuerke, Toward energy-efficient 5G wireless communications technologies: tools for decoupling the scaling of networks from the growth of operating power. IEEE Signal Process. Mag. **31**(6), 24 (2014)
324. M.S. Bazaraa, H.D. Sherali, C.M. Shetty, *Convex Optimization*, 3rd edn. (Wiley, New York, 2006)
325. E. Telatar, Capacity of multi-antenna Gaussian channels. Eur. Trans. Telecommun. **10**(6), 585 (1999)

Index

A. Gründinger, *Statistical Robust Beamforming for Broadcast Channels and Applications in Satellite Communication*, Foundations in Signal Processing, Communications and Networking 22, https://doi.org/10.1007/978-3-030-29578-3